# Recent Advances in Understanding of the Role of Synuclein Family Members in Health and Disease

# Recent Advances in Understanding of the Role of Synuclein Family Members in Health and Disease

Editor

**Natalia Ninkina**

Basel • Beijing • Wuhan • Barcelona • Belgrade • Novi Sad • Cluj • Manchester

*Editor*
Natalia Ninkina
Division of Biomedicine
Cardiff University
Cardiff, UK

*Editorial Office*
MDPI
St. Alban-Anlage 66
4052 Basel, Switzerland

This is a reprint of articles from the Special Issue published online in the open access journal *Biomedicines* (ISSN 2227-9059) (available at: https://www.mdpi.com/journal/biomedicines/special_issues/Synuclein_Family_Members).

For citation purposes, cite each article independently as indicated on the article page online and as indicated below:

Lastname, A.A.; Lastname, B.B. Article Title. *Journal Name* **Year**, *Volume Number*, Page Range.

**ISBN 978-3-0365-8938-1 (Hbk)**
**ISBN 978-3-0365-8939-8 (PDF)**
**doi.org/10.3390/books978-3-0365-8939-8**

# Contents

# About the Editor

**Natalia Ninkina**

Graduated from Pirogov Russian National Research Medical University, Moscow and obtained a PhD degree from the Institute of Molecular Biology, Russian Academy of Sciences. In 1990 she moved to the UK and worked on neuroscience projects in the Sandoz (now Novartis) Institute for Medical Research in London. She moved to the University of St Andrews to carry out research on developmental neurobiology and continue her studies as Principal Investigator (PI) in the University of Edinburgh. From 2004, she held a permanent Senior Research Fellow position at the Cardiff University School of Biosciences headed by Sir Martin Evans, a Nobel Prize winner. There, the focus of her research was on the creation of new mouse models for studying molecular and cellular mechanisms of neurodegenerative diseases. During that time, she led a number of collaborative projects with the Russian Academy of Science.

*Editorial*

# Editorial of the Special Issue: Recent Advances in Understanding of the Role of Synuclein Family Members in Health and Disease

Natalia Ninkina [1,2,3,*] and Michail S. Kukharsky [2,4]

1 School of Biosciences, Cardiff University, Cardiff CF10 3AX, UK
2 Institute of Physiologically Active Compounds at Federal Research Center of Problems of Chemical Physics and Medicinal Chemistry, Russian Academy of Sciences, 142432 Chernogolovka, Russia; kukharskym@gmail.com
3 Department of Pharmacology and Clinical Pharmacology, Belgorod State National Research University, 308015 Belgorod, Russia
4 Department of General and Cell Biology, Faculty of Medical Biology, Pirogov Russian National Research Medical University, 117997 Moscow, Russia
* Correspondence: ninkinan@cardiff.ac.uk; Tel.: +44-(0)29-2087-9068

**Citation:** Ninkina, N.; Kukharsky, M.S. Editorial of the Special Issue: Recent Advances in Understanding of the Role of Synuclein Family Members in Health and Disease. *Biomedicines* **2023**, *11*, 2330. https://doi.org/10.3390/biomedicines11092330

Received: 15 August 2023
Accepted: 18 August 2023
Published: 22 August 2023



Extensive studies of α-synuclein function and dysfunction revealed its involvement in multiple normal and aberrant molecular processes and, consequently, numerous and diverse effects on the neuronal cell biology. The other two members of this family, namely β-synuclein and γ-synuclein, have also drawn increasing attention from researchers and clinicians over the past few years, but their roles in homeostasis and pathology of the nervous system are still poorly understood. Due to a high level of amino acid sequence homology, all three members of the synuclein family share many physicochemical properties. They also have overlapping patterns of intracellular localisation and expression throughout the nervous system. Taken together, these similarities point towards a possible functional redundancy within this family. However, each synuclein has its unique, i.e., unshared with the other two synucleins, functions, and in the context of certain cellular mechanisms and pathways, functions of these family members could be antagonistic rather than synergistic [1–3]. Therefore, it is important to contemplate the synuclein family as a conjoint protein system with a dynamic mode of functioning that requires fine tuning and balance to efficiently modulate the physiological processes these proteins are involved in. Moreover, dysfunction of the coordinated network of synucleins could trigger or critically contribute to the formation of pathomechanisms that lead to the development of certain neurological disorders. The number of research papers on synucleins that have been published since their discovery around 25 years ago has already exceeded 16,000. The articles published in the Special Issue "Recent Advances in Understanding of the Role of Synuclein Family Members in Health and Disease" add further value to this vast collection by addressing several aspects of synuclein biology and pathology, some of which are not in the mainstream of current research. The involvement of synucleins, particularly α-synuclein, in the pathogenesis of protein misfolding diseases (proteinopathies) has been the most thoroughly studied and well documented stream of data. The comprehensive review by A. Surguchov et al. [4] summarises the latest progress and main trends in research on the molecular mechanisms of synucleinopathies, a group of neurodegenerative diseases associated with the malfunction of synucleins. The current focus remains on the scrutinising processes of the pathogenic α-synuclein accumulation, aggregation, and formation of transmissible species that drive the aggregation pathology spreading across the nervous system. The authors of this review also covered the relatively recent studies that demonstrated the critical of role of epigenetics in the development of a synuclein-driven pathology. In vitro cellular models of synuclenopathies still remains a powerful tool for studying pathogenic fibrillation due to their throughput and fewer ethical issues faced. The capacity of this approach has been significantly reinforced through the advent of pluripotent stem cells and the ability to create mini brains in a dish. Katrina Albert and colleagues describe, in their review, how data that have been produced on various cellular models, such as immortalised cell lines, primary neurons, human-induced pluripotent stem cells (hiPSCs), blood–brain barrier models, and brain organoids, have substantially widened our knowledge of the mechanisms underlying synuclein aggregation and confirmed the crucial role of certain types of pathogenic fibrils in the triggering and transmission of proteinopathy from cell to cell, thereby allowing disease spread [5]. Several potentially important factors that could affect α-synuclein aggregation propensity have been recently identified. For example, α-synuclein fibrillation could be promoted through the receptor-binding domain of SARS-CoV-2 proteins [6]. Researchers have confirmed the formation of complexes between the receptor-binding domain (RBD) of the viral S-protein and the α-synuclein monomer following the immobilisation of RBD on its specific receptor ACE2. Several spectral approaches were used to produce novel data on the kinetics of amyloid fibril formation, which strongly suggested that the RBD prevents the amyloid transformation of α-synuclein. Moreover, the fibrils obtained in the presence of the RBD exhibited a significantly lower level of cytotoxicity on SH-SY5Y neuroblastoma cells.

It is well known that synucleins are involved in the regulation of dopamine homeostasis [3,7–10]. The latest research revealed and delineated their role in the optimisation of dopamine uptake through synaptic vesicles and suggested that β-synuclein, rather than α-synuclein, potentiates this uptake [1,11]. These were discussed in connection with the different sensitivity of various synuclein deficient mouse models to MPTP toxicity [12]. In this review, the authors analysed available published data on how single, double, and triple knockouts of synuclein family members affect animal sensitivity to MPTP toxicity and attempted to explain the impact of α-, β-, and γ-synuclein on the mechanisms of MPTP and its active derivate, MPP+, toxicity on dopaminergic neurones. These authors also proposed a mechanistic explanation of why and how α-synuclein knockout mice demonstrated an increased resistance to MPTP, whereas the absence of β-synuclein attenuated this effect.

Consequences of the deletion of one or more synuclein genes and the resulting disbalance of these synucleins could manifest themselves in varied ways and modify certain physiological functions in model mice. For instance, Vorobyov et al. observed changes in electrical activity in the brains of synuclein knockout mice as their frequency spectra of electroencephalograms (EEGs) were being recorded [13]. EEGs were recorded from the motor cortex, the putamen, the ventral tegmental area, and the substantia nigra in mice lacking α-, β-, and γ-synucleins in all possible combinations, including completely synuclein-free triple knockout animals. Additionally, changes in the EEGs of these mice were assessed following the systemic injection of a DA receptor agonist, apomorphine (APO). The obtained data clearly demonstrated that it was not the absence of any particular synuclein, but rather that a disbalance of synucleins caused widespread changes in the EEG spectral profiles. Research by Kokhan et al. focused on studying the engagement of α-, β-, and γ-synucleins in mechanisms of craving for alcohol and developing alcohol dependence [14]. They described changes in the levels of expression in genes encoding for synucleins in the hippocampus and midbrain of alcohol-consuming male and female rats. Moreover, these authors revealed the sex-related differences in α-synuclein levels in the brains of adult rats that have been exposed to alcohol prenatally.

Due to the direct involvement of α-synuclein in the aetiology and pathogenesis of Parkinson's disease, changes in the brain dopamine system are the main target for studies of pathological conditions caused by the malfunction of synucleins. However, the brain serotonin system is also affected by lesions in the synuclein network, and this could further exacerbate the patient's health decline caused by problems with dopaminergic transmission. The serotonin pathways regulate mood and emotions, and therefore dysfunction of this system is crucial for the occurrence of certain non-motor symptoms in Parkinson's disease patients at all stages of the disease progression, including in its very early stages, when a confident diagnosis is problematic. Consequently, better insight into how synuclein pathology affects the function of serotoninergic neurons is important for the development of novel therapeutic approaches for the treatment of non-motor symptoms that significantly affect patients' quality of life. The most recent review by Miquel-Rio et al. summarised the latest progress and current knowledge about the involvement of $\alpha$-synuclein in regulating serotonin system function in the context of health and disease [15]. The importance underlying the careful evaluation of non-motor symptoms, along with a multimodal MRI analysis, for assessing changes in brain function and managing patients with Parkinson's disease and dementia with Lewy bodies, the two most common synucleinopathies, was also emphasised in the multimodal imaging study by Lucas-Jiménez et al. [16].

$\gamma$-synuclein is a well-known marker of certain types of malignant tumours; although, its association with neuronal pathology is less obvious than for the other two members of the family. However, increased levels of $\gamma$-synuclein expression in neurons causes its pathological aggregation and the development of severe neuronal pathology in transgenic mice [17–19]. Thus, it is feasible that a compensatory increase in $\gamma$-synuclein expression following the functional depletion of $\alpha$- or/and $\beta$-synucleins could contribute to the nervous system malfunctioning. This may be linked to the appearance of autoantibodies against this protein in the CSF and peripheral blood of patients with certain disorders. For example, its expression and localisation are changed in the retina and optic nerves in patients with glaucoma. Moreover, Pavlenko et al. assessed the presence of autoantibodies to $\gamma$-synuclein in the blood serum of patients with primary open-angle glaucoma (POAG), and they have detected them in 20% of patients. Using $\gamma$-synuclein knockout mice as a model, they confirmed that $\gamma$-synuclein dysfunction contributes to pathological processes in glaucoma, including the dysregulation of intraocular pressure [20]. This is in accordance with the results of earlier studies that suggested the particular vulnerability of the visual system to malfunction of $\gamma$-synuclein due to the physiologically high levels of its expression in the retina and optic nerves [21].

Current therapeutic approaches to combat synucleinopathies are limited. Sai Sriram et al. have suggested, in their review, that $\alpha$-synuclein aggregation pathology could be an actual target of deep brain stimulation (DBS), a surgical method that is currently in use for the treatment of three synucleinopathies: Parkinson's disease (PD), dementia with Lewy bodies (DLB), and multiple system atrophy (MSA). In their review, these authors also discussed the usefulness and benefits of other surgical approaches, including focused ultrasound (FUS), for the management of these diseases [22].

The past few decades have been characterised by a sharp increase in the number of experimental and clinical studies of a group of severe and eventually fatal neurodegenerative diseases known as synucleinopathies due to the direct involvement of members of the synuclein family of proteins, primarily $\alpha$-synuclein, in the aetiology and pathogenesis of these diseases. Although substantial progress has been achieved in understanding the molecular and cellular events associated with pathological changes in the nervous system typical to each of the synucleinopathies, our knowledge is still insufficient for the designing and creating of therapies that are capable of either preventing or halting the progression of these diseases. Therefore, the principal aim of current and future studies is to find out which components of these pathomechanisms are crucially responsible for the triggering and progression of these synucleinopathies, which would facilitate the development of efficient treatments for these diseases. Revealing an interplay of function and dysfunction of the synuclein family members in all its complexity and diversity would be an important future step in this direction.

**Author Contributions:** Conceptualization, N.N. and M.S.K.; writing—original draft preparation, N.N.; writing—review and editing, M.S.K.; supervision, N.N.; funding acquisition, N.N. and M.S.K. All authors have read and agreed to the published version of the manuscript.

**Funding:** N.N. and M.S.K. research on modelling of synuclein pathology in mice was funded by Ministry of Science and Higher Education of the Russian Federation (grant agreement No. 075-15-2020-795, state contract No. 13.1902.21.0027 of 29.09.2020 unique project ID: RF-190220X0027).

**Conflicts of Interest:** The authors declare no conflict of interest.

## References


1. Ninkina, N.; Millership, S.J.; Peters, O.M.; Connor-Robson, N.; Chaprov, K.; Kopylov, A.T.; Montoya, A.; Kramer, H.; Withers, D.J.; Buchman, V.L. beta-synuclein potentiates synaptic vesicle dopamine uptake and rescues dopaminergic neurons from MPTP-induced death in the absence of other synucleins. *J. Biol. Chem.* **2021**, *297*, 101375. [CrossRef] [PubMed]
2. Ninkina, N.; Peters, O.M.; Connor-Robson, N.; Lytkina, O.; Sharfeddin, E.; Buchman, V.L. Contrasting effects of alpha-synuclein and gamma-synuclein on the phenotype of cysteine string protein alpha (CSPalpha) null mutant mice suggest distinct function of these proteins in neuronal synapses. *J. Biol. Chem.* **2012**, *287*, 44471–44477. [CrossRef] [PubMed]
3. Connor-Robson, N.; Peters, O.M.; Millership, S.; Ninkina, N.; Buchman, V.L. Combinational losses of synucleins reveal their differential requirements for compensating age-dependent alterations in motor behavior and dopamine metabolism. *Neurobiol. Aging* **2016**, *46*, 107–112. [CrossRef] [PubMed]
4. Surguchov, A.; Surguchev, A. Synucleins: New Data on Misfolding, Aggregation and Role in Diseases. *Biomedicines* **2022**, *10*, 3241. [CrossRef]
5. Albert, K.; Kalvala, S.; Hakosalo, V.; Syvanen, V.; Krupa, P.; Niskanen, J.; Peltonen, S.; Sonninen, T.M.; Lehtonen, S. Cellular Models of Alpha-Synuclein Aggregation: What Have We Learned and Implications for Future Study. *Biomedicines* **2022**, *10*, 2649. [CrossRef]
6. Stroylova, Y.; Konstantinova, A.; Stroylov, V.; Katrukha, I.; Rozov, F.; Muronetz, V. Does the SARS-CoV-2 Spike Receptor-Binding Domain Hamper the Amyloid Transformation of Alpha-Synuclein after All? *Biomedicines* **2023**, *11*, 498. [CrossRef]
7. Senior, S.L.; Ninkina, N.; Deacon, R.; Bannerman, D.; Buchman, V.L.; Cragg, S.J.; Wade-Martins, R. Increased striatal dopamine release and hyperdopaminergic-like behaviour in mice lacking both alpha-synuclein and gamma-synuclein. *Eur. J. Neurosci.* **2008**, *27*, 947–957. [CrossRef]
8. Al-Wandi, A.; Ninkina, N.; Millership, S.; Williamson, S.J.; Jones, P.A.; Buchman, V.L. Absence of alpha-synuclein affects dopamine metabolism and synaptic markers in the striatum of aging mice. *Neurobiol. Aging* **2010**, *31*, 796–804. [CrossRef] [PubMed]
9. Venda, L.L.; Cragg, S.J.; Buchman, V.L.; Wade-Martins, R. alpha-Synuclein and dopamine at the crossroads of Parkinson's disease. *Trends Neurosci.* **2010**, *33*, 559–568. [CrossRef]
10. Anwar, S.; Peters, O.; Millership, S.; Ninkina, N.; Doig, N.; Connor-Robson, N.; Threlfell, S.; Kooner, G.; Deacon, R.M.; Bannerman, D.M.; et al. Functional alterations to the nigrostriatal system in mice lacking all three members of the synuclein family. *J. Neurosci. Off. J. Soc. Neurosci.* **2011**, *31*, 7264–7274. [CrossRef]
11. Carnazza, K.E.; Komer, L.E.; Xie, Y.X.; Pineda, A.; Briano, J.A.; Gao, V.; Na, Y.; Ramlall, T.; Buchman, V.L.; Eliezer, D.; et al. Synaptic vesicle binding of alpha-synuclein is modulated by beta- and gamma-synucleins. *Cell Rep.* **2022**, *39*, 110675. [CrossRef] [PubMed]
12. Goloborshcheva, V.V.; Kucheryanu, V.G.; Voronina, N.A.; Teterina, E.V.; Ustyugov, A.A.; Morozov, S.G. Synuclein Proteins in MPTP-Induced Death of Substantia Nigra Pars Compacta Dopaminergic Neurons. *Biomedicines* **2022**, *10*, 2278. [CrossRef] [PubMed]
13. Vorobyov, V.; Deev, A.; Chaprov, K.; Ustyugov, A.A.; Lysikova, E. Age-Related Modifications of Electroencephalogram Coherence in Mice Models of Alzheimer's Disease and Amyotrophic Lateral Sclerosis. *Biomedicines* **2023**, *11*, 1151. [CrossRef] [PubMed]
14. Kokhan, V.S.; Chaprov, K.; Ninkina, N.N.; Anokhin, P.K.; Pakhlova, E.P.; Sarycheva, N.Y.; Shamakina, I.Y. Sex-Related Differences in Voluntary Alcohol Intake and mRNA Coding for Synucleins in the Brain of Adult Rats Prenatally Exposed to Alcohol. *Biomedicines* **2022**, *10*, 2163. [CrossRef]
15. Miquel-Rio, L.; Sarries-Serrano, U.; Pavia-Collado, R.; Meana, J.J.; Bortolozzi, A. The Role of alpha-Synuclein in the Regulation of Serotonin System: Physiological and Pathological Features. *Biomedicines* **2023**, *11*, 541. [CrossRef]
16. Lucas-Jimenez, O.; Ibarretxe-Bilbao, N.; Diez, I.; Pena, J.; Tijero, B.; Galdos, M.; Murueta-Goyena, A.; Del Pino, R.; Acera, M.; Gomez-Esteban, J.C.; et al. Brain Degeneration in Synucleinopathies Based on Analysis of Cognition and Other Nonmotor Features: A Multimodal Imaging Study. *Biomedicines* **2023**, *11*, 573. [CrossRef]
17. Ninkina, N.; Peters, O.; Millership, S.; Salem, H.; van der Putten, H.; Buchman, V.L. Gamma-synucleinopathy: Neurodegeneration associated with overexpression of the mouse protein. *Hum. Mol. Genet.* **2009**, *18*, 1779–1794. [CrossRef]
18. Peters, O.M.; Millership, S.; Shelkovnikova, T.A.; Soto, I.; Keeling, L.; Hann, A.; Marsh-Armstrong, N.; Buchman, V.L.; Ninkina, N. Selective pattern of motor system damage in gamma-synuclein transgenic mice mirrors the respective pathology in amyotrophic lateral sclerosis. *Neurobiol. Dis.* **2012**, *48*, 124–131. [CrossRef]
19. Pavia-Collado, R.; Rodriguez-Aller, R.; Alarcon-Aris, D.; Miquel-Rio, L.; Ruiz-Bronchal, E.; Paz, V.; Campa, L.; Galofre, M.; Sgambato, V.; Bortolozzi, A. Up and Down gamma-Synuclein Transcription in Dopamine Neurons Translates into Changes in Dopamine Neurotransmission and Behavioral Performance in Mice. *Int. J. Mol. Sci.* **2022**, *23*, 1807. [CrossRef]
20. Pavlenko, T.A.; Roman, A.Y.; Lytkina, O.A.; Pukaeva, N.E.; Everett, M.W.; Sukhanova, I.S.; Soldatov, V.O.; Davidova, N.G.; Chesnokova, N.B.; Ovchinnikov, R.K.; et al. Gamma-Synuclein Dysfunction Causes Autoantibody Formation in Glaucoma Patients and Dysregulation of Intraocular Pressure in Mice. *Biomedicines* **2022**, *11*, 60. [CrossRef]

21.  Surgucheva, I.; Ninkina, N.; Buchman, V.L.; Grasing, K.; Surguchov, A. Protein aggregation in retinal cells and approaches to cell protection. *Cell. Mol. Neurobiol.* **2005**, *25*, 1051–1066. [CrossRef] [PubMed]
22.  Sriram, S.; Root, K.; Chacko, K.; Patel, A.; Lucke-Wold, B. Surgical Management of Synucleinopathies. *Biomedicines* **2022**, *10*, 2657. [CrossRef] [PubMed]

 *biomedicines*

*Article*

# Brain Degeneration in Synucleinopathies Based on Analysis of Cognition and Other Nonmotor Features: A Multimodal Imaging Study

Olaia Lucas-Jiménez [1,*], Naroa Ibarretxe-Bilbao [1], Ibai Diez [2], Javier Peña [1], Beatriz Tijero [3,4], Marta Galdós [5], Ane Murueta-Goyena [3,6], Rocío Del Pino [3], Marian Acera [3], Juan Carlos Gómez-Esteban [3,4,6], Iñigo Gabilondo [3,4,7,†] and Natalia Ojeda [1,†]

[1]  Department of Psychology, Faculty of Health Sciences, University of Deusto, 48007 Bilbao, Spain
[2]  Gordon Center for Medical Imaging, Department of Radiology, Massachusetts General Hospital, Harvard Medical School, Boston, MA 02114-1107, USA
[3]  Neurodegenerative Diseases Group, Biocruces Bizkaia Health Research Institute, 48903 Barakaldo, Spain
[4]  Department of Neurology, Cruces University Hospital, 48903 Barakaldo, Spain
[5]  Ophthalmology Department, Cruces University Hospital, 48903 Barakaldo, Spain
[6]  Department of Neurosciences, University of the Basque Country (UPV/EHU), 48940 Leioa, Spain
[7]  IKERBASQUE, The Basque Foundation for Science, 48009 Bilbao, Spain
*   Correspondence: olaia.lucas@deusto.es; Tel./Fax: +34-944-139000 (ext. 3231)
†   These authors contributed equally to this work.

**Citation:** Lucas-Jiménez, O.; Ibarretxe-Bilbao, N.; Diez, I.; Peña, J.; Tijero, B.; Galdós, M.; Murueta-Goyena, A.; Del Pino, R.; Acera, M.; Gómez-Esteban, J.C.; et al. Brain Degeneration in Synucleinopathies Based on Analysis of Cognition and Other Nonmotor Features: A Multimodal Imaging Study. *Biomedicines* **2023**, *11*, 573. https://doi.org/10.3390/biomedicines11020573

Academic Editor: Natalia Ninkina

Received: 27 December 2022
Revised: 9 February 2023
Accepted: 10 February 2023
Published: 15 February 2023



**Abstract:** Background: We aimed to characterize subtypes of synucleinopathies using a clustering approach based on cognitive and other nonmotor data and to explore structural and functional magnetic resonance imaging (MRI) brain differences between identified clusters. Methods: Sixty-two patients ($n$ = 6 E46K-SNCA, $n$ = 8 dementia with Lewy bodies (DLB) and $n$ = 48 idiopathic Parkinson's disease (PD)) and 37 normal controls underwent nonmotor evaluation with extensive cognitive assessment. Hierarchical cluster analysis (HCA) was performed on patients' samples based on nonmotor variables. T1, diffusion-weighted, and resting-state functional MRI data were acquired. Whole-brain comparisons were performed. Results: HCA revealed two subtypes, the mild subtype ($n$ = 29) and the severe subtype ($n$ = 33). The mild subtype patients were slightly impaired in some nonmotor domains (fatigue, depression, olfaction, and orthostatic hypotension) with no detectable cognitive impairment; the severe subtype patients (PD patients, all DLB, and the symptomatic E46K-SNCA carriers) were severely impaired in motor and nonmotor domains with marked cognitive, visual and bradykinesia alterations. Multimodal MRI analyses suggested that the severe subtype exhibits widespread brain alterations in both structure and function, whereas the mild subtype shows relatively mild disruptions in occipital brain structure and function. Conclusions: These findings support the potential value of incorporating an extensive nonmotor evaluation to characterize specific clinical patterns and brain degeneration patterns of synucleinopathies.

**Keywords:** synucleinopathies; Lewy bodies diseases; Parkinson's disease; E46K-SNCA; cognition; nonmotor; clustering analysis; multimodal MRI

## 1. Introduction

The interest in the cognition of Parkinson's disease (PD) has grown considerably over the years [1,2]. Only 15% of PD patients remain cognitively intact in the long-term [3]; although 20% of PD patients will fulfill criteria for mild cognitive impairment (MCI) [4], up to 46% of patients with PD and MCI will progress to dementia in 10 years [5,6]. These alterations in cognition also vary depending on whether the PD case is idiopathic or genetic. Apart from cognitive dysfunction, PD is a complex and heterogenic disease in terms of clinical presentation. The heterogeneity of this disease has led to increased interest in patient subtyping based on motor and nonmotor manifestations, and it is only now starting to be understood [7,8]. What it is clear is that good subtyping at baseline study selection is crucial for future clinical trial designs. Data-driven approaches and cross-sectional studies [8–12] have hypothesized that there are different PD subtypes. Few studies on PD subtypes consider a complete assessment of nonmotor symptoms as well as an extensive cognitive evaluation. A recent study [10] found four clusters replicated in two independent cohorts (Tracking Parkinson's and Discovery) of newly diagnosed patients with PD. However, in this recent study [10], cognition was only recorded by using the Montreal Cognitive Assessment (MoCA) adjusted for education years and by using semantic verbal fluency (animals). The classification of MCI in PD (PDMCI) as established by the Movement Disorders Society (MDS) Task Force criteria [4] defines two levels of assessment. Level I is based on a global cognitive scale, whereas Level II is based on a comprehensive assessment that includes two tests per cognitive domain. It is therefore important to ascertain which subtypes of PD exist based on motor and nonmotor symptoms, but one must bear in mind the specific differences in the entire cognitive profile based on Level II of the PDMCI.

Gray matter (GM) and white matter (WM) data obtained from magnetic resonance imaging (MRI) also helped classify PD patients with cluster analysis [13–16]. However, there are very few studies investigating PD subtyping based on whole-brain resting-state functional connectivity (FC). A recent study showed heterogeneous subtypes of PD patients in which depression symptoms had a considerable impact on brain damage affecting FC in patients [17]. In addition, in some PD patients with more aggressive phenotypes, cognitive impairment occurred in early phases of the disease, when, neurobiologically, the cause of cognitive fluctuations is likely to originate from alterations in the functional network rather than from structural alterations [18]. These patients with aggressive phenotypes of PD share clinical and pathological characteristics with two less common diffuse synucleinopathies: PD associated with the E46K mutation of the alpha-synuclein gene (E46K- SNCA) [19,20] and dementia with Lewy bodies (DLB) [21–23]. In 2004, our group described a mutation in the SNCA gene (E46K substitution in SNCA) in a family with autosomal dominant PD and DLB [20]. The mutation produced glutamic acid substitution with lysine in position 46 of the alpha-synuclein gene (E46K-SNCA). Mutation carriers showed extensive Lewy bodies and Lewy neurites in subcortical and cortical structures that met the pathological criteria for DLB, and it induced a Lewy body disease in the brain with an aggressive clinical phenotype, including motor and nonmotor alterations (mood disorders, early cognitive impairment, and visuospatial disorders). In fact, one of the strengths of our work is that we tried to investigate the brain mechanisms of synucleinopathies while differentiating between specific clinical subtypes and while using an excellent genetic model of idiopathic PD. We sought to know whether, in addition to a different clinical profile, the described brain alterations are specific to clinical subtypes or are shared across different subtypes.

Therefore, we aimed to characterize patients using a clustering approach based on cognitive and other nonmotor data, and we involved idiopathic PD patients, E46K-SNCA carriers, and DLB patients. Additionally, we explored whole-brain structural (T1 and diffusion-weighted) and resting-state functional differences between the clusters identified and compared them to normal controls. A good definition of these clusters will be important for understanding the etiology of the disease, for discovering biomarkers related to prognosis, and even for making different interventions that are much more appropriate to the clinical subtype and to the specific cognitive profile.

## 2. Materials and Methods

### 2.1. Participants

Sixty-two patients (6 E46K-SNCA, 8 DLB, 48 idiopathic PD) and thirty-seven normal control patients were included in this study. Participants were recruited at Cruces University Hospital (Department of Neurology) and at the PD Biscay Association (ASPARBI). Patients with idiopathic PD fulfilled the Parkinson's UK Brain Bank criteria for the diagnosis of PD, and patients with DLB fulfilled the diagnosis of probable DLB by revised criteria for the clinical diagnosis of DLB. All patients were evaluated in on-medication states (more information in [24]). For the MRI part of the study, further exclusion criteria included problems with the pre-processing of MRI data or with whole-group analysis. From the initial sample of 62 patients and 37 controls, one patient refused to attend MRI acquisition, four patients were excluded from the T1-weighted structural MRI analysis, and one control was excluded from the resting-state functional MRI analysis. Hence, MRI analyses were carried out on 57 patients and 36 controls. No significant differences were found between the included and the excluded patients.

## 2.2. Demographic and PD-Related Features Assessment

Age, sex, and years of education were registered for all participants. PD-related features were also recorded (see Supplementary Material Table S1).

## 2.3. Nonmotor Assessment
### 2.3.1. Cognitive and Clinical Assessment

Cognition was assessed with MoCA as a test of cognitive screening and with a broad range of standardized neuropsychological tests. Five cognitive domains with the tests recommended by the MDS criteria for diagnosis of PDMCI (Level II) were created [4]: attention and working memory, executive functions, language, memory, and visuospatial functions. Single-domain MCI (SDMCI) was categorized as when abnormalities in two tests within a single cognitive domain were present, and multiple-domain MCI (MDMCI) was categorized as when abnormalities in at least one test in two or more cognitive domains were present. Patients who did not meet these specific criteria were classified as noMCI. In addition, processing speed and theory of mind were measured. Depression, apathy, fatigue symptoms, quality of life, and activities of daily living were also recorded (see Supplementary Material Table S1).

### 2.3.2. Dysautonomia, Olfaction, and Visual Assessment

Orthostatic hypotension (OHT), blood pressure recovery time (PRT), heart rate response (variability) to deep breathing (HRVdb) (more information in [25] and Supplementary Material Table S1), olfaction (BSIT), visual functioning (VFQ-25), binocular low-contrast visual acuity (LCVA), and photopic contrast sensitivity (PCS) were measured (more information in [24] and Supplementary Material Table S1).

## 2.4. Selection of Variables and Clustering Analysis

To simplify the model and reduce the bias in clustering algorithms due to highly correlated variables, we used a random forest feature selection technique. This method allowed us to select the combination of cognitive and other nonmotor variables that best differentiated between patients and controls. These variables were chosen for hierarchical clustering analysis (HCA): attention and working memory, executive functions, language, memory, visuospatial functions, and processing speed (cognition); GDS-15 (depressive symptoms); OHT, PRT, and HRVdb (dysautonomia); BSIT (olfaction); LCVA and PCS (visual). Variables were converted to z scores to conduct the HCA, which was performed including only patients (synucleinopathies). Features related to the disease were not included in the HCA. The HCA was based on a bottom-up approach. A complete linkage criterion was used to minimize the maximum distance between observations of pairs of clusters. Using the silhouette method, we found k = 2 clusters to be the optimal partition. For both clusters, we obtained the average z score of each variable to perform the HCA. Scikit-Learning running under Python version 3.6.5 was employed.

## 2.5. Neuroimaging Preprocessing and Analysis

Structural and functional neuroimaging brain data were acquired using a 3–T MRI scanner (Philips Achieva TX, USA) at OSATEK, Hospital of Galdakao. All sequences were acquired during a single session. The neuroimaging acquisition parameters' descriptions are included in Supplementary Material 1 (S2: Neuroimage acquisition).

### 2.5.1. Structural and Diffusion MRI Preprocessing and Analysis

Voxel-based morphometry (VBM) [26] preprocessing was carried out using the FMRIB Software Library (FSL) tools version 6.0 [27] (for more information, see Supplementary Material 1, S3: Structural, diffusion and resting-state functional MRI preprocessing). Whole-brain GM analyses were performed with a randomized tool (5000 permutations) and with the threshold-free cluster enhancement (TFCE) methodology. The statistical threshold for analysis was set at $p < 0.05$ and was corrected for multiple comparisons by using the family-wise error (FWE) rate and by including sex, age, years of education, and total intracranial volume (TIV) as covariates.

The FSL [27] version 6.0 and Brain Extraction Tool (BET) [28] was used for the preprocessing and analysis of diffusion data (for more information, see Supplementary Material 1, S3: Structural, diffusion and resting-state functional MRI preprocessing). Whole-brain WM analyses were performed with a randomized tool (5000 permutations) and with TFCE methodology. The statistical threshold for analysis was set at $p < 0.05$ and was corrected for multiple comparisons by using the FWE rate and by including sex, age, and years of education as covariates.

### 2.5.2. Resting-State Functional MRI Preprocessing and Analysis

Resting-state functional MRI (rs-fMRI) data were preprocessed (band-pass filtering was performed with a frequency window of 0.008 to 0.09 Hz [29]) and analyzed using Conn Functional Connectivity Toolbox version 20.0 [30] (for more information, see Supplementary Material 1 S3: Structural, diffusion and resting-state functional MRI preprocessing). FC differences were assessed with the region of interest (ROI)-to-ROI methodology, and the statistical threshold was set at $p < 0.05$ and was corrected corrected for multiple comparisons by using false discovery rate (FDR) and by including sex, age, and years of education as covariates. In addition, LEDD data were also included in the FC analysis as covariates [31]. The ROIs selected for FC analysis were based on the FC atlas networks of the CONN toolbox: Default Mode Network, Sensorimotor, Visual, Salience/Cingulo-Opercular, Dorsal Attention, Fronto Parietal/Central Executive, Language, and Cerebellar. For specific network information, see CONN network cortical ROIs HCP-ICA [30].

### 2.6. Data Analysis

The normality of the data was tested using the Shapiro–Wilk test. Significant differences in demographic, cognitive, clinical, and other nonmotor variables were assessed with analysis of variance (ANOVA) and LSD post-hoc tests. PD-related features' differences between identified clusters were assessed with a two-tailed $t$-test. Categorical data were analyzed with a chi-squared ($\chi2$) test. Statistical analyses were performed using the statistical package SPSS program (IBM SPSS Statistics v 27.0).

## 3. Results

### 3.1. Subtypes of Synucleinopathies Based on Cluster Analysis

We identified two cluster subtypes in synucleinopathies (see Figure 1). Twenty-six idiopathic PD patients and the three less affected and younger E46K-SNCA carriers comprised the mild subtype ($n = 29$), whereas all DLB patients ($n = 8$), 22 idiopathic PD patients and the three more affected E46K-SNCA carriers comprised the severe subtype ($n = 33$).

The mild subtype patients showed no significant differences in demographic variables when compared to normal controls. Mild subtype patients scored significantly higher in orthostatic hypotension, fatigue, and depression, and they had lower scores in visual acuity and olfaction than controls (see Figure 2). Additionally, mild subtype patients were younger, had more education years, were younger at disease onset, and scored significantly higher in all cognitive domains, dysautonomia, and PD-related features as well as in ADL compared to the severe subtype. The severe subtype patients showed significantly higher ages and fewer years of education compared to controls and to mild subtype. Additionally, severe subtype patients showed more nonmotor alterations compared to controls and had more severe motor symptoms (UPDRS III) compared to mild subtype patients. Specifically, more marked cognitive, visual and bradykinesia alterations were found. Severe subtype patients had the greatest average age of disease onset and the oldest ages, but they did not differ significantly in disease duration compared to the mild subtype. The demographic, cognition, PD-related features, dysautonomia, visual, and clinical differences between all patients (mild subtype and severe subtype) and controls are shown in Table 1. Specifically, in demographic variables, controls showed significant differences from the severe subtype patients in age and years of education; therefore, these variables were used as covariates in neuroimaging analysis.

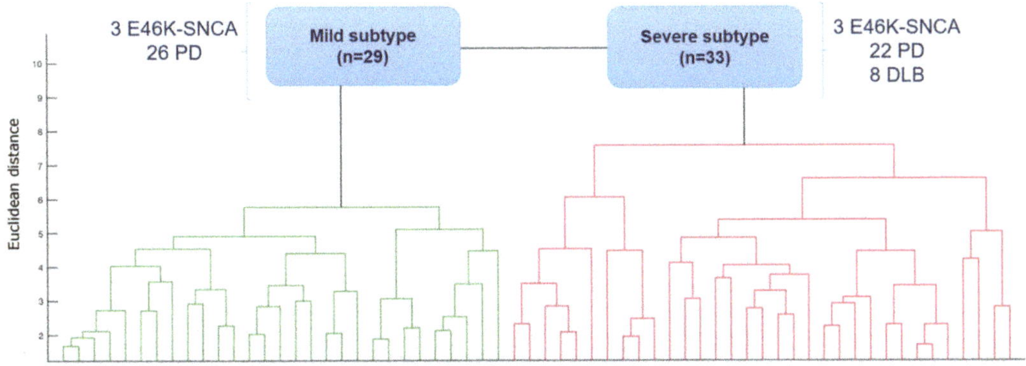


**Figure 1.** Dendrogram of patients clustered according to nonmotor data. Abbreviations: E46K-SNCA = E46K mutation of the alpha-synuclein gene; PD = Parkinson's disease; DLB = Dementia with Lewy bodies.

(a) Differences between subtypes (mild or severe) compared to controls

| | Mild subtype vs controls | Severe subtype vs controls |
|---|---|---|
| Age | <60 years | >60 years |
| Education (years) | <10 years | >10 years |
| Sex (m/f) | 18/11 | 22/11 |
| Global cognition (MoCA) | | |
| Apathy | | |
| Fatigue | | |
| Depression | | |
| Olfaction | | |
| Activities of daily living | | |
| Orthostatic hypotension | | |
| Blood pressure recovery | | |
| Heart rate response (variability) | | |
| Binocular low contrast visual | | |
| Photopic contrast sensitivity | | |
| Visual Function Questionnaire | | |

(b) PD-related features differences among subtypes

| | Mild subtype | Severe subtype |
|---|---|---|
| Age at disease onset (years) | <50 years | >50 years |
| Disease duration (years) | <5 years | >5 years |
| LEDD | | |
| UPDRS III L or R | | |
| Bradykinesia L or R | | |
| Tremor L or R | | |
| Rigidity L or R | | |
| PDQ-39 Mobility | | |
| PDQ-39 AVDL | | |
| PDQ-39 Emotional well-being | | |
| PDQ-39 Stigma | | |
| PDQ-39 Social support | | |
| PDQ-39 Cognition | | |
| PDQ-39 Communication | | |
| PDQ-39 Bodily discomfort | | |

**Figure 2.** Demographic, nonmotor, and PD-related features of subtypes. (a) Graphical representation of demographic and nonmotor differences between the two identified clusters compared to controls. Red color indicates significantly ($p < 0.05$) lower scores compared to controls; green color indicates no

significant ($p > 0.05$) differences compared to controls. (**b**) Graphical representation of PD-related features' differences between clusters identified (mild subtype vs. severe subtype). Red color indicates significantly ($p < 0.05$) lower scores from the severe subtype compared to the mild subtype; green color indicates significantly ($p < 0.05$) higher scores from the mild subtype compared to the severe subtype; blue color indicates no significant ($p > 0.05$) differences between clusters. Abbreviations: m = male; f = female; MoCA = Montreal Cognitive Assessment; LEDD = Levodopa Equivalent Daily Dose; UPDRS = Unified Parkinson's Disease Rating Scale; L = left; R = right; PDQ-39 = Parkinson's Disease Questionnaire 39 items; AVDL = Activities of Daily Living.

**Table 1.** Demographic, PD-related, and nonmotor characteristics of the sample.

| Variables | Synucleinopathies (*n*:62) | | | Comparisons | | |
| | Mild (*n*:29) | Severe (*n*:33) | Controls (*n*:37) | Among Subtypes (*p* Value) | Subtypes vs. Controls (*p* Value) | |
| | Mean (SD) | Mean (SD) | Mean (SD) | | Mild vs. Controls | Severe vs. Controls |
| **Demographics** | | | | | | |
| Age, years | 53.60 (8.78) | 67.55 (6.73) | 53.48 (12.74) | **0.000** | 0.960 | **0.000** |
| Sex (m/f) | 18/11 | 22/11 | 21/16 | 0.793 | 0.802 | 0.465 |
| Education, years | 13.24 (3.45) | 8.61 (4.03) | 13.84 (5.05) | **0.000** | 0.577 | **0.000** |
| **Cognition** | | | | | | |
| MoCA | 26.51 (1.76) | 20.54 (5.02) | 27.24 (3.14) | **0.000** | 0.419 | **0.000** |
| Attention and WM | 0.44 (0.52) | −0.81 (0.56) | 0.36 (0.78) | **0.000** | 0.601 | **0.000** |
| Memory | 0.37 (0.62) | −0.96 (0.64) | 0.53 (0.58) | **0.000** | 0.302 | **0.000** |
| Executive functions | 0.39 (0.56) | −0.92 (0.87) | 0.48 (0.44) | **0.000** | 0.375 | **0.000** |
| Language | 0.41 (0.49) | −0.94 (0.48) | 0.46 (0.82) | **0.000** | 0.161 | **0.000** |
| Visuospatial abilities | 0.29 (0.55) | −0.76 (0.87) | 0.47 (0.55) | **0.000** | 0.439 | **0.000** |
| Processing speed | 0.40 (0.51) | −0.99 (0.50) | 0.53 (0.80) | **0.000** | 0.423 | **0.000** |
| Theory of mind | 0.37 (0.74) | −0.87 (0.96) | 0.44 (0.70) | **0.000** | 0.720 | **0.000** |
| **PD-related features** | | | | | | |
| Disease duration, years | 5.80 (3.52) | 7.85 (4.86) | - | 0.078 | - | - |
| Age of onset, years | 48.34 (7.29) | 59.29 (8.05) | - | **0.000** | - | - |
| UPDRS III R no midline | 7.14 (5.27) | 11.28 (3.89) | - | **0.001** | - | - |
| UPDRS III L no midline | 8.61 (4.49) | 12.21 (5.11) | - | **0.007** | - | - |
| Bradykinesia R | 4.30 (2.83) | 7.07 (1.98) | - | **0.000** | - | - |
| Bradykinesia L | 5.52 (2.46) | 7.66 (2.72) | - | **0.003** | - | - |
| Rigidity | 2.11 (1.55) | 2.93 (1.71) | - | 0.066 | - | - |
| Rigidity L | 2.44 (1.58) | 3.14 (1.66) | - | 0.116 | - | - |
| Tremor R | 0.89 (2.23) | 1.45 (1.76) | - | 0.300 | - | - |
| Tremor L | 0.93 (1.47) | 1.59 (1.97) | - | 0.159 | - | - |
| LEDD | 618.76 (369.41) | 736.22 (454.39) | - | 0.295 | - | - |
| PDQ-39 Mobility | 9.40 (8.88) | 11.79 (9.08) | - | 0.325 | - | - |
| PDQ-39 AVDL | 4.51 (5.42) | 8.10 (6.93) | - | **0.035** | - | - |
| PDQ-39 EW | 7.11 (5.43) | 6.31 (6.89) | - | 0.633 | - | - |
| PDQ-39 Stigma | 2.59 (2.91) | 2.20 (4.91) | - | 0.725 | - | - |
| PDQ-39 SS | 1.25 (2.56) | 1.59 (2.18) | - | 0.609 | - | - |
| PDQ-39 Cognition | 3.25 (2.83) | 5.38 (4.73) | - | **0.046** | - | - |
| PDQ-39 Com | 1.22 (1.88) | 3.20 (3.57) | - | **0.012** | - | - |
| PDQ-39 BD | 4 (3.44) | 5.41 (3.81) | - | 0.152 | - | - |
| **Dysautonomia** | | | | | | |
| OHT | 0.66 (0.87) | 0.58 (0.84) | 0.12 (0.33) | 0.663 | **0.012** | **0.028** |
| Valsalva PRT | 3.69 (2.05) | 7.61 (4.28) | 2.64 (1.74) | **0.000** | 0.177 | **0.000** |
| HRVdb | 1.01 (0.09) | 0.90 (0.05) | 1.03 (0.09) | **0.000** | 0.334 | **0.000** |

**Table 1.** *Cont.*

| Variables | Synucleinopathies (*n*:62) | | Controls (*n*:37) | Comparisons | | |
| | Mild (*n*:29) | Severe (*n*:33) | | Among Subtypes (*p* Value) | Subtypes vs. Controls (*p* Value) | |
| | Mean (SD) | Mean (SD) | Mean (SD) | | Mild vs. Controls | Severe vs. Controls |
| **Vision** | | | | | | |
| Binocular LCVA | 32.93 (6.11) | 18.19 (13.21) | 37.10 (6.38) | **0.000** | **0.031** | **0.000** |
| Photopic CS | 2.0 (0.11) | 1.85 (0.14) | 2.05 (0.13) | **0.000** | 0.086 | **0.000** |
| VFQ-25 | 92.22 (11.99) | 84.32 (15.01) | 96.08 (4.66) | **0.008** | 0.187 | **0.000** |
| **Clinical** | | | | | | |
| Apathy | 27.89 (5.13) | 22.62 (8.61) | 29.5 (4.04) | 0.057 | 0.551 | **0.004** |
| Fatigue | 29.76 (17.25) | 34.50 (16.57) | 20.81 (10.96) | 0.218 | **0.017** | **0.000** |
| Depression | 2.59 (2.33) | 3.75 (3.52) | 1.11 (1.62) | 0.082 | **0.023** | **0.000** |
| Olfaction | 7.68 (2.30) | 5.48 (2.61) | 10.48 (1.21) | **0.000** | **0.000** | **0.000** |
| AVDL | 7.83 (0.75) | 6.19 (2.28) | 8 (0) | **0.000** | 0.612 | **0.000** |

Abbreviations: SD = standard deviation; m = male; f = female; MoCA = Montreal Cognitive Assessment; WM = working memory; UPDRS = Unified Parkinson's Disease Rating Scale; III = motor part; R = right; L = left; LEDD = Levodopa Equivalent Daily Dose; PDQ-39 = Parkinson's Disease Questionnaire 39 items; AVDL = Activities of Daily Living; EW = emotional wellbeing; SS = social support; Com = communication; BD = bodily discomfort; OHT = orthostatic hypotension; PRT = blood pressure recovery time; HRVdb = heart rate variability to deep breathing; LCVA = low-contrast visual acuity; CS = contrast sensitivity; VFQ-25 = Visual Functionary Questionnaire 25 items.

### 3.2. Cognitive Profile of Subtypes According to PDMCI Level II Criteria

Mild subtype patients scored lower in executive functions, memory, visuospatial abilities, processing speed, and theory of mind, and they scored higher in language compared to controls and higher in all cognitive domains compared to severe subtype patients (see Figure 3a). Moreover, mean differences between mild subtype patients and controls were higher in the memory, executive functions, and theory of mind domains (see Figure 3b). Severe subtype patients scored lower than mild subtype patients and controls in all domains. Memory, language, processing speed, and theory of mind were the domains with the lowest means based on z scores (see Figure 3a) and based on the mean differences between groups (see Figure 3b).

The percentage of MCI in each subtype showed two patterns. The severe subtype comprised 94% of patients with MDMCI (*n* = 31), 3% with noMCI (*n* = 1), and 3% of patients with SDMCI (*n* = 1, memory domain), whereas in the mild subtype, 86% of patients were categorized as noMCI (*n* = 25), 10% as MDMCI (*n* = 3), and 4% of patients as SDMCI (*n* = 1, executive functions domain).

### 3.3. Structural and Functional Brain Degeneration in Synucleinopathies Based on Clusters

Structural and functional brain results in patients based on clusters are shown in Tables A1–A3 in the Appendix A and in Figure 4. Mild subtype patients exhibited marked left occipital and precuneus GM alterations and lower FCs between visual occipital and language networks (left inferior frontal gyrus) compared to controls, but they exhibited no significant WM alterations compared to controls. Severe subtype patients showed bilateral frontotemporal (including left hippocampus) and occipital GM alterations compared to controls as well as lower FA WMs in bilateral anterior thalamic radiation and right longitudinal fasciculus and higher MD WMs in bilateral anterior thalamic radiation, left cingulum, right longitudinal, and body of corpus callosum compared to controls. In addition, visual, dorsal-attentional, salience, language, and sensorimotor FC alterations compared to controls were found. When both subtypes were compared, similar GM patterns were found when we compared the mild subtype patients with normal controls. Severe subtype patients exhibited GM alterations in the left hippocampus, right fronto-orbital, left occipital, and left precuneus compared to those with the mild subtype. WM FA differences were also found between subtypes; specifically, the severe subtype presented lower FA WM in right anterior thalamic radiation, left cingulum, right longitudinal, and body of corpus callosum compared to the mild subtype. Finally, marked lower FC in the posterior-lateral DMN was found in severe subtype patients compared to those with the mild subtype.

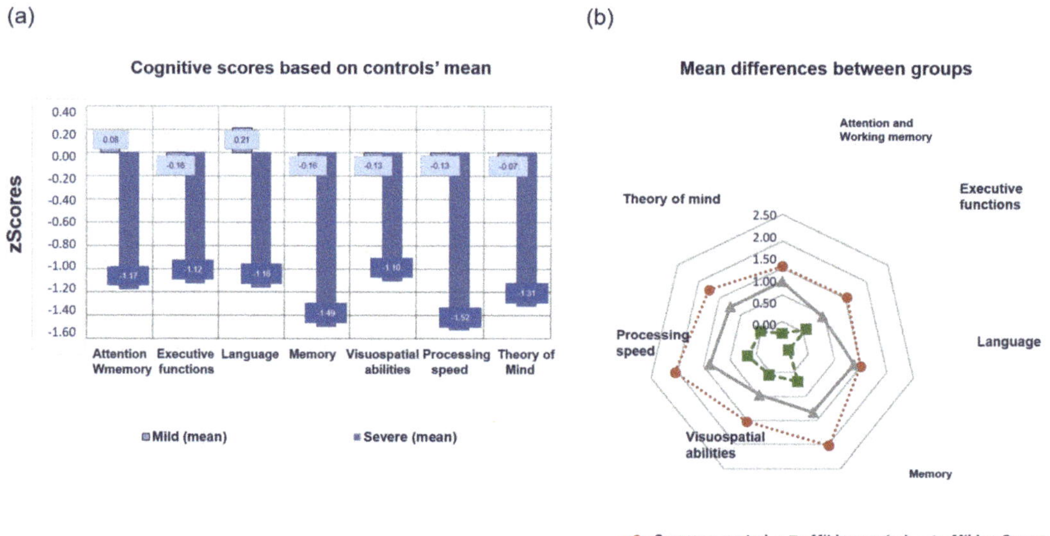

**Figure 3.** Cognitive profile of subtypes. (**a**) Means of cognitive domains of patients compared to controls (z = 0 is the baseline according to the mean of controls). Data are presented as z scores, and in all cases, lower z scores indicate worse performance. (**b**) Mean differences of ANOVA post-hoc analysis of the sample between subtypes and controls. Data are presented as z scores, and in all cases, lower z scores indicate less prominent differences between groups.

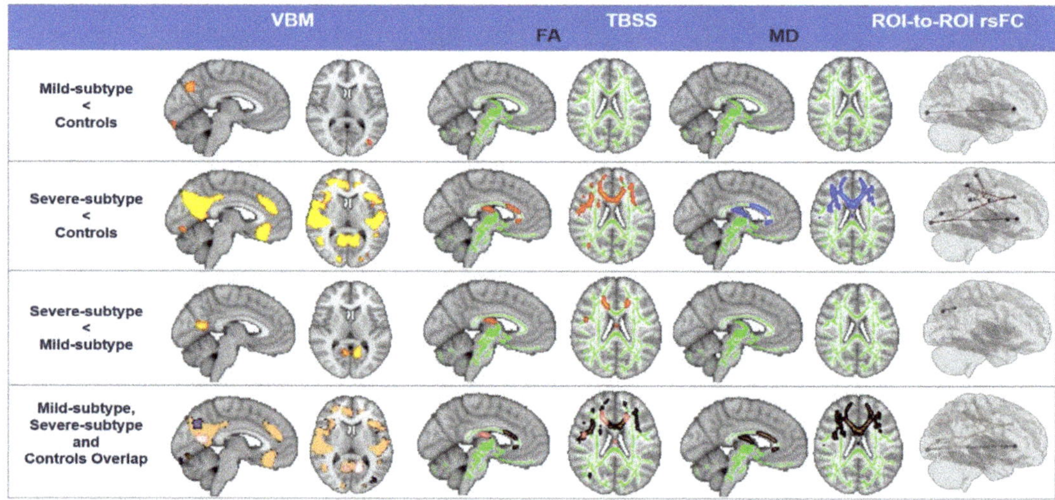

**Figure 4.** Structural and functional brain degeneration in synucleinopathies compared to controls. Voxel-based morphometry (VBM), tract-based spatial statistics (TBSS), and region of interest (ROI)-

to-ROI functional connectivity (FC) analyses of the sample. VBM analysis: regions with less gray matter volume are shown in red–yellow shades ($p < 0.05$, FWE-corrected). Results were adjusted by age, sex, years of education, and TIV. TBSS analysis: regions with less FA are shown in red–yellow shades, and those with higher MD are shown in blue to light blue shades. FA skeleton mask (green) ($p < 0.05$, FWE-corrected). Results were adjusted by age, sex, and years of education. FC analysis: regions with less FC are shown in red ($p < 0.05$, FDR-corrected). Results were adjusted by age, sex, years of education, TIV, and Levodopa Equivalent Daily Dose (LEDD). Abbreviations: FA = fractional anisotropy; MD = medial diffusivity.

## 4. Discussion

To the best of our knowledge, this is the first study that investigated the heterogeneity of synucleinopathies, classified them using clustering analysis, and included an extensive neuropsychological battery of tests and a broad spectrum of other nonmotor features. Additionally, we explored the specific cognitive profiles of patients based on PDMCI Level II criteria, and we investigated with multimodal MRI the brain's structural and functional alterations behind these subtypes. The main findings were as follows: (a) clustering analysis using nonmotor features defined two subtypes with very different clinical profiles; (b) the severe subtype was characterized by motor and nonmotor alterations with marked cognitive impairment, whereas the mild subtype only presented nonmotor alterations in visual acuity, olfaction, dysautonomia, fatigue, and depression; (c) the severe subtype showed MDMCI with memory, processing speed, language, and theory of mind as the most affected domains; (d) older E46K-SNCA carriers and DLB patients showed MDMCI with similar cognitive profiles and marked visuospatial alterations, whereas PD patients showed heterogeneity with the PD severe subtype showing MDMCI and amnestic SDMCI, and the PD mild subtype showing noMCI and executive SMDMCI; (e) the severe subtype revealed widespread disruptions in function and structure in the fronto-temporal and occipital areas, whereas the mild subtype showed relatively mild brain abnormalities that were mainly in occipital areas.

### 4.1. Distinct Clinically Relevant Patterns in Synucleinopathies

In this study, we considered several nonmotor features together, and we revealed two clinically relevant subtypes that were most associated with specific nonmotor symptoms. The main results of this study are in line with previous studies in the literature. According to Van Rooden [8], the majority of studies reported two clusters with very similar profiles in terms of age-at-onset and rates of disease progression. In fact, a recent study [32] also showed two clusters of PD patients, which were mild-motor predominant and severe cluster (diffuse malignant); apart from motor and nonmotor symptoms [33], MRI studies also showed two subgroups of PD patients [13] with differences in cortical atrophy. However, other studies showed three or four different clusters [14,34]. In the study by Mu and colleagues [34], four clusters were identified: mild, nonmotor-dominant, motor-dominant, and severe. In another structural MRI study [14], three subtypes of PD were found with prominent differences in GM patterns and little WM involvement. Finally, regarding rs-fMRI, one study evaluated the FC of networks in PD, and, as in our study, two significant patterns were found: "motor-related pattern" and "depression-related pattern".

It is widely known that nonmotor symptoms of PD may have a greater impact on quality of life than motor features, and they may even precede overt signs and symptoms of motor disturbances [35,36]. In our study, severe subtype patients exhibited a clear deterioration of their cognitive and clinical profiles with alterations in motor features, the more marked symptom being bradykinesia alterations. Interestingly, we observed in the mild subtype mild motor alterations but significant disturbances in olfaction (hyposmia), dysautonomia (orthostatic hypotension), visual acuity, fatigue, and depression. In addition, these results were found when patients did not have significantly different disease durations, although the mean age and age at onset were lower in the mild subtype. Therefore, and according to a similar study [14], these findings reinforce the idea that later onset of the disease is associated with rapid disease progression [37]. Furthermore, recent reviews in animal models [38] highlight the importance of the presence of Lewy bodies (made predominantly of alpha-synuclein) in relation to motor and nonmotor symptomatology, such as olfactory dysfunction, anxiety, depression, and cognitive dysfunction.

One hypothesis about the neuropathological stages of PD suggests that Lewy body pathology in the nigrostriatal system only develops after lower brainstem areas and the olfactory system are affected [39,40]. In our study, one of the prominent alterations of the mild subtype was related to olfaction. Therefore, this could imply that hyposmia is present from the start of the disease. Recent studies suggest that hyposmia may be an early preclinical sign, is related to an increased risk to develop overt PD in asymptomatic first-degree relatives of PD patients, and is a risk marker for the general population [41,42]. Orthostatic hypotension is a frequent nonmotor symptom in alpha-synucleinopathies including PD and DLB, and it occurs in 20–50% of PD cases and 30–70% of DLB cases [43,44]. This alteration was also found in mild subtype patients. In fact, a previous study of our group showed that dysautonomia was also related to neuropsychological performance and to depression and apathy symptoms in Lewy body diseases [25]. LCVA (low-contrast visual acuity) was another alteration in the mild subtype. Visual disturbances in general are relatively common in PD, with some studies reporting that up to 78% of patients are affected, including reduced contrast sensitivity, impaired color discrimination, convergence insufficiency, and dry eye syndrome [45]. In a previous study, we identified that primary visual function was significantly worse in patients than in controls, mainly expressed as LCVA, which was severely impaired in the E46K–SNCA and DLB patients and moderately impaired in idiopathic PD patients [46]. Depression could appear in the prodromal phase of PD and was shown to nearly double an individual's risk of subsequently developing PD [47]. In addition, fatigue often appears in the early motor stage, and it is identified by persons with PD as one of their most disabling symptoms with the greatest impact on their quality of life [47,48]. Fatigue can be present in 50% of people with PD. In some patients, fatigue is also related to depression and autonomic symptoms; however, whereas depression and its brain correlates are highly investigated in PD, very little is known about the mechanisms of fatigue in PD. Finally, these alterations (hyposmia, orthostatic hypotension, visual deficits, depression, and fatigue) are usually related to future cognitive decline in PD. Thus, the five nonmotor alterations identified in the mild subtype of this study mark the increasingly important role of nonmotor symptoms in synucleinopathies' heterogeneity and the importance to take steps toward early identification and subtype-specific treatment packages.

### 4.2. Distinct Cognitive Profiles in Synucleinopathies

The severe subtype was 94% composed of patients with MDMCI, whereas in the mild subtype, 86% of patients were categorized as noMCI, 10% as MDMCI, and 4% of patients as SDMCI (executive functions domain). Interestingly, mild cognitive impairment is also a common in early PD, and the main feature is usually impairment in executive functions [47]. One study showed that the worst cognitive domains among converters to PD dementia compared with nonconverters to PD dementia were the executive function/working memory domains [49]. In addition, the severe subtype group, apart from MDMCI, also showed more bradykinetic symptoms; patients with those cognitive alterations plus patients with the bradykinetic-rigid form of PD are more at risk of developing dementia [47].

Additionally, the results showed that young E46K-SNCA carriers' cognitive profiles were very similar to the controls' cognitive profiles, demonstrating noMCI (100% of noMCI) in the sample. However, all aggressive E46K-SNCA carriers exhibited MDMCI, as all DLB patients did, with more marked visuospatial alterations and even lower scores in all domains evaluated compared to the DLB group. This concept is related to previous studies in the literature suggesting that early visual-cognitive dysfunction is one of the main predictors for the development of cognitive disability in PD [50].

Apart from PD-MCI Level II criteria domains, other cognitive domains, such as processing speed or theory of mind, are frequently altered in PD. When we compared the cognitive profiles of the patients, the mild subtype group scored lower in executive functions, memory, visuospatial abilities, processing speed, and theory of mind, and they scored higher in language compared to controls. One study revealed that the unique domain that did not worsen among converters compared with non-converters PD dementia was the language domain [49]. However, other studies showed that semantic fluency was shown to be a predictor of dementia in PD [50]. The severe subtype group scored lower than the mild subtype group and controls in all domains with the lowest mean scores, those being memory, language, processing speed, and theory of mind. Processing speed is a cognitive domain that is not included in the Level II of PD-MCI, but processing speed disturbances are widely present in PD, and they usually affect daily living activities [51]. In fact, cognitive alterations in processing speed are correlated with fatigue in PD [52] and clinical symptoms which usually appear in the early phases of the disease. In the case of social cognition and theory of mind, one recent study [53] suggested that deficits in the DMN may be contributing to theory of mind deficits in amnestic MCI, highlighting the importance of including measures of social cognition in the clinical routine to detect MCI. Moreover, neurodegenerative disorders frequently present with social cognitive, memory, and executive impairments, offering an opportunity to explore the intersection between theory of mind and cognitive functions. Interestingly, when executive function performance is controlled for, specific WM alterations were found, implying a domain-specific theory of mind impairment in PD [54].

*4.3. Brain Degeneration in Synucleinopathies Based on Neuroimaging: Towards a Pathophysiological Explanation*

Severe subtype patients showed bilateral frontotemporal and occipital GM and WM alterations compared to controls as well as visual, dorsal attentional, salience, language, and sensorimotor FC alterations. However, the mild subtype only showed marked left occipital and precuneus GM alterations, no significant WM alterations, and lower FC between visual occipital and language networks (left inferior frontal gyrus) compared to normal controls. When both subtypes were compared, the severe subtype group exhibited GM alterations in the left hippocampus, right fronto-orbital, left occipital, and left precuneus compared to the mild subtype group as well as lower FA WM in the right anterior thalamic radiation, left cingulum, right longitudinal fasciculus, and body of corpus callosum compared to the mild subtype group. Finally, markedly lower FC in the posterior-lateral DMN was found in severe subtype patients compared to mild subtype patients.

It is also worth noting that even though both WM and GM contributed to explain the different subtypes, GM degeneration patterns were more relevant in the characterization of PD groups than WM alterations. The majority of whole-brain studies evidenced the involvement of fronto-occipital WM tracts [14], such as the corpus callosum, cingulum, and other major association tracts in PD patients with MCI [16,55,56]. Indeed, in this study, WM FA differences were found between subtypes specifically in the cingulum, corpus callosum, and longitudinal fasciculus, suggesting that WM impairment in PD might be a sign preceding the neuronal loss in associated GM areas.

What it is clear is that there are extensive atrophic changes as well as FC changes in bilateral fronto-temporo-occipital regions in the severe subtype, whereas marked fronto-occipital alterations appear in the mild subtype. This dissociation in the severe subtype could be related not only to cognitive impairment but also to other nonmotor symptoms. The frontal and medial occipital lobes are speculated to be associated with nonmotor symptoms [57], and in previous studies in the literature, prefrontal disturbances were also associated with depression [17] and hyposmia [58] in PD. In addition, mild subtype patients showed depressive symptoms. A recent study indicated that the presence of depressive symptoms in patients with early PD is associated with a higher risk of progression to MCI, and that early depression may reflect subsequent cortical atrophy [59]. Moreover, visual disturbances were also found in the mild subtype; given that there are important nodes associated with visual information processing in occipital region, this suggests that abnormal FC in occipital lobe underpins both primary visual perceptual and intrinsic visual integration. In fact, visual disturbances are associated with more pronounced cognitive deterioration in PD [24]. Therefore, incorporating resting-state FC measures other than structure into these studies involves moving one step closer to multimodal MRI approaches.

However, some limitations of this study should be noted. First, our data-driven cluster analysis needs to be validated in independent cohorts. Second, longitudinal studies on large cohorts of patients will be crucial to confirm our results and to accurately follow brain modifications from synucleinopathies' progression. Third, although the inclusion of synucleinopathies and, specifically, E46K mutation is one of the highlights of the present work, the sample size was small, and future studies with larger samples are needed to deeply explore the structural and functional brain differences across stages of synucleinopathies.

## 5. Conclusions

The current study opens new perspectives by being the first to assess brain network degeneration based on cognition and other nonmotor features in synucleinopathies by using a multimodal neuroimaging approach. Moreover, it is important to remark that SNCA-linked mutations are limited to specific families and series worldwide, and that their study is a unique opportunity to improve our understanding of different synucleinopathies' phenotypes. These results shed light on different phenotypes in synucleinopathies, which not only differ in cognitive performance but also in other nonmotor symptoms (visual acuity, olfaction, dysautonomia, fatigue, and depression) and brain degeneration. Thus, the hypothesis of distinct disease courses and treatments is supported.

**Supplementary Materials:** The following supporting information can be downloaded at: https://www.mdpi.com/article/10.3390/biomedicines11020573/s1, Table S1. PD-related features and non-motor assessment; S2. Neuroimaging acquisition; S3. Structural, diffusion and resting-state functional MRI preprocessing.

**Author Contributions:** Conceptualization, O.L.-J. and J.C.G.-E.; methodology, O.L.-J., I.D. and I.G.; software, O.L.-J. and I.D.; validation, O.L.-J., N.I.-B., J.P., I.G. and N.O.; formal analysis, O.L.-J., I.D. and I.G.; data curation, O.L.-J., R.D.P., M.A., M.G., A.M.-G., B.T., J.C.G.-E. and I.G.; writing—original draft preparation, O.L.-J.; writing—review and editing, N.I.-B., I.D., J.P., B.T., M.G., A.M.-G., R.D.P., M.A., J.C.G.-E., I.G. and N.O.; supervision, N.I.-B., J.P., I.G. and N.O.; project administration, J.C.G.-E. and I.G.; funding acquisition, J.C.G.-E. and I.G. All authors have read and agreed to the published version of the manuscript.

**Funding:** This study was co-funded by the Michael J. Fox Foundation [RRIA 2014 (Rapid Response Innovation Awards) Program (Grant ID: 10189)] and Instituto de Salud Carlos III through the project "PI14/00679" and Juan Rodes grant "JR15/00008" (IG) (Co-funded by European Regional Development Fund/European Social Fund-"Investing in your future").

**Institutional Review Board Statement:** The study was conducted in accordance with the Declaration of Helsinki and approved by the Ethics Committee at the Helath Department of the Basque Mental Helath in Spain (CEIC-E PI2014154), attached to the Health Research and Innovation Directorate of the Basque Government Department of Health.

**Informed Consent Statement:** All participants gave written informed consent prior to their participation in the study in accordance with the tenets of the Declaration of Helsinki.

**Data Availability Statement:** The data supporting the findings of this study are available from the corresponding author upon reasonable request. They are not publicly available due to ethical restrictions.

**Acknowledgments:** We would like to thank ASPARBI and all of the participants in the study.

**Conflicts of Interest:** The authors declare no conflict of interest.

## Appendix A

**Table A1.** Structural gray matter differences.

| Regions | Cluster Size (Voxels) | Stats | *p* Value (FWE-Corrected) | MNI Coordinates X Y Z |
|---|---|---|---|---|
| Mild < Controls | | | | |
| Inferior Temporooccipital Left | 522 | 3.98 | 0.020 | −50 −54 −10 |
| Precuneus Left | 318 | 4.29 | 0.020 | −2 −70 −34 |
| Lateral Occipital Left | 223 | 3.96 | 0.031 | −36 −82 −4 |
| Occipital Pole Right | 105 | 4.21 | 0.033 | 2 −90 −24 |
| Severe < Controls | | | | |
| Superior Temporal Right | 3972 | 4.47 | <0.001 | 52 −22 0 |
| Superior Temporal Left | 2408 | 4.56 | 0.005 | 54 −4 −12 |
| Superior Lateral Occipital Right | 2151 | 5.10 | 0.007 | 14 −84 24 |
| Temporal Fusiform Right | 1769 | 4.54 | 0.011 | 40 −34 −16 |
| Hippocampus Left | 430 | 4.58 | 0.022 | −28 −32 −14 |
| Occipital Pole Left | 313 | 3.84 | 0.020 | −22 −90 6 |
| Occipital Pole Right | 185 | 3.86 | 0.025 | 18 −88 4 |
| Frontal Orbital Left | 174 | 4.31 | 0.036 | −32 30 −2 |
| Frontal Medial Left | 147 | 4.73 | 0.038 | −2 30 −20 |
| Temporal Pole Left | 113 | 3.91 | 0.037 | −22 6 −24 |
| Frontal Orbital Right | 106 | 4.76 | 0.032 | 32 28 −6 |
| Severe < Mild | | | | |
| Precuneus Left | 1036 | 4.91 | 0.002 | −12 −64 8 |
| Frontal Orbital Right | 82 | 4.34 | 0.035 | 36 30 4 |
| Occipital Left | 32 | 3.69 | 0.044 | −18 −84 −14 |
| Hippocampus Left | 16 | 4.45 | 0.044 | −36 −32 −6 |

Abbreviations: MNI = Montreal Neurological Institute.

**Table A2.** White matter fractional anisotropy and mean diffusivity differences.

| Regions | Cluster Size (Voxels) | Stats | *p* Value (FWE-Corrected) | MNI Coordinates X Y Z |
|---|---|---|---|---|
| Severe < Controls | | | | |
| FA | | | | |
| Anterior Thalamic Radiation Right | 13,377 | 3.51 | <0.001 | 19 42 −4 |
| Anterior Thalamic Radiation Left | 366 | 5.51 | 0.007 | −7 −24 14 |
| Inferior Longitudinal Right | 65 | 5.41 | 0.009 | 47 −21 1 |
| Superior Longitudinal Right | 56 | 4.39 | 0.010 | 33 4 40 |
| Severe > Controls | | | | |
| MD | | | | |
| Cingulum Left | 8361 | 3.52 | 0.003 | −16 32 16 |
| Anterior Thalamic Radiation Right | 764 | 4.24 | 0.009 | 21 12 10 |
| Superior Longitudinal Right | 719 | 2.31 | 0.010 | 35 2 21 |
| Anterior Thalamic Radiation Left | 657 | 6.24 | 0.005 | −11 −31 12 |
| Body of Corpus Callosum | 545 | 3.57 | 0.009 | 14 −29 28 |
| Cingulum Left | 239 | 4.71 | 0.009 | −9 25 −5 |
| Inferior Fronto-Occipital Right | 127 | 3.43 | 0.010 | 22 51 −9 |
| Longitudinal Right | 103 | 2.29 | 0.010 | 25 −47 25 |
| Severe < Mild | | | | |
| FA | | | | |
| Body of Corpus Callosum | 1866 | 4.21 | <0.010 * | 19 −17 38 |
| Anterior Thalamic Radiation Right | 515 | 5.68 | 0.046 | 12 −29 13 |
| Cingulum Left | 482 | 4.29 | <0.010 * | −16 31 21 |
| Superior Longitudinal Right | 66 | 3.59 | <0.010 * | 47 0 25 |

Abbreviations: MNI = Montreal Neurological Institute; FA = fractional anisotropy; MD = medial diffusivity.
* Uncorrected results with *p* < 0.01.

**Table A3.** Resting-state functional connectivity differences in networks.

| Seed | Target | Stats | *p* Value (FDR-Corrected) |
|---|---|---|---|
| Mild < Controls | | | |
| Visual Occipital (Visual) | IFG Left (Language) | 3.75 | <0.001 |
| Severe < Controls | | | |
| IPS Left (DAN) | SMG Right (Salience) | 4.19 | 0.024 |
| Visual Lateral Left (Visual) | ACC (Salience) | 3.93 | 0.028 |
| Visual Occipital (Visual) | ACC (Salience) | 3.61 | 0.031 |
| Visual Occipital (Visual) | IFG Left (Language) | 3.58 | 0.031 |
| IPS Left (DAN) | ACC (Salience) | 3.57 | 0.031 |
| Sensorimotor Superior (Sensorimotor) | Sensorimotor Lateral Right (Sensorimotor) | 3.55 | 0.031 |
| Sensorimotor Superior (Sensorimotor) | Sensorimotor Lateral Left (Sensorimotor) | 3.48 | 0.033 |
| SMG Left (Salience) | aInsula Left (Salience) | 3.43 | 0.034 |
| SMG Left (Salience) | ACC (Salience) | 3.33 | 0.041 |
| Visual Occipital (Visual) | aInsula Left (Salience) | 3.24 | 0.048 |
| Severe < Mild | | | |
| PCC (Default Mode) | LP Left (Default Mode) | 4.19 | 0.028 |

Abbreviations: DAN = Dorsal Attention Network; IPS = Intraparietal Sulcus; IFG = Inferior Frontal Gyrus; SMG = Supramarginal Gyrus; ACC = Anterior Cingulate Cortex; A = Anterior; PCC = Posterior Cingulate Cortex.

## References


1. Aarsland, D.; Brønnick, K.; Larsen, J.P.; Tysnes, O.B.; Alves, G. Cognitive impairment in incident, untreated parkinson disease: The norwegian parkwest study. *Neurology* **2009**, *72*, 1121–1126. [CrossRef] [PubMed]
2. Muslimović, D.; Post, B.; Speelman, J.D.; Schmand, B. Cognitive profile of patients with newly diagnosed Parkinson's disease. *Neurology* **2005**, *65*, 1239–1245. [CrossRef] [PubMed]
3. Aarsland, D.; Bronnick, K.; Williams-Gray, C.; Weintraub, D.; Marder, K.; Kulisevsky, J.; Burn, D.; Barone, P.; Pagonabarraga, J.; Allcock, L.; et al. Mild cognitive impairment in Parkinson disease: A multicenter pooled analysis. *Neurology* **2010**, *75*, 1062–1069. [CrossRef] [PubMed]
4. Litvan, I.; Goldman, J.G.; Tröster, A.I.; Ben, A.; Weintraub, D.; Petersen, R.C.; Mollenhauer, B.; Adler, C.H.; Marder, K.; Williams-gray, C.H. Diangostic Criteria for Mild Cognitive Impairment in Parkinson's disease:Movement Disorder Society Task Force Guidelines. *Mov. Disord.* **2012**, *27*, 349–356. [CrossRef] [PubMed]
5. Hely, M.A.; Reid, W.G.J.; Adena, M.A.; Halliday, G.M.; Morris, J.G.L. The Sydney Multicenter Study of Parkinson's disease: The inevitability of dementia at 20 years. *Mov. Disord.* **2008**, *23*, 837–844. [CrossRef]
6. Emre, M.; Aarsland, D.; Brown, R.; Burn, D.J.; Duyckaerts, C.; Mizuno, Y.; Broe, G.A.; Cummings, J.; Dickson, D.W.; Gauthier, S.; et al. Clinical diagnostic criteria for dementia associated with Parkinson's disease. *Mov. Disord.* **2007**, *22*, 1689–1707. [CrossRef]
7. Foltynie, T.; Brayne, C.; Barker, R.A. The heterogeneity of idiopathic Parkinson's disease. *J. Neurol.* **2002**, *249*, 138–145. [CrossRef]
8. Van Rooden, S.M.; Heiser, W.J.; Kok, J.N.; Verbaan, D.; Van Hilten, J.J.; Marinus, J. The identification of Parkinson's disease subtypes using cluster analysis: A systematic review. *Mov. Disord.* **2010**, *25*, 969–978. [CrossRef]
9. Erro, R.; Picillo, M.; Vitale, C.; Palladino, R.; Amboni, M.; Moccia, M.; Pellecchia, M.T.; Barone, P. Clinical clusters and dopaminergic dysfunction in de-novo Parkinson disease. *Park. Relat. Disord.* **2016**, *28*, 137–140. [CrossRef]
10. Lawton, M.; Ben-shlomo, Y.; May, M.T.; Baig, F.; Barber, T.R.; Klein, J.C.; Swallow, D.M.A.; Malek, N.; Grosset, K.A.; Bajaj, N.; et al. Developing and validating Parkinson's disease subtypes and their motor and cognitive progression. *J. Neurol. Neurosurg. Psychiatry* **2018**, *89*, 1279–1287. [CrossRef]
11. Campbell, M.C.; Myers, P.S.; Weigand, A.J.; Foster, E.R.; Cairns, N.J.; Jackson, J.J.; Lessov-schlaggar, C.N.; Perlmutter, J.S. Parkinson disease clinical subtypes: Key features & clinical milestones. *Ann. Clin. Transl. Neurol.* **2020**, *7*, 1272–1283. [CrossRef]
12. Fereshtehnejad, S.-M.; Romenets, S.R.; Anang, J.B.M.; Latreille, V.; Gagnon, J.-F.; Postuma, R.B. New Clinical Subtypes of Parkinson Disease and Their Longitudinal Progression. *JAMA Neurol.* **2015**, *72*, 863. [CrossRef] [PubMed]
13. Uribe, C.; Segura, B.; Cesar, H.; Abos, A.; Garcia-diaz, A.I.; Campabadal, A.; Jose, M. Parkinsonism and Related Disorders Cortical atrophy patterns in early Parkinson's disease patients using hierarchical cluster analysis. *Park. Relat. Disord.* **2018**, *50*, 3–9. [CrossRef] [PubMed]
14. Inguanzo, A.; Sala-llonch, R.; Segura, B.; Erostarbe, H.; Abos, A. Parkinsonism and Related Disorders Hierarchical cluster analysis of multimodal imaging data identifies brain atrophy and cognitive patterns in Parkinson's disease. *Park. Relat. Disord.* **2021**, *82*, 16–23. [CrossRef] [PubMed]

15. Cigdem, O.; Demirel, H.; Unay, D. The Performance of Local-Learning Based Clustering Feature Selection Method on the Diagnosis of Parkinson's Disease Using Structural MRI. In Proceedings of the 2019 IEEE International Conference on Systems, Man and Cybernetics (SMC), Bari, Italy, 6–9 October 2019; pp. 1286–1291.

16. Abbasi, N.; Fereshtehnejad, S.; Zeighami, Y.; Larcher, K.M.; Postuma, R.B.; Dagher, A. Predicting severity and prognosis in Parkinson's disease from brain microstructure and connectivity. *NeuroImage Clin.* **2019**, *25*, 102111. [CrossRef] [PubMed]

17. Guo, T.; Guan, X.; Zhou, C.; Gao, T.; Wu, J.; Song, Z.; Xuan, M.; Gu, Q.; Huang, P.; Pu, J.; et al. Clinically relevant connectivity features define three subtypes of Parkinson's disease patients. *Hum. Brain Mapp.* **2020**, *41*, 4077–4092. [CrossRef]

18. Taylor, J.; Colloby, S.J.; Mckeith, I.G.; Brien, J.T.O. Covariant perfusion patterns provide clues to the origin of cognitive fluctuations and attentional dysfunction in Dementia with Lewy bodies. *Int. Psychogeriatr.* **2013**, *25*, 1917–1928. [CrossRef]

19. Somme, J.H.; Gomez-Esteban, J.C.; Molano, A.; Tijero, B.; Lezcano, E.; Zarranz, J.J. Initial neuropsychological impairments in patients with the E46K mutation of the α-synuclein gene (PARK 1). *J. Neurol. Sci.* **2011**, *310*, 86–89. [CrossRef]

20. Zarranz, J.J.; Alegre, J.; Gómez-Esteban, J.C.; Lezcano, E.; Ros, R.; Ampuero, I.; Vidal, L.; Hoenicka, J.; Rodriguez, O.; Atarés, B.; et al. The new mutation, E46K, of alpha-synuclein causes Parkinson and Lewy body dementia. *Ann. Neurol.* **2004**, *55*, 164–173. [CrossRef]

21. Tröster, A.I. Neuropsychological Characteristics of Dementia with Lewy Bodies and Parkinson's Disease with Dementia: Differentiation, Early Detection, and Implications for "Mild Cognitive Impairment" and Biomarkers. *Neuropsychol. Rev.* **2008**, *18*, 103–119. [CrossRef]

22. Peraza, L.R.; Kaiser, M.; Firbank, M.; Graziadio, S.; Bonanni, L.; Onofrj, M.; Colloby, S.J.; Blamire, A.; Brien, J.O.; Taylor, J. NeuroImage: Clinical fMRI resting state networks and their association with cognitive fluctuations in dementia with Lewy bodies. *NeuroImage Clin.* **2014**, *4*, 558–565. [CrossRef] [PubMed]

23. Cagnin, A.; Gardini, S.; Guzzo, C.; Gnoato, F.; Mitolo, M.; Caffarra, P.; Jelcic, N.; Ermani, M. Clinical and Cognitive Phenotype of Mild Cognitive Impairment Evolving to Dementia with Lewy Bodies. *Dement. Geriatr. Cogn. Disord. Extra* **2015**, *5*, 442–449. [CrossRef] [PubMed]

24. Murueta-Goyena, A.; Del Pino, R.; Galdós, M.; Arana, B.; Acera, M.; Carmona-Abellán, M.; Fernández-Valle, T.; Tijero, B.; Lucas-Jiménez, O.; Ojeda, N.; et al. Retinal Thickness Predicts the Risk of Cognitive Decline in Parkinson Disease. *Ann. Neurol.* **2021**, *89*, 165–176. [CrossRef] [PubMed]

25. Del Pino, R.; Murueta-Goyena, A.; Acera, M.; Carmona-Abellan, M.; Tijero, B.; Lucas-Jiménez, O.; Ojeda, N.; Ibarretxe-Bilbao, N.; Peña, J.; Gabilondo, I.; et al. Autonomic dysfunction is associated with neuropsychological impairment in Lewy body disease. *J. Neurol.* **2020**, *267*, 1941–1951. [CrossRef] [PubMed]

26. Douaud, G.; Smith, S.; Jenkinson, M.; Behrens, T.; Johansen-Berg, H.; Vickers, J.; James, S.; Voets, N.; Watkins, K.; Matthews, P.M. Anatomically related grey and white matter abnormalities in adolescent-onset schizophrenia. *Brain* **2007**, *130*, 2375–2386. [CrossRef] [PubMed]

27. Smith, S.M.; Jenkinson, M.; Woolrich, M.W.; Beckmann, C.F.; Behrens, T.E.J.; Johansen-Berg, H.; Bannister, P.R.; De Luca, M.; Drobnjak, I.; Flitney, D.E.; et al. Advances in functional and structural MR image analysis and implementation as FSL. *Neuroimage* **2004**, *23*, S208–S219. [CrossRef]

28. Jones, D.K.; Cercignani, M. Twenty-five pitfalls in the analysis of diffusion MRI data. *NMR Biomed.* **2010**, *23*, 803–820. [CrossRef]

29. Weissenbacher, A.; Kasess, C.; Gerstl, F.; Lanzenberger, R.; Moser, E.; Windischberger, C. Correlations and anticorrelations in resting-state functional connectivity MRI: A quantitative comparison of preprocessing strategies. *Neuroimage* **2009**, *47*, 1408–1416. [CrossRef]

30. Whitfield-Gabrieli, S.; Nieto-Castanon, A. A Functional Connectivity Toolbox for Correlated and Anticorrelated Brain Networks. *Brain Connect.* **2012**, *2*, 125–141. [CrossRef]

31. Mattay, V.S.; Tessitore, A.; Callicott, J.H.; Bertolino, A.; Goldberg, T.E.; Chase, T.N.; Hyde, T.M.; Weinberger, D.R. Dopaminergic modulation of cortical function in patients with Parkinson's disease. *Ann. Neurol.* **2002**, *51*, 156–164. [CrossRef]

32. Belvisi, D.; Fabbrini, A.; De Bartolo, M.I.; Costanzo, M.; Manzo, N.; Fabbrini, G.; Defazio, G.; Conte, A.; Berardelli, A. The Pathophysiological Correlates of Parkinson's Disease Clinical Subtypes. *Mov. Disord.* **2021**, *36*, 370–379. [CrossRef]

33. Yang, H.J.; Kim, Y.E.; Yun, J.Y.; Kim, H.J.; Jeon, B.S. Identifying the clusters within nonmotor manifestations in early Parkinson's disease by using unsupervised cluster analysis. *PLoS ONE* **2014**, *9*, e91906. [CrossRef] [PubMed]

34. Mu, J.; Chaudhuri, K.R.; Bielza, C.; De Pedro-Cuesta, J.; Larrañaga, P.; Martinez-Martin, P. Parkinson's disease subtypes identified from cluster analysis of motor and non-motor symptoms. *Front. Aging Neurosci.* **2017**, *9*, 301. [CrossRef] [PubMed]

35. Poewe, W.; Seppi, K.; Tanner, C.M.; Halliday, G.M.; Brundin, P.; Volkmann, J.; Schrag, A.-E.; Lang, A.E. Parkinson disease. *Nat. Rev.* **2017**, *3*, 17013. [CrossRef]

36. Pont-Sunyer, C.; Hotter, A.; Gaig, C.; Seppi, K.; Compta, Y.; Katzenschlager, R.; Mas, N.; Hofeneder, D.; Brücke, T.; Bayés, A.; et al. The Onset of Nonmotor Symptoms in Parkinson's disease (The ONSET PDStudy). *Mov. Disord.* **2015**, *30*, 229–237. [CrossRef] [PubMed]

37. Fereshtehnejad, S.M.; Zeighami, Y.; Dagher, A.; Postuma, R.B. Clinical criteria for subtyping Parkinson's disease: Biomarkers and longitudinal progression. *Brain* **2017**, *140*, 1959–1976. [CrossRef] [PubMed]

38. Deng, I.; Corrigan, F.; Zhai, G.; Zhou, X.; Bobrovskaya, L. Brain, Behavior, & Immunity—Health Lipopolysaccharide animal models of Parkinson's disease: Recent progress and relevance to clinical disease. *Brain Behav. Immun. Health* **2020**, *4*, 100060. [CrossRef]

39. Braak, H.; Del Tredici, K.; Rüb, U.; De Vos, R.A.I.; Steur, E.N.H.J.; Braak, E. Staging of brain pathology related to sporadic Parkinson's disease. *Neurobiol. Aging* **2003**, *24*, 197–211. [CrossRef]

40. Del Tredici, K.; Rüb, U.; De Vos, R.A.I.; Bohl, J.R.E.; Braak, H. Where does Parkinson disease pathology begin in the brain? *J. Neuropathol. Exp. Neurol.* **2002**, *61*, 413–426. [CrossRef]

41. Ponsen, M.M.; Stoffers, D.; Booij, J.; Van Eck-Smit, B.L.F.; Wolters, E.C.; Berendse, H.W. Idiopathic hyposmia as a preclinical sign of Parkinson's disease. *Ann. Neurol.* **2004**, *56*, 173–181. [CrossRef]

42. Sui, X.; Zhou, C.; Li, J.; Chen, L.; Yang, X.; Li, F. Hyposmia as a predictive marker of Parkinson's disease: A systematic review and meta-analysis. *Biomed Res. Int.* **2019**, *2019*, 23–27. [CrossRef] [PubMed]

43. Coon, E.A.; Cutsforth-Gregory, J.K.; Benarroch, E.E. Neuropathology of autonomic dysfunction in synucleinopathies. *Mov. Disord.* **2018**, *33*, 349–358. [CrossRef] [PubMed]

44. Goedert, M.; Jakes, R.; Spillantini, M.G. The Synucleinopathies: Twenty Years on. *J. Parkinsons. Dis.* **2017**, *7*, S53–S71. [CrossRef] [PubMed]

45. Pfeiffer, R.F. Non-motor symptoms in Parkinson's disease. *Park. Relat. Disord.* **2016**, *22*, S119–S122. [CrossRef]

46. Murueta-Goyena, A.; Del Pino, R.; Reyero, P.; Galdós, M.; Arana, B.; Lucas-Jiménez, O.; Acera, M.; Tijero, B.; Ibarretxe-Bilbao, N.; Ojeda, N.; et al. Parafoveal thinning of inner retina is associated with visual dysfunction in Lewy body diseases. *Mov. Disord.* **2019**, *34*, 1315–1324. [CrossRef]

47. Schapira, A.H.V.; Chaudhuri, K.R.; Jenner, P. Non-motor features of Parkinson disease. *Nat. Rev. Neurosci.* **2017**, *18*, 435–450. [CrossRef]

48. Marras, C.; Chaudhuri, K.R. Nonmotor features of Parkinson's disease subtypes. *Mov. Disord.* **2016**, *31*, 1095–1102. [CrossRef]

49. Weintraub, D.; Chahine, L.M.; Hawkins, K.A.; Siderowf, A.; Eberly, S.; Oakes, D.; Seibyl, J.; Stern, M.B.; Marek, K. Cognition and the Course of Prodromal Parkinson's Disease Methods Study Description. *Mov. Disord.* **2017**, *32*, 1640–1645. [CrossRef]

50. Williams-Gray, C.H.; Foltynie, T.; Brayne, C.E.G.; Robbins, T.W.; Barker, R.A. Evolution of cognitive dysfunction in an incident Parkinson's disease cohort. *Brain* **2007**, *130*, 1787–1798. [CrossRef]

51. Rosenthal, E.; Brennan, L.; Xie, S.; Hurtig, H.; Milber, J.; Weintraub, D.; Karlawish, J.; Siderowf, A. Association between cognition and function in patients with Parkinson disease with and without dementia. *Mov. Disord.* **2010**, *25*, 1170–1176. [CrossRef]

52. Kluger, B.M.; Freddy, K.; Tysnes, O.; Ongre, S.O.; Øygarden, B.; Herlofson, K. Is fatigue associated with cognitive dysfunction in early Parkinson's disease? *Park. Relat. Disord.* **2017**, *37*, 87–91. [CrossRef] [PubMed]

53. Michaelian, J.C.; Duffy, S.L.; Mckinnon, A.C.; Naismith, S.L.; Mowszowski, L.; Guastella, A. Theory of Mind in mild cognitive impairment: Relationship with the default mode network (DMN). *Alzheimer's Dement.* **2020**, *16*, 39164. [CrossRef]

54. Díez-Cirarda, M.; Ojeda, N.; Peña, J.; Cabrera-Zubizarreta, A.; Gómez-Beldarrain, M.Á.; Gómez-Esteban, J.C.; Ibarretxe-Bilbao, N. Neuroanatomical correlates of theory of mind deficit in Parkinson's disease: A multimodal imaging study. *PLoS ONE* **2015**, *10*, e0142234. [CrossRef]

55. Rektor, I.; Svátková, A.; Vojtíšek, L.; Zikmundová, I.; Vaníček, J.; Király, A.; Szabó, N. White matter alterations in Parkinson's disease with normal cognition precede grey matter atrophy. *PLoS ONE* **2018**, *13*, e0187939. [CrossRef] [PubMed]

56. Auning, E.; Kjærvik, V.K.; Selnes, P.; Aarsland, D.; Haram, A.; Bjørnerud, A.; Hessen, E.; Esnaashari, A. White matter integrity and cognition in Parkinson's disease: A cross-sectional study. *BMJ Open* **2014**, *4*, e003976. [CrossRef]

57. Zhang, J.; Bi, W.; Zhang, Y.; Zhu, M.; Zhang, Y.; Hua, F.; Wang, J.; Zhang, Y.; Jiang, T. Abnormal functional connectivity density in Parkinson's disease. *Behav. Brain Res.* **2015**, *280*, 113–118. [CrossRef]

58. Su, M.; Wang, S.; Fang, W.; Zhu, Y.; Li, R.; Sheng, K.; Zou, D.; Han, Y.; Wang, X.; Cheng, O. Parkinsonism and Related Disorders Alterations in the limbic/paralimbic cortices of Parkinson's disease patients with hyposmia under resting-state functional MRI by regional homogeneity and functional connectivity analysis. *Park. Relat. Disord.* **2015**, *21*, 698–703. [CrossRef]

59. Park, J.H.; Lee, S.H.; Kim, Y.; Park, S.; Byeon, G.H. Depressive symptoms are associated with worse cognitive prognosis in patients with newly diagnosed idiopathic Parkinson disease. *Psychogeriatrics* **2020**, *20*, 880–890. [CrossRef]

 *biomedicines*

*Article*

# Does the SARS-CoV-2 Spike Receptor-Binding Domain Hamper the Amyloid Transformation of Alpha-Synuclein after All?


Yulia Stroylova [1,2], Anastasiia Konstantinova [3], Victor Stroylov [4,5], Ivan Katrukha [6,7], Fedor Rozov [7] and Vladimir Muronetz [1,*]

1. Belozersky Institute of Physico-Chemical Biology, Lomonosov Moscow State University, Moscow 119991, Russia
2. Institute of Molecular Medicine, Sechenov First Moscow State Medical University, Moscow 119991, Russia
3. Faculty of Biotechnology, Lomonosov Moscow State University, Moscow 119991, Russia
4. Zelinsky Institute of Organic Chemistry, Russian Academy of Sciences, Moscow 119991, Russia
5. Chemical Faculty, National Research University Higher School of Economics (HSE), Moscow 101000, Russia
6. Department of Biochemistry, School of Biology, Lomonosov Moscow State University, Moscow 119991, Russia
7. HyTest Ltd., 20520 Turku, Finland
* Correspondence: vimuronets@belozersky.msu.ru; Tel.: +7-495-939-14-56



**Abstract:** Interactions of key amyloidogenic proteins with SARS-CoV-2 proteins may be one of the causes of expanding and delayed post-COVID-19 neurodegenerative processes. Furthermore, such abnormal effects can be caused by proteins and their fragments circulating in the body during vaccination. The aim of our work was to analyze the effect of the receptor-binding domain of the coronavirus S-protein domain (RBD) on alpha-synuclein amyloid aggregation. Molecular modeling showed that the predicted RBD complex with monomeric alpha-synuclein is stable over 100 ns of molecular dynamics. Analysis of the interactions of RBD with the amyloid form of alpha-synuclein showed that during molecular dynamics for 200 ns the number of contacts is markedly higher than that for the monomeric form. The formation of the RBD complex with the alpha-synuclein monomer was confirmed immunochemically by immobilization of RBD on its specific receptor ACE2. Changes in the spectral characteristics of the intrinsic tryptophans of RBD and hydrophobic dye ANS indicate an interaction between the monomeric proteins, but according to the data of circular dichroism spectra, this interaction does not lead to a change in their secondary structure. Data on the kinetics of amyloid fibril formation using several spectral approaches strongly suggest that RBD prevents the amyloid transformation of alpha-synuclein. Moreover, the fibrils obtained in the presence of RBD showed significantly less cytotoxicity on SH-SY5Y neuroblastoma cells.

**Keywords:** SARS-CoV-2 spike RBD domain; alpha-synuclein; amyloid aggregation; post-COVID-19 neurodegeneration; SARS-CoV-2 vaccine; Parkinson's disease



**Citation:** Stroylova, Y.; Konstantinova, A.; Stroylov, V.; Katrukha, I.; Rozov, F.; Muronetz, V. Does the SARS-CoV-2 Spike Receptor-Binding Domain Hamper the Amyloid Transformation of Alpha-Synuclein after All? *Biomedicines* **2023**, *11*, 498. https://doi.org/10.3390/biomedicines11020498

Academic Editor: Natalia Ninkina

Received: 21 December 2022
Revised: 30 January 2023
Accepted: 3 February 2023
Published: 9 February 2023

## 1. Introduction

Numerous data have been accumulated about the effects of coronavirus infection and vaccination by vector and RNA vaccines on the nervous system, from the common complication on the olfactory receptors to a variety of cognitive disorders [1–4]. However, there is relatively little data about the relationship between coronavirus infection and vaccination and socially significant neurodegenerative diseases, especially Alzheimer's and Parkinson's disease, due to the short period since the onset of the coronavirus pandemic [5–9]. The SARS-CoV-2 virus is thought to exacerbate Parkinson's disease by affecting pathological transformation of alpha-synuclein, stimulation of mitochondrial dysfunction, and depletion of dopamine [9]. Several studies have also emphasized the increased vulnerability of patients with Parkinson's disease to COVID-19 [10,11].

In order for the coronavirus to invade the cell, proteolysis of the site between the S1 and S2 subunits of the spike protein by cellular proteases, primarily furin, must occur, resulting in the release of the S1 subunit [12]. It has been shown that in mice the S1 protein is able to permeate the blood–brain barrier [13]. At present, the question of the interaction of alpha-synuclein with the full-length S-protein has been analyzed in two papers [14,15], but in the context of the effect on synuclein aggregation, the differences between the full-length S-protein and the RBD-containing S1 protein can be very crucial. There is some evidence from molecular modeling that the RBD fragment released as a part of the S1 subunit of the spike protein during proteolysis can interact with major amyloidogenic proteins alpha-synuclein, the beta-amyloid peptide precursor, or with the peptide itself as well as with the prion protein [16]. During vaccination, both the intact C-protein and its fragments can circulate in the organism, including RBD, formed during proteolysis. Moreover, there are vaccines in which the RBD fragment is the only active component [17,18]. Thus, studying the effect of the RBD fragment on the pathological transformation of alpha-synuclein has implications both for elucidating the mechanisms of synucleinopathies in COVID-19infections and for understanding approaches to prevent undesired effects of vaccination.

Molecular modeling techniques have shown that binding of alpha-synuclein to the RBD domain of the spike protein of coronavirus may occur [15,16]. However, the experimental evidence for this interaction is still not at all clear. An attempt to pull the full-length S-protein expressed in HEK293 cells by immunoprecipitation was unsuccessful, although its co-localization with alpha-synuclein has been shown [15]. In addition, it was shown that the full-length S-protein had no noticeable effect on alpha-synuclein aggregation [14].

It should also be noted that molecular modeling of interactions between the RBD domain spike protein of coronavirus and alpha-synuclein is complicated by the fact that in the monomeric form alpha-synuclein is completely unstructured, while in the oligomeric or fibrillar form it may exist in different conformational states. Currently, there is only one publication whose authors applied protein–protein docking to study the interaction of synuclein with RBD [15]. However, this method does not allow us to consider the conformational dynamics of the proteins that form the complex, and in the case of synuclein this can be very important. To study the structure of potential complexes between synuclein and RBD, it is necessary to use the molecular dynamics approach to investigate the stability of the complexes over time, as well as to clarify the structures of the complexes.

Thus, the aim of this work was to perform both experimental and more advanced molecular modeling of the interaction of the RBD domain of the spike protein of coronavirus with various forms of alpha-synuclein, as well as to test experimentally the possible influence of the individual RBD domain of the spike protein of coronavirus on the amyloid transformation of alpha-synuclein.

## 2. Materials and Methods

### 2.1. Materials

In the current work, we used the following chemicals: Thioflavin T, Congo Red, ANS (1-anilinonaphthalene-8-sulfonic Acid), o-phenylenediamine and MTT (Sigma, Burlington, MA, USA); inorganic salts (Panreac, Barcelona, Spain) and buffers (Amresco, Solon, OH, USA); fetal bovine serum (FBS) (HyClone, Logan, UT, USA); GlutaMAX (Gibco, Waltham, MA, USA); Dulbecco's Modified Eagle Medium (DMEM), trypsin, and penicillin/streptomycin (Paneco, Moscow, Russia). Human neuroblastoma SH-SY5Y cell culture (ATCC) was kindly gifted by Dr. Irina Naletova (University of Catania, Catania, Italy) [19]. Recombinant human ACE2-Fc (the extracellular domain of human angiotensin-converting enzyme 2 (1-740 a.a. of human angiotensin-converting enzyme 2, UniProtKB Q9BYF1), containing at the C-terminus of Fc fragment of human immunoglobulin G, SARS-CoV-2 Spike RBD (a fragment Arg319-Phe541 of SARS-CoV-2 Spike surface glycoprotein, GenBank: QHD43416.1) and anti-RBD antibodies (5308) were from HyTest LLC (Russia); anti-alpha-synuclein antibodies [LB 509] were from Abcam (Cambridge, UK). HRP-conjugated goat anti-mouse secondary polyclonal antibodies were purchased from Jackson Immunoresearch (West Grove, PA, USA).

### 2.2. Purification of Recombinant Alpha-Synuclein

Full-length wild type alpha-synuclein without additional motifs was expressed in E. coli and purified as previously described [16], with minor modifications. A codon 136 mutation from TAC to TAT (Tyr->Tyr) was introduced into the gene to avoid erroneous inclusion of cysteine in the bacterial expression system [20,21]. In the first step, acid precipitation of unwanted proteins was performed by adding 9% HCl to pH 2.8 in the cell extract and by centrifugation at $15,000\times g$ (5 min, 4 °C). The pH of the solution was then quickly corrected to 7.6 using 1 M potassium phosphate solution, pH 11. The pH value of the supernatant was adjusted to 7.6 using 1 M potassium phosphate solution, pH 11. Alpha-synuclein was purified by salting out ammonium sulfate to 40% saturation and stored as a concentrated suspension at 4 °C. The final protein concentration in the suspension was 1.5 mg/mL.

Protein concentration of alpha-synuclein solutions was determined spectrophotometrically at 280 nm, using extinction coefficient for 0.1% solution equal to 0.412.

### 2.3. Preparation of Alpha-Synuclein Fibrils

The ammonium sulfate precipitate of alpha-synuclein was collected by centrifugation and dialyzed against PBS buffer, pH 7.4. The sample was then diluted to a final concentration of 0.4 mg/mL (28 μM) with or without 2.8 or 5.6 μM RBD in the same buffer. Glass tubes (0.3 mL) were used to avoid sorption to plastic, pipetted (0.3 mL) into glass tubes and incubated at 37 °C with constant shaking at 600 rpm for 52 h. During incubation at 37 °C with constant shaking at 600 rpm for 52 h, 10 l aliquots were taken for thioflavin T fluorescence analysis to monitor the formation of amyloid aggregates. Simultaneously, the turbidity of the samples was measured as absorbance at 400 nm to detect the process of protein aggregation.

### 2.4. Thioflavin T Fluorescence Assay

Thioflavin T (ThT) fluorescence assay was performed in 96-well FLUOTRAC 200 black immunology plates (Greiner) using excitation wavelength set to 430 nm and the fluorescence was registered close to its maximum at 485 nm. Aliquots of 10 μL were taken from the tested samples of growing fibrils and added to 100 μL of 35 μM ThT solution. After 10 min of incubation at 20 °C, fluorescence intensity was measured on the plate reader PerkinElmer 2030 Multilabel Reader Victor X5. Each point was obtained by three separate measurements of the same protein sample.

### 2.5. Fluorescence Spectroscopy

Fluorescence spectra were measured using FluoroMax-3 spectrofluorimeter (Horiba Jobin Yvon, Longjumeau, France) in a 100-μL quartz fluorimetric cuvette. The slit widths were both set at 5 nm in emission and in excitation pathways. Tested samples were placed in quartz cuvette with optical pathway of 3 mm. Emission spectra were recorded between 300 nm and 400 nm, every 1 nm, with excitation wavelength set at 295 nm at 20 °C. The fluorescence emission of tryptophanyl residue shows a blue shift when its local environment becomes more hydrophobic.

### 2.6. ANS Fluorescence

For the ANS experiments, 10 μL of tested samples were incubated at 20 °C with a 50-fold molar excess of freshly prepared ANS for 60 min in the dark before the analysis. For the acquisition, the excitation was fixed at 365 nm and the emission was collected between 400 and 600 nm at 20 °C, using a 3 mm path-length cuvette. Spectra of ANS fluorescence were acquired with a FluoroMax-3 spectrofluorimeter (Horiba Jobin Yvon, Longjumeau, France).

### 2.7. Congo Red Binding Assay

Freshly prepared water solution of Congo Red was added to tested samples in PBS buffer, pH 4.0 at a molar ratio of 10:1. Congo Red was incubated for 10 min with protein samples before the measurements. Spectra of Congo Red absorption were acquired using Implen NanoPhotometer® NP80 (Munich, Germany) at 20 °C with a 3 mm path-length cuvette. The buffer spectrum was used as blank and subtracted from all other spectra.

### 2.8. Modified ELISA for Studying Protein–Protein Interactions

Solution of ACE2-Fc (10 μg/mL in 100 μL of PBS) were applied into wells of an ELISA plate. The proteins were adsorbed during 1 h incubation on a shaker at 20 °C, then the plate was washed with PBST (PBS with 0.05 % Tween 20). The next step was the addition of the mixture (10 μg/mL in 100 μL of PBS) of RBD and alpha-synuclein in PBS, previously incubated in 1:1 ratio (1.1 mg/mL of each protein) during 1 h at 20 °C, and incubation for 1 h on a shaker at 20 °C. Then the plate was washed with PBST and incubated with primary monoclonal antibodies against RBD or alpha-synuclein in PBST for 1 h at 20 °C. After washing with PBST, the plate was incubated with peroxidase-conjugated secondary anti-mouse antibodies (1 μg/mL in PBST) for 1 h. After washing, the plate was stained with a solution of o-phenylenediamine with hydrogen peroxide according to the standard procedure. The optical density in the wells was determined at 492 nm using a Stat Fax 2100 plate reader.

### 2.9. Circular Dichroism

Circular dichroism (CD) spectra of 20 μM RBD, 20 μM alpha-synuclein and their equimolar mixture were recorded in the far UV region (190–250 nm) at 20 °C using a 0.1 mm path-length cuvette on Applied Photophysics Chirascan CD spectrometer (Applied Photophysics, Leatherhead, UK). CD spectra were recorded after dialysis against 10 mM potassium phosphate buffer, pH 7.4. Each spectrum was the average of five scans.

### 2.10. Cell Viability of Neuroblastoma SH-SY5Y Cells by MTT-Test

SH-SY5Y cells were grown in DMEM/F12 (PanEco, Moscow, Russia), 10% (*v/v*) fetal bovine serum (Hyclone, Logan, UT, USA), 1% (*v/v*) L-glutamax (Sigma), and 1% (*v/v*) antibiotic solution (penicillin, streptomycin) (PanEco, Moscow, Russia). Cell line was used in experiments at passages 25–28. Cells were maintained at 37 °C in 5% $CO_2$, and culture medium was changed every 3–4 days. For cell viability, test cells were seeded at a density of 15,000 cells/well (in 100 μL of medium) onto 96-well cell culture treated plates (Eppendorf, Hamburg, Germany). Experiments were performed 24 h post seeding. Amounts of 0.26 μM RBD, 0.8 μM alpha-synuclein monomers, 0.8 μM alpha-synuclein fibrils or mixture (0.26 μM RBD, 0.8 μM alpha-synuclein) after 24 h of fibrillization under described protocol and further dilution were added. Cells were exposed to proteins for 24 h. Cells exposed to full DMEM/F12 were used as controls. Cell viability was estimated by MTT-reagent, based on the reduction of the tetrazolium dye MTT 3-(4,5-dimethylthiazol-2-yl)-2,5-diphenyltetrazolium bromide to formazan, which can be measured spectrophotometrically [22]. Cell medium was changed for DMEM/F12 with tetrazolium dye (final concentration 0.375 mg/mL) and after 4 h of incubation cells were dissolved in 100 mL DMSO. After 10 min at 37 °C absorbance at 570 nm was registered (VersaMax microplate reader, Molecular Devices, San Jose, CA, USA). The reference wavelength at 630 nm was used.

### 2.11. Docking and Molecular Dynamics Simulations

Structures of alpha-synuclein bound to sodium dodecyl sulfate micelles (PDB ID: 1XQ8) and alpha-synuclein fibril obtained by solid state NMR (PDB ID:2N0A) were used for modelling. Protein–protein docking of SARS-CoV-2 S-RBD (PDB ID: 6M0J) with alpha-synuclein was performed on the HDOCK server (http://hdock.phys.hust.edu.cn/, accessed on 11 April 2022), which is based on a hybrid algorithm of template-based modeling and ab initio free docking [23]. The binding modes of macromolecules were evaluated visually using VMD and by the value of binding energy. Then, the interactions within protein complexes were analyzed through the PDBSUM server (http://www.ebi.ac.uk/pdbsum, accessed on 19 April 2022), which is a web server that provides structural information including protein secondary structure, protein–ligand, and protein–protein interactions [24]. Dissociation constant (Kd) of a putative aSyn-RBD complex was estimated using the PPA-Pred server [25].

Molecular dynamics simulations were carried out using the GROMACS 2018 simulation package [26]. Intramolecular interactions are described for the fibril by the CHARMM 36m all-atom force field [27] and the TIP3P water model [28]. Prior to the production MD, each model was energy minimized using the steepest descent algorithm, and equilibrated for 200 ps in an NVT ensemble of 310 K, and then for 200 ps in an NPT ensemble at 310 K and 1 atm, with the heavy atoms restrained with a force constant of 1000 kJ mol$^{-1}$ nm$^{-2}$.

Production MD simulations were commenced from the final frame of equilibration trajectory and run at 310 K and 1 atm. Thermal stabilization was achieved by a v-rescale thermostat with a coupling constant of 0.1 ps, and the pressure was kept constant using the Parrinello–Rahman barostat with a relaxation time of 2 ps. Rigidity of water molecules was kept using the SETTLE algorithm 28, and other covalent bonds involving hydrogen atoms were restrained at their equilibrium lengths using LINCS29 algorithm, with a time step of 2.

Long-range electrostatic interactions were computed with the Particle-mesh Ewald approach, using cutoff of 12 Å and a Fourier grid spacing of 1.6 Å. Short-range van der Waal interactions were truncated at 12 Å, with smoothing starting at 10.5 Å.

### 3. Results

*3.1. Docking and Molecular Dynamics of Protein–Protein Interaction between SARS-CoV-2 Spike Receptor-Binding Domain (SARS-CoV-2 RBD) and Alpha-Synuclein (SYN)*

The appearance of protein–protein interactions is the first and key step for many biological interactions involved in cell signaling, immunity, and cellular transport [29]. Molecular modeling techniques are well established for studying protein structure and ligand activity [30,31]. Potential interactions between the RBD domain of SARS-CoV-2 spike protein and alpha-synuclein were studied using the HDOCK server. Model 1 with the highest docking energy and the lowest RMSD of the ligand was chosen. For this model, a molecular dynamic trajectory of 100 ns was performed to clarify the structure of the complex. The results of the docking and molecular dynamics are shown in Figure 1. The PDBSum server was used to determine the interacting residues of protein complexes; such surfaces and residues are shown in Figure 1.

Docking results showed that the interaction between alpha-synuclein and RBD was mediated by three hydrogen bonds via Lys21, Lys80, and Thr22 residues to Val503, Gln498, and Phe 374 of the RBD (Figure 1B,C) and 127 unlinked contacts, which corresponds essentially to the result described by Wu et al. [15]. Interestingly, after molecular dynamics, salt bridges between Lys32 and Lys60 with Asp389 and Asp427 RBD residues, which were absent before, appeared in the structure of the complex. The number of hydrogen bonds and nonpolar contacts increased to 9 and 155, respectively.

The binding affinity of the docking structures, represented by the dissociation constant (Kd), was obtained using PPA-Pred [25]. The predicted value of the dissociation constant was 7.26 $\times$ 10$^{-10}$ M, which is about two orders of magnitude lower than previously reported (1.46 $\times$ 10$^{-8}$ M) [25].

Thus, we can say with reasonable certainty that an interaction between RBD and the monomeric form of alpha-synuclein is possible.

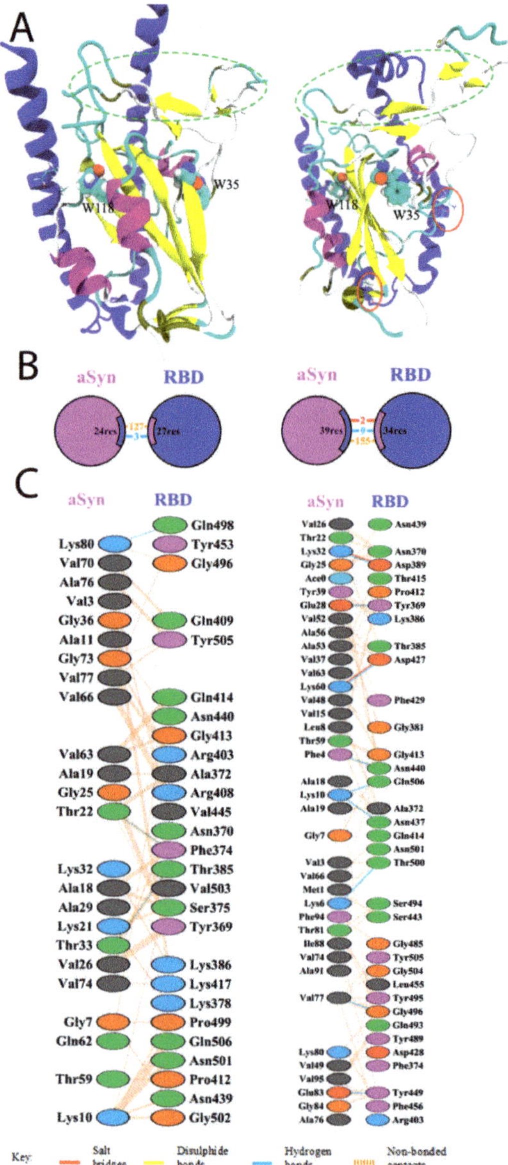

**Figure 1.** Interaction of SARS-CoV-2 RBD with alpha-synuclein by docking (**left**) or after 100 ns of molecular dynamics (**right**). (**A**) General view of binding. Alpha-synuclein is stained blue. RBD: α-helices are stained violet; β-sheets are yellow; the unstructured part is cyan. Tryptophan residues are highlighted on the RBD; the interaction site with ACE2 is highlighted with green ovals. On the alpha-synuclein as Licorice, residues within 5 Å of tryptophan residues are marked. Red ovals indicate the salt bridges formed during the dynamic simulation. (**B**) Schematic representation of the interaction of RBD with alpha-synuclein. Numbers indicate the amount of salt bridges (red), non-bonded contacts (orange), and hydrogen bonds (blue) that were discovered by docking (**left**) and molecular dynamics (**right**). (**C**) Residues involved in the interaction.(**left**): docking; (**right**): molecular dynamics.

Potential interactions between the RBD domain of the SARS-CoV-2 S-protein and the alpha-synuclein oligomer/fibril (PDB ID 2n0a) were studied using the HDOCK server. Model 2 was chosen, featuring both high interaction energy and the involvement of the RBD site, which is characterized by interaction with ACE2, in the protein–protein contact. For this model a molecular dynamic trajectory of 200 ns was calculated to clarify the structure of the complex. The results of docking and molecular dynamics are shown in Figure 2. The PDBSum server was used to determine the interacting residues of the protein complexes. The interacting surfaces and binding residues are shown in Figure 2.

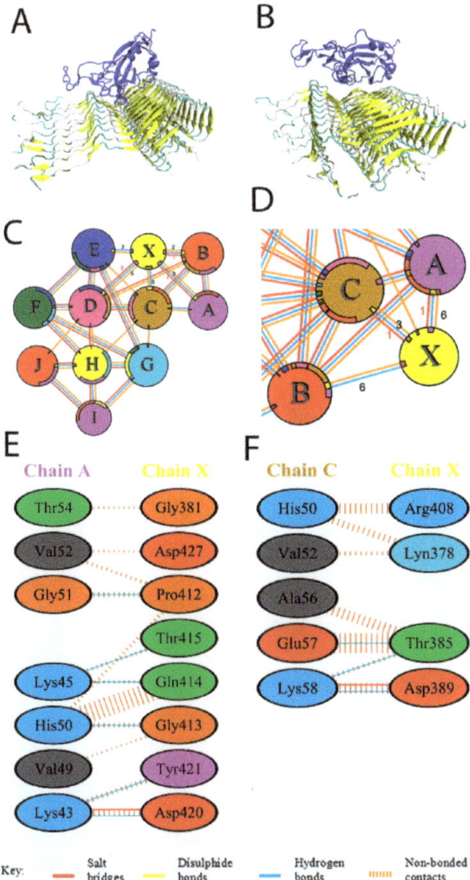

**Figure 2.** Interaction of SARS-CoV-2 RBD with the oligomer of alpha-synuclein by docking (**left**) or after 200 ns of molecular dynamics (**right**). Chains from A to I correspond to alpha-synuclein; chain X corresponds to RBD. (**A,C**) General view of binding and scheme of interactions according to docking results. Alpha-synuclein is colored yellow; RBD is colored blue. (**A,B**) General view. (**C,D**) Schematic representation of the interaction of RBD with alpha-synuclein after molecular dynamics. (**E,F**) Residues with the most significant contribution to the binding according to calculations. (**left**) docking; (**right**) molecular dynamics.

The docking results showed that RBD is able to form a fairly tight contact with the oligomer of alpha-synuclein, which involves the site responsible for the interaction with ACE2. However, as a result of molecular dynamics, a change in the orientation of RBD relative to the micelle is observed, which leads to a shift of the contact region to the lateral surface of the receptor fragment. This shift, however, does not hamper the binding to the full-length spike protein. Two salt bridges and 15 hydrogen bonds are observed in the resulting structure, which exceeds the parameters obtained for the alpha-synuclein monomer. Lys43 and Lys58 residues are involved in the formation of salt bridges on the alpha-synuclein side, and Asp389 and Asp420 residues on the RBD side (numbering shifted).

It is particularly noteworthy that the oligomer/fibril structure of alpha-synuclein is considerably compacted with an increase in the number of interchain interactions during the process of molecular dynamics (Figure 2E,F).

### 3.2. Interaction of the Monomeric Form of Alpha-Synuclein with the SARS-CoV-2 RBD

First, the possibility of binding of the monomeric form of alpha-synuclein to RBD was checked. For this purpose, an immunochemical method was used to identify the complex of the two proteins. To obtain the complex, the monomeric form of alpha-synuclein was incubated for one hour with RBD and then injected into wells of a plate containing pre-sorbed ACE2-Fc, which binds to the RBD of spike protein upon virus entry into the cell. This approach mimics immunoprecipitation, in which ACE2-Fc acts as an antibody by interacting with RBD. As follows from the data in Figure 3, using specific monoclonal antibodies to these two proteins, we showed that not only RBD but also the associated alpha-synuclein is sorbed onto ACE2-Fc, indicating formation of a complex between the two proteins.

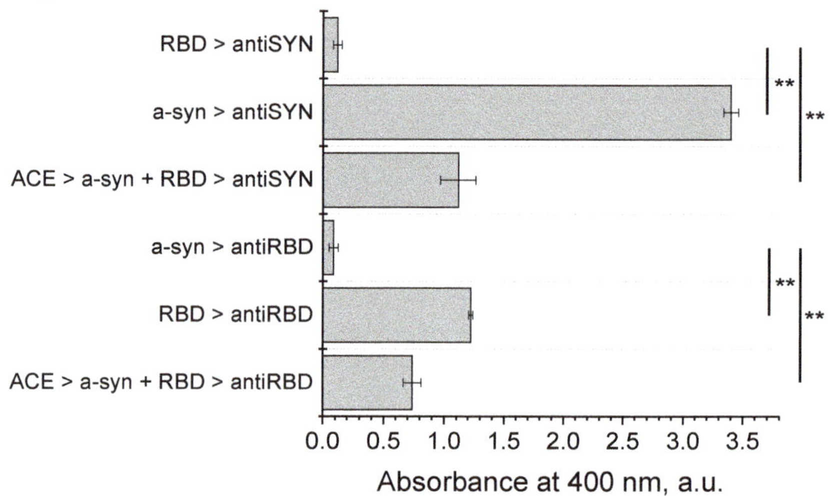


**Figure 3.** ACE2 interacts with pre-incubated SARS-CoV-2 RBD and alpha-synuclein by modified ELISA. Pre-incubated alpha-synuclein and SARS-CoV-2 S1 RBD during 1 h at 20 °C was applied to ACE2-Fc and detected by anti-RBD (5308) and anti-alpha-synuclein (LB509) antibodies. The poles represent the mean value of three independent measurements and error bars show the standard deviation. P-value represents results of one-way ANOVA; ** $p < 0.01$.

To confirm the formation of a complex between the monomeric form of alpha-synuclein and RBD, we also analyzed the change in the spectral characteristics of the proteins after their mixing. Figure 4A show the fluorescence spectra of RBD's own tryptophans before and after its binding to alpha-synuclein, which has no tryptophans. The shift in the fluorescence spectrum of RBD tryptophans may indicate their partial transition from the hydrophobic to the aqueous environment. The increased intensity and shift of the fluorescence maximum of 1,8-ANS interacting with hydrophobic sites when it binds to the complex of two proteins (Figure 4B) indicates a change in the conformation of alpha-synuclein and/or RBD during their interaction. However, the secondary structure of the proteins does not change during complex formation. This is evidenced by the CD spectra shown in Figure 5. As follows from these data, summing the spectra of the original proteins gives a spectrum characteristic of their complex.

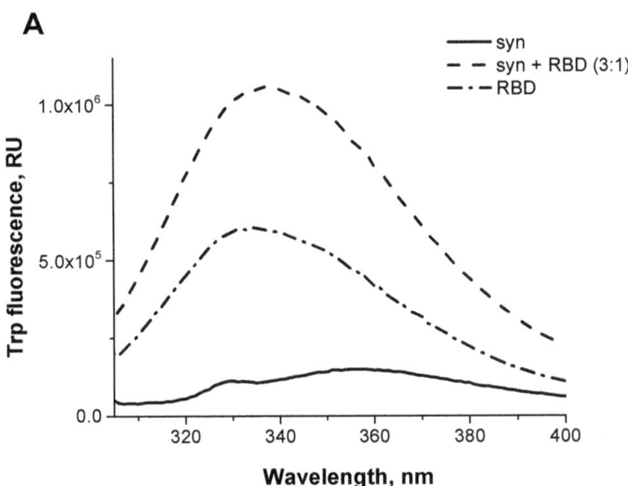


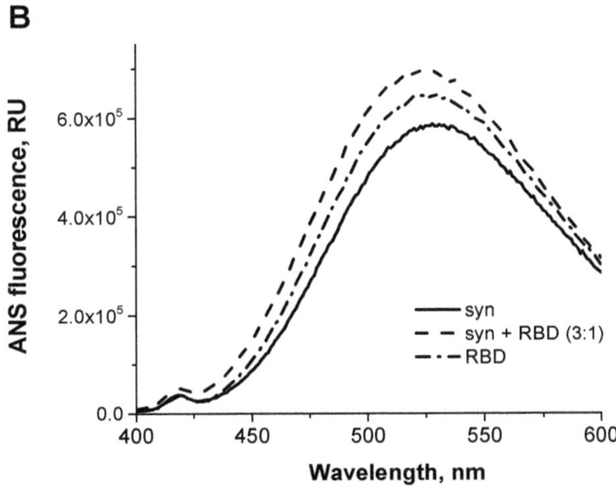


| Sample | Fluorescence Maximum Wavelength, nm | |
| --- | --- | --- |
| | **Trp** | **ANS** |
| syn | n/d | 528 |
| syn − RBD (3:1) | 336 | 522 |
| RBD | 332 | 528 |

**Figure 4.** The structural change of the SARS-CoV-2 RBD observed in the presence of alpha-synuclein by Trp (**A**) and ANS (**B**) fluorescence. Trp (**A**) and ANS (**B**) fluorescence emission spectra of 5.6 µM SARS-CoV-2 RBD (dot-dashed line), 28 µM alpha-synuclein (solid line), 28 µM alpha-synuclein mixed with 9.3 µM SARS-CoV-2 RBD (molar ratio 3:1, dashed line). Please note that alpha-synuclein does not contain tryptophans (solid line).

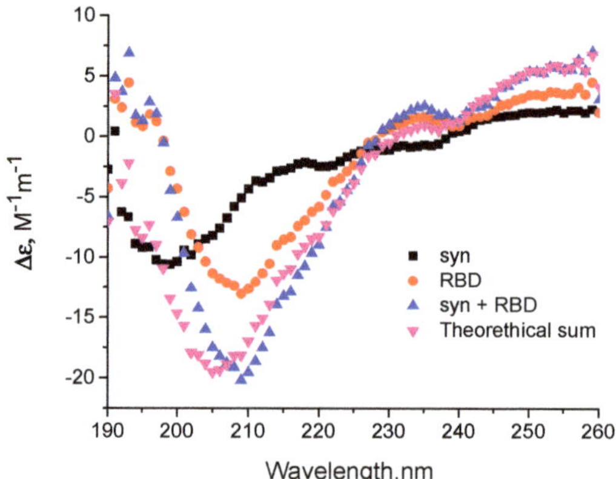

**Figure 5.** Influence of alpha-synuclein on circular dichroism spectra of SARS-CoV-2 RBD. CD spectra of 20 µM alpha-synuclein (squares) and 20 µM SARS-CoV-2 RBD (circles), equimolar mixture of Syn and RBD (blue triangles pointed up), and theoretical sum (violet triangles pointed down) are shown. CD spectra of the investigated samples were recorded after buffer exchange to 10 mM potassium phosphate buffer, pH 7.4. Each spectrum is an average of five records.

Thus, it was shown that monomeric alpha-synuclein is bound by RBD, causing changes in some spectral properties of the proteins, indicating a change in their conformation while keeping the secondary structure intact.

### 3.3. Effect of SARS-CoV-2 RBD on the Amyloid Transformation of Alpha-Synuclein

As results of thioflavin T fluorescence on Figure 6 clearly demonstrate, RBD of the spike protein completely blocks the amyloid transformation of alpha-synuclein. However, the RBD does not prevent disordered aggregation of alpha-synuclein, based on changes in the turbidity of the solution of the mixture of the two proteins (Figure 7). Some differences in the turbidity of the mixture of the two proteins probably depend both on their total concentration and on the characteristics of aggregation at different ratios of alpha-synuclein to RBD.

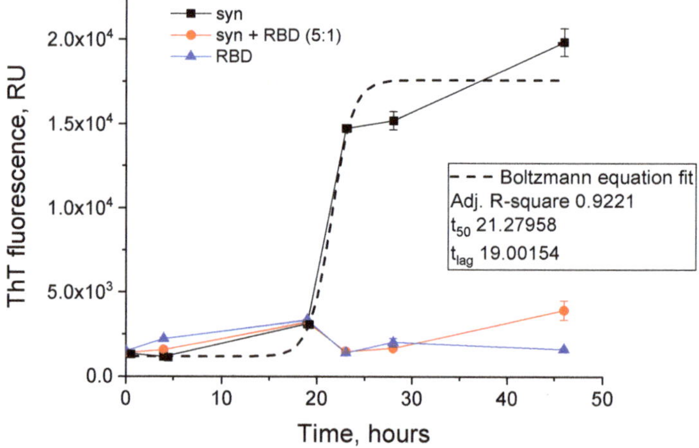

**Figure 6.** ThT fluorescence fibril formation kinetics of alpha-synuclein in the absence and presence of SARS-CoV-2 RBD. ThT fluorescence of 5.6 µM SARS-CoV-2 RBD (triangles), 28 µM alpha-synuclein (squares), 28 µM alpha-synuclein and 5.6 µM SARS-CoV-2 RBD (molar ratio 5:1, circles).

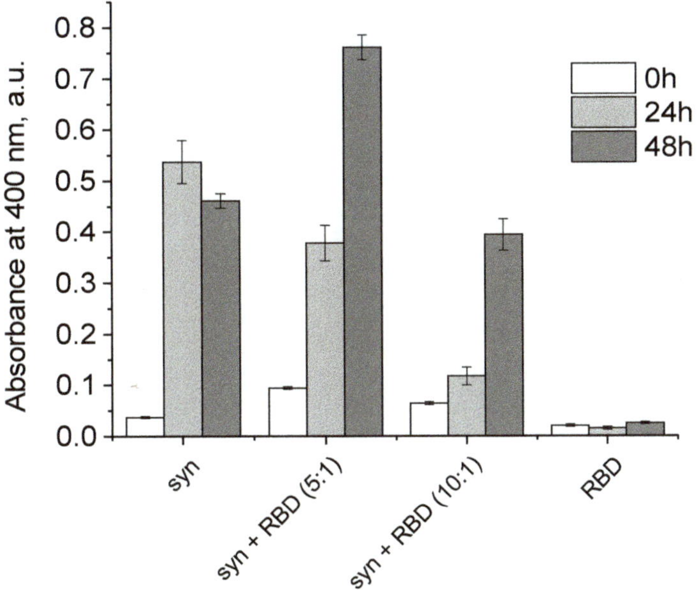

**Figure 7.** Aggregation of the alpha-synuclein, fibrillized in the presence of SARS-CoV-2 RBD. Absorbance at 400 nm of 5.6 µM SARS-CoV-2 RBD, 28 µM alpha-synuclein, 28 µM alpha-synuclein and 5.6 µM SARS-CoV-2 RBD (molar ratio 5:1), and 28 µM alpha-synuclein and 2.8 µM SARS-CoV-2 RBD (molar ratio 10:1) for 0 h (white), 24 h (light gray), and 48 h (dark gray).

The prevention of amyloid transformation of alpha-synuclein in the presence of the RBD is also evidenced by the spectra of Congo Red shown in Figure 8. During the interaction of Congo Red with alpha-synuclein fibrils, a shift of the dye absorption spectrum by almost 20 nm is observed. The spectra of the dye upon addition of the aggregates obtained in the presence of the RBD practically do not change. RBD also prevents changes in the ANS fluorescence spectra characteristic of dye interactions with amyloid forms of alpha-synuclein, primarily the amplitude (Figure 9). The spectra of tryptophan residues of the RBD fragment in interaction with monomeric forms and in aggregates are almost identical (Figures 4 and 9).

*3.4. Changes in Cytotoxicity of Synuclein Fibrils during Their Formation in the Presence of RBD*

The influence of RBD on the formation of alpha-synuclein amyloid fibrils was also tested in cytotoxicity experiments. Alpha-synuclein fibrils were produced by incubating 28 µM monomeric alpha-synuclein for 24 h at pH 4.0 as well as by incubating a mixture of alpha-synuclein monomers in the presence of 5.6 µM RBD. The resulting fibril preparations were diluted in DMEM/F12 to final concentrations of 0.8 µM for alpha-synuclein and 0.26 µM for RBD. The fibril solution was added to the neuroblastoma cells and cell viability was assessed after 24 h of incubation. As follows from the Figure 10, amyloid fibrils reduce cell viability by 50 percent, while RBD itself has no cytotoxic effect. The cytotoxic effect of alpha-synuclein fibrils formed in the presence of RBD is significantly lower than that of conventional alpha-synuclein fibrils, indicating their diminished amyloidogenic properties.

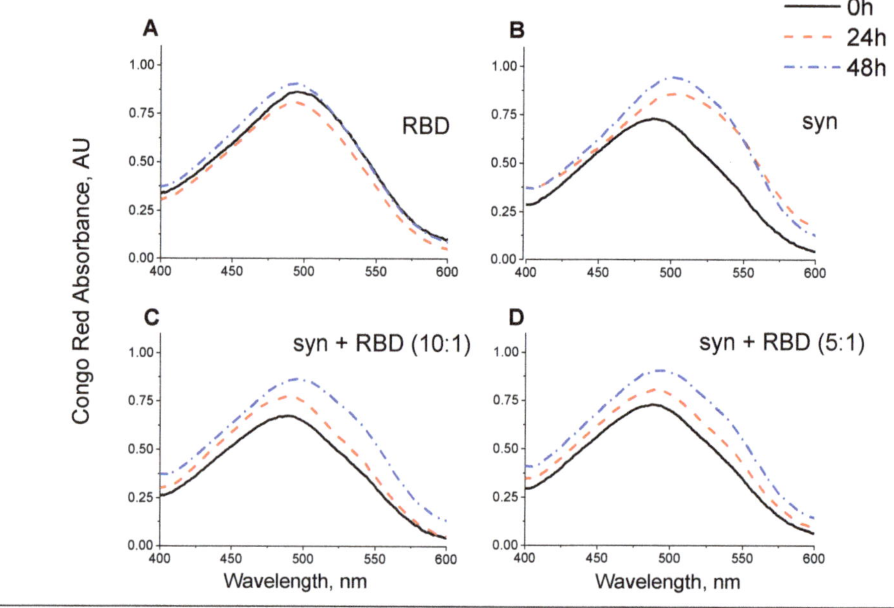



| Sample | λmax, nm | | |
| --- | --- | --- | --- |
| | 0 h | 24 h | 48 h |
| syn | 487 | 506 | 504 |
| syn + RBD (5:1) | 488 | 489 | 491 |
| syn + RBD (10:1) | 488 | 489 | 498 |
| RBD | 494 | 496 | 492 |

**Figure 8.** Congo Red absorbance changes of the alpha-synuclein, fibrillized in the presence of SARS-CoV-2 RBD. Congo Red absorbance spectra of SARS-CoV-2 RBD (**A**), 28 μM alpha-synuclein (**B**), 28 μM alpha-synuclein and 5.6 μM SARS-CoV-2 RBD (molar ratio 5:1, (**C**)), 28 μM alpha-synuclein and 2.8 μM SARS-CoV-2 RBD (molar ratio 10:1, (**D**)) for 0 h (solid line), 24 h (dashed line), and 48 h (dash-dotted line).

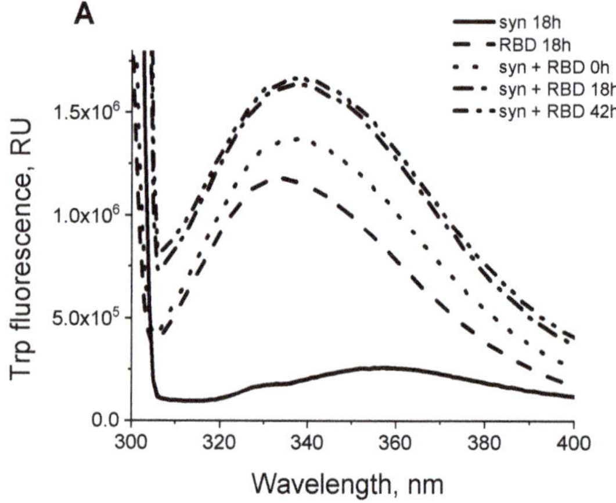



**Figure 9.** *Cont.*

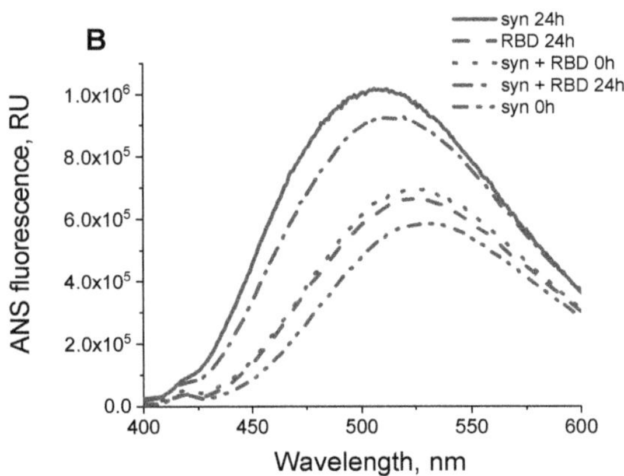



**Figure 9.** Structural changes of the SARS-CoV-2 RBD observed under fibrillization of alpha-synuclein by Trp and ANS fluorescence. Trp (**A**) and ANS (**B**) fluorescence emission spectra SARS-CoV-2 RBD (dashed line), 28 μM alpha-synuclein and 5.6 μM SARS-CoV-2 RBD (molar ratio 5:1) for 0 h (dash-dotted line) and 18–24 h (dotted line). Alpha-synuclein does not contain tryptophans (solid line for 18-24 h and dash-double-dot line for 0 h).

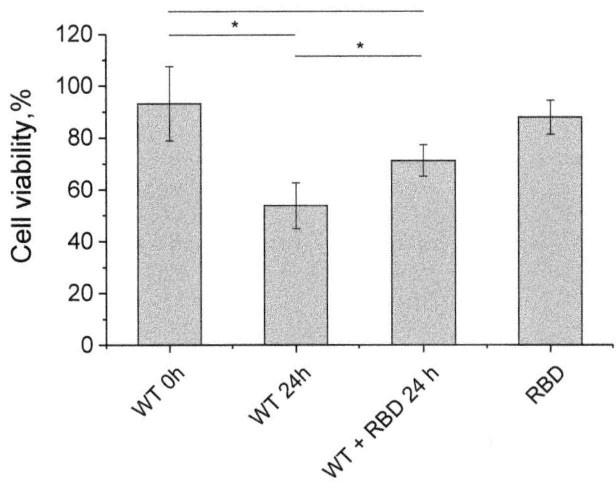



**Figure 10.** The effect of the alpha-synuclein fibrils, obtained in the presence of SARS-CoV-2 RBD, on neuroblastoma SH-SY5Y viability by MTT-test. The 28 μM alpha-synuclein fibrils (with or without 5.6 μM RBD) were produced for 24 h (as described in the Materials and Methods section). After dilution, 0.8 μM synuclein, 0.26 μM RBD, and 0.8 μM synuclein with 0.26 μM RBD (molar ratio 3:1) were added to SH-SY5Y neuroblastoma cells, and cell viability was determined after 24 h using the MTT test. $p$-value represents results of one-way ANOVA; * $p < 0.05$; ** $p < 0.01$.

## 4. Discussion

The data of our molecular simulations are in good agreement with the experiments performed. According to our molecular simulation, the binding site of the RBD fragment to alpha-synuclein is nearly almost unaffected by the regions involved in the interaction with the cellular ACE2 receptor (labeled with ovals in Figure 1A). We tested this interaction in an experiment by binding a mixture of RBD fragment and alpha-synuclein to immobilized ACE2, followed by the detection of the developed complex with anti-RBD and anti-Syn antibodies. Indeed, judging from the results of the experiment, the binding of alpha-synuclein to RBD does not interfere with the interaction of RBD with ACE. Moreover, according to docking, two tryptophan residues in the RBD molecule are located quite close to the alpha-helices of alpha-synuclein (marked in Figure 1A); therefore, these residues can serve as sensitive indicators of interaction, as we see in the observed change in fluorescence in the experiments.

In contrast to amorphous aggregation, the formation of amyloid is a complex organized process with a number of specific features, such as changes in the initial conformation of the protein, ability of "seeding", formation of intermediate oligomers, and growth of an unbranched fibril in a strictly determined mode. Therefore, among the possible explanations for how amyloid aggregation is inhibited when interacting with RBD, we can assume the difficulty of changing conformation during binding, inhibition of the formation of intermediate oligomers, and "capping" of the fibril ends to prevent further growth. Indeed, interaction with other molecules may both prevent [32,33] and stimulate amyloid transformation of alpha-synuclein [34,35].

The S-protein was chosen for immunization because it is a surface penetration protein abundantly present in the viral particle, and a strategy aimed at producing neutralizing antibodies that prevent the virus from penetrating cells is most effective in terms of preventing/suppressing disease. The question of whether circulating S-protein and its fragments will influence the pathological transformation of amyloid proteins seems important, since this interaction might lead to side effects of vaccination. Evidence of worsening the course of infection in Parkinson's disease patients [10] indicated the possibility of stimulation of amyloid transformation of alpha-synuclein by coronavirus proteins. However, in vitro it has been shown that full-length coronavirus S-protein did not bind alpha-synuclein during immunoprecipitation and had no effect on its aggregation [15]. Yet, given the release of the S1 subunit during proteolysis of the protein in the cell and the use of truncated proteins with the RBD fragment for coronavirus vaccine development, we performed a study of the RBD fragment. It was shown that the RBD fragment not only does not stimulate amyloid transformation of alpha-synuclein, but instead prevents it by stimulating amorphous aggregation. The combination of these data shows that the use of vaccines causing RBD of S-protein to circulate in the blood should not stimulate the formation of amyloid aggregates, at least in synucleinopathies. However, such stimulation can occur with the COVID-19 disease itself or with the use of whole virus particles containing both S- and N-proteins. At the same time, work has appeared showing that N-protein causes amyloid transformation of alpha-synuclein [14]. So we can presume that the pro-Parkinson manifestations of SARS-CoV-2 could be attributed to the action of N- but not S-protein.

Given that Parkinson's disease and other amyloid diseases usually have a very prolonged (years!) preclinical stage, when vaccinating with an agent capable of contacting amyloidogenic proteins, the effect of vaccines on the pathological transformation of proteins must be taken into account. At the same time, different types of vaccines can have different effects on the development of amyloidosis. For example, in the case of Parkinson's disease, the RBD fragment of the S-protein may even act as an anti-amyloid agent, whereas N-protein-containing vaccines, on the contrary, may provoke the disease. It is likely that clinical trials of new vaccines should include similar studies with major amyloidogenic in vitro proteins, since traditional animal and human safety trials are not conducted for a sufficient amount of time, and the consequences can be more than serious. In this case, the in vitro approach allows us to simulate the accelerated interaction of proteins and the possibility of amyloid transformation. Of course, based on the data obtained on the effect of only one RBD fragment on the pathological transformation of alpha-synuclein, it is difficult to talk about the possible mechanisms of the development of postvaccine complications. However, these data may already be useful for the development of vaccines against this infection. At the current stage of development of this field of science, in our opinion, it is preferable to use strictly defined proteins or their fragments produced by both RNA and vector vaccines for vaccination in order to reduce the possibility of unpredictable consequences due to the circulation of a whole set of proteins in the body. Moreover, the use of vaccines with only the RBD fragment, which does not stimulate the amyloid transformation of alpha-synuclein, as opposed to the full-length S-protein capable of forming amyloidogenic proteolytic fragments, may be sufficient.

Nystrom et al. recently showed that the S-protein contains several peptides capable of forming amyloid fibrils, and its coincubation with protease neutrophil elastase can result in the accumulation of one of the amyloidogenic peptides 194–203 [36]. As the amino acid sequence of RBD (319–541) does not include those S-protein peptides (192–211, 601–620, and 1166–1185) for which maximum amyloid criteria were shown, as well as amyloidogenic peptide 194–203 derived from cleavage of full-length S-protein by protease neutrophil elastase. This, in turn, indicates the likely benefits and safety of using RBD-based vaccines.

### 5. Conclusions

Using molecular modeling methods and biochemical approaches, we showed that the RBD of the coronavirus S-protein is able to interact with both monomeric and amyloid forms of alpha-synuclein. According to the molecular modeling data, the binding to the amyloid of alpha-synuclein is stronger than to the original monomer of the protein. It is also likely that this interaction makes further formation of amyloid fibrils more difficult. Although aggregates do form in the presence of RBD, they not only have less pronounced amyloid properties but also possess less cytotoxicity. In regards of post-COVID-19 neurodegeneration and provocation of Parkinson's disease manifestation, our results refute the initial assumption that the RBD of S-protein provokes the amyloid transformation of alpha-synuclein.

Thus, our results support the possibility of using the RBD of S-protein for vaccination and immune response production as sufficiently safe in the context of the development of neurodegenerative complications. At the very least, the presence of RBD in the body should not provoke the development of synucleinopathies.

**Author Contributions:** Conceptualization, V.M. and Y.S.; methodology, Y.S.; software and modelling, V.S.; validation, I.K., Y.S. and V.M.; formal analysis, Y.S.; investigation, Y.S. and A.K.; resources, I.K. and F.R.; data curation, Y.S. and V.S.; writing—original draft preparation, Y.S and V.M.; writing—review and editing, I.K.; visualization, Y.S. and V.S.; supervision, V.M.; project administration, Y.S.; funding acquisition, V.M. All authors have read and agreed to the published version of the manuscript.

**Funding:** This research was funded by Russian Scientific Foundation, project NO 21-14-00037.

**Institutional Review Board Statement:** Not applicable.

**Informed Consent Statement:** Not applicable.

**Data Availability Statement:** Not applicable.

**Acknowledgments:** The molecular dynamics simulations were supported by the Supercomputing Center of Lomonosov Moscow State University.

**Conflicts of Interest:** The authors declare no conflict of interest.

## References

1. Mukaetova-Ladinska, E.B.; Kronenberg, G.; Raha-Chowdhury, R. COVID-19 and neurocognitive disorders. *Curr. Opin. Psychiatry* **2021**, *34*, 149–156. [CrossRef]
2. De Luca, P.; Di Stadio, A.; Colacurcio, V.; Marra, P.; Scarpa, A.; Ricciardiello, F.; Cassandro, C.; Camaioni, A.; Cassandro, E. Long COVID, audiovestibular symptoms and persistent chemosensory dysfunction: A systematic review of the current evidence. *Acta Otorhinolaryngol. Ital.* **2022**, *42*, S87–S93. [CrossRef]
3. De Felice, F.G.; Tovar-Moll, F.; Moll, J.; Munoz, D.P.; Ferreira, S.T. Severe Acute Respiratory Syndrome Coronavirus 2 (SARS-CoV-2) and the Central Nervous System. *Trends Neurosci.* **2020**, *43*, 355–357. [CrossRef]
4. Chatterjee, A.; Chakravarty, A. Neurological Complications Following COVID-19 Vaccination. *Curr. Neurol. Neurosci. Rep.* **2023**, *23*, 1–14. [CrossRef]
5. Lingor, P.; Demleitner, A.F.; Wolff, A.W.; Feneberg, E. SARS-CoV-2 and neurodegenerative diseases: What we know and what we don't. *J. Neural. Transm.* **2022**, *129*, 1155–1167. [CrossRef]

6.  Krey, L.; Huber, M.K.; Hoglinger, G.U.; Wegner, F. Can SARS-CoV-2 Infection Lead to Neurodegeneration and Parkinson's Disease? *Brain Sci.* **2021**, *11*, 1654. [CrossRef]
7.  Silva, J.; Patricio, F.; Patricio-Martinez, A.; Santos-Lopez, G.; Cedillo, L.; Tizabi, Y.; Limon, I.D. Neuropathological Aspects of SARS-CoV-2 Infection: Significance for Both Alzheimer's and Parkinson's Disease. *Front. Neurosci.* **2022**, *16*, 867825. [CrossRef]
8.  Rahmani, B.; Ghashghayi, E.; Zendehdel, M.; Baghbanzadeh, A.; Khodadadi, M. Molecular mechanisms highlighting the potential role of COVID-19 in the development of neurodegenerative diseases. *Physiol. Int.* **2022**, *109*, 135–162. [CrossRef]
9.  Fu, Y.W.; Xu, H.S.; Liu, S.J. COVID-19 and neurodegenerative diseases. *Eur. Rev. Med. Pharmacol. Sci.* **2022**, *26*, 4535–4544.
10. Sinha, S.; Mittal, S.; Roy, R. Parkinson's Disease and the COVID-19 Pandemic: A Review Article on the Association between SARS-CoV-2 and alpha-Synucleinopathy. *J. Mov. Disord.* **2021**, *14*, 184–192. [CrossRef]
11. Zenesini, C.; Vignatelli, L.; Belotti, L.M.B.; Baccari, F.; Calandra-Buonaura, G.; Cortelli, P.; Descovich, C.; Giannini, G.; Guaraldi, P.; Guarino, M.; et al. Risk of SARS-CoV-2 infection, hospitalization and death for COVID-19 in people with Parkinson's disease or parkinsonism over a 15-month period: A cohort study. *Eur. J. Neurol.* **2022**, *29*, 3205–3217. [CrossRef]
12. Shang, J.; Wan, Y.; Luo, C.; Ye, G.; Geng, Q.; Auerbach, A.; Li, F. Cell entry mechanisms of SARS-CoV-2. *Proc. Natl. Acad. Sci. USA* **2020**, *117*, 11727–11734. [CrossRef]
13. Rhea, E.M.; Logsdon, A.F.; Hansen, K.M.; Williams, L.M.; Reed, M.J.; Baumann, K.K.; Holden, S.J.; Raber, J.; Banks, W.A.; Erickson, M.A. The S1 protein of SARS-CoV-2 crosses the blood-brain barrier in mice. *Nat. Neurosci.* **2021**, *24*, 368–378. [CrossRef]
14. Semerdzhiev, S.A.; Fakhree, M.A.A.; Segers-Nolten, I.; Blum, C.; Claessens, M. Interactions between SARS-CoV-2 N-Protein and alpha-Synuclein Accelerate Amyloid Formation. *ACS Chem. Neurosci.* **2022**, *13*, 143–150. [CrossRef]
15. Wu, Z.; Zhang, X.; Huang, Z.; Ma, K. SARS-CoV-2 Proteins Interact with Alpha Synuclein and Induce Lewy Body-like Pathology In Vitro. *Int. J. Mol. Sci.* **2022**, *23*, 3394. [CrossRef]
16. Idrees, D.; Kumar, V. SARS-CoV-2 spike protein interactions with amyloidogenic proteins: Potential clues to neurodegeneration. *Biochem. Biophys. Res. Commun.* **2021**, *554*, 94–98. [CrossRef]
17. Salimian, J.; Ahmadi, A.; Amani, J.; Olad, G.; Halabian, R.; Saffaei, A.; Arabfard, M.; Nasiri, M.; Nazarian, S.; Abolghasemi, H.; et al. Safety and immunogenicity of a recombinant receptor-binding domain-based protein subunit vaccine (Noora vaccine) against COVID-19 in adults: A randomized, double-blind, placebo-controlled, Phase 1 trial. *J. Med. Virol.* **2022**, *95*. [CrossRef]
18. Yang, J.; Liu, M.Q.; Liu, L.; Li, X.; Xu, M.; Lin, H.; Liu, S.; Hu, Y.; Li, B.; Liu, B.; et al. A triple-RBD-based mucosal vaccine provides broad protection against SARS-CoV-2 variants of concern. *Cell. Mol. Immunol.* **2022**, *19*, 1279–1289. [CrossRef]
19. Greco, V.; Naletova, I.; Ahmed, I.M.M.; Vaccaro, S.; Messina, L.; La Mendola, D.; Bellia, F.; Sciuto, S.; Satriano, C.; Rizzarelli, E. Hyaluronan-carnosine conjugates inhibit Abeta aggregation and toxicity. *Sci. Rep.* **2020**, *10*, 15998. [CrossRef]
20. Barinova, K.V.; Kuravsky, M.L.; Arutyunyan, A.M.; Serebryakova, M.V.; Schmalhausen, E.V.; Muronetz, V.I. Dimerization of Tyr136Cys alpha-synuclein prevents amyloid transformation of wild type alpha-synuclein. *Int. J. Biol. Macromol.* **2017**, *96*, 35–43. [CrossRef]
21. Masuda, M.; Dohmae, N.; Nonaka, T.; Oikawa, T.; Hisanaga, S.; Goedert, M.; Hasegawa, M. Cysteine misincorporation in bacterially expressed human alpha-synuclein. *FEBS Lett.* **2006**, *580*, 1775–1779. [CrossRef]
22. Niks, M.; Otto, M. Towards an optimized MTT assay. *J. Immunol. Methods* **1990**, *130*, 149–151. [CrossRef]
23. Yan, Y.; Zhang, D.; Zhou, P.; Li, B.; Huang, S.Y. HDOCK: A web server for protein-protein and protein-DNA/RNA docking based on a hybrid strategy. *Nucleic Acids Res.* **2017**, *45*, W365–W373. [CrossRef]
24. Laskowski, R.A.; Jablonska, J.; Pravda, L.; Varekova, R.S.; Thornton, J.M. PDBsum: Structural summaries of PDB entries. *Protein Sci. A Publ. Protein Soc.* **2018**, *27*, 129–134. [CrossRef]
25. Yugandhar, K.; Gromiha, M.M. Protein-protein binding affinity prediction from amino acid sequence. *Bioinformatics* **2014**, *30*, 3583–3589. [CrossRef]
26. Abraham, M.J.; Murtola, T.; Schulz, R.; Páll, S.; Smith, J.C.; Hess, B.; Lindahl, E. GROMACS: High performance molecular simulations through multi-level parallelism from laptops to supercomputers. *SoftwareX* **2015**, *1–2*, 19–25. [CrossRef]
27. Huang, J.; Rauscher, S.; Nawrocki, G.; Ran, T.; Feig, M.; de Groot, B.L.; Grubmuller, H.; MacKerell, A.D., Jr. CHARMM36m: An improved force field for folded and intrinsically disordered proteins. *Nat. Methods* **2017**, *14*, 71–73. [CrossRef]
28. Lee, S.; Tran, A.; Allsopp, M.; Lim, J.B.; Henin, J.; Klauda, J.B. CHARMM36 united atom chain model for lipids and surfactants. *J. Phys. Chem. B* **2014**, *118*, 547–556. [CrossRef]
29. Yan, Y.; Huang, S.Y. Modeling Protein-Protein or Protein-DNA/RNA Complexes Using the HDOCK Webserver. *Methods Mol. Biol.* **2020**, *2165*, 217–229.
30. Stroylov, V.S.; Svitanko, I.V.; Maksimenko, A.S.; Kislyi, V.P.; Semenova, M.N.; Semenov, V.V. Computational modeling and target synthesis of monomethoxy-substituted o-diphenylisoxazoles with unexpectedly high antimitotic microtubule destabilizing activity. *Bioorganic Med. Chem. Lett.* **2020**, *30*, 127608. [CrossRef]
31. Agapova, Y.K.; Altukhov, D.A.; Timofeev, V.I.; Stroylov, V.S.; Mityanov, V.S.; Korzhenevskiy, D.A.; Vlaskina, A.V.; Smirnova, E.V.; Bocharov, E.V.; Rakitina, T.V. Structure-based inhibitors targeting the alpha-helical domain of the Spiroplasma melliferum histone-like HU protein. *Sci. Rep.* **2020**, *10*, 15128. [CrossRef] [PubMed]
32. Medvedeva, M.; Barinova, K.; Melnikova, A.; Semenyuk, P.; Kolmogorov, V.; Gorelkin, P.; Erofeev, A.; Muronetz, V. Naturally occurring cinnamic acid derivatives prevent amyloid transformation of alpha-synuclein. *Biochimie* **2020**, *170*, 128–139. [CrossRef]
33. Priss, A.; Afitska, K.; Galkin, M.; Yushchenko, D.A.; Shvadchak, V.V. Rationally Designed Protein-Based Inhibitor of alpha-Synuclein Fibrillization in Cells. *J. Med. Chem.* **2021**, *64*, 6827–6837. [CrossRef]

34. Leisi, E.V.; Barinova, K.V.; Kudryavtseva, S.S.; Moiseenko, A.V.; Muronetz, V.I.; Kurochkina, L.P. Effect of bacteriophage-encoded chaperonins on amyloid transformation of alpha-synuclein. *Biochem. Biophys. Res. Commun.* **2022**, *622*, 136–142. [CrossRef] [PubMed]

35. Golts, N.; Snyder, H.; Frasier, M.; Theisler, C.; Choi, P.; Wolozin, B. Magnesium inhibits spontaneous and iron-induced aggregation of alpha-synuclein. *J. Biol. Chem.* **2002**, *277*, 16116–16123. [CrossRef]

36. Nystrom, S.; Hammarstrom, P. Amyloidogenesis of SARS-CoV-2 Spike Protein. *J. Am. Chem. Soc.* **2022**, *144*, 8945–8950. [CrossRef]

 *biomedicines*

*Article*

# Gamma-Synuclein Dysfunction Causes Autoantibody Formation in Glaucoma Patients and Dysregulation of Intraocular Pressure in Mice

Tatiana A. Pavlenko [1], Andrei Y. Roman [2], Olga A. Lytkina [2], Nadezhda E. Pukaeva [2,3], Martha W. Everett [3], Iuliia S. Sukhanova [2,3], Vladislav O. Soldatov [4], Nina G. Davidova [1], Natalia B. Chesnokova [1], Ruslan K. Ovchinnikov [2,3] and Michail S. Kukharsky [2,3,*]

[1] Helmholtz Moscow Research Institute of Eye Diseases, Ministry of Health of the Russian Federation, 105062 Moscow, Russia
[2] Institute of Physiologically Active Compounds at Federal Research Center of Problems of Chemical Physics and Medicinal Chemistry, Russian Academy of Sciences, 142432 Chernogolovka, Russia
[3] Department of General and Cell Biology, Faculty of Medical Biology, Pirogov Russian National Research Medical University, 117997 Moscow, Russia
[4] Department of Pharmacology and Clinical Pharmacology, Belgorod State National Research University, 308015 Belgorod, Russia
* Correspondence: kukharskym@gmail.com; Tel.: +7-9164859560

**Abstract:** Dysregulation of intraocular pressure (IOP) is one of the main risk factors for glaucoma. γ-synuclein is a member of the synuclein family of widely expressed synaptic proteins within the central nervous system that are implicated in certain types of neurodegeneration. γ-synuclein expression and localization changes in the retina and optic nerve of patients with glaucoma. However, the mechanisms by which γ-synuclein could contribute to glaucoma are poorly understood. We assessed the presence of autoantibodies to γ-synuclein in the blood serum of patients with primary open-angle glaucoma (POAG) by immunoblotting. A positive reaction was detected for five out of 25 patients (20%) with POAG. Autoantibodies to γ-synuclein were not detected in a group of patients without glaucoma. We studied the dynamics of IOP in response to IOP regulators in knockout mice (γ-KO) to understand a possible link between γ-synuclein dysfunction and glaucoma-related pathophysiological changes. The most prominent decrease of IOP in γ-KO mice was observed after the instillation of 1% phenylephrine and 10% dopamine. The total protein concentration in tear fluid of γ-KO mice was approximately two times higher than that of wild-type mice, and the activity of neurodegeneration-linked protein α2-macroglobulin was reduced. Therefore, γ-synuclein dysfunction contributes to pathological processes in glaucoma, including dysregulation of IOP.

**Keywords:** γ-synuclein; autoantibodies; glaucoma; dopamine; α2-macroglobulin; tear fluid; intraocular pressure

**Citation:** Pavlenko, T.A.; Roman, A.Y.; Lytkina, O.A.; Pukaeva, N.E.; Everett, M.W.; Sukhanova, I.S.; Soldatov, V.O.; Davidova, N.G.; Chesnokova, N.B.; Ovchinnikov, R.K.; et al. Gamma-Synuclein Dysfunction Causes Autoantibody Formation in Glaucoma Patients and Dysregulation of Intraocular Pressure in Mice. *Biomedicines* **2023**, *11*, 60. https://doi.org/10.3390/biomedicines11010060

Academic Editor: Andrei Surguchov

Received: 3 December 2022
Revised: 18 December 2022
Accepted: 23 December 2022
Published: 27 December 2022

## 1. Introduction

Glaucoma is a multifactorial disease with various mechanisms of development and is defined as chronic progressive optic neuropathy with morphological changes in the optic nerve head and the layer of retinal nerve fibers associated with the death of retinal ganglion cells and visual field defects [1]. It is the second-leading cause of blindness worldwide and the most common cause of irreversible blindness [2,3]. Pathological processes in glaucoma have common features with neurodegenerative diseases, including the death of a specific nerve cell type—retinal ganglion cells (RGCs). An increase in intraocular pressure (IOP) is the most common cause of the development of this pathology. Drug therapy for glaucoma is mainly aimed at reducing IOP [4]. However, a form of glaucoma occurring with normal IOP (normal-tension glaucoma) emphasizes the complex nature of pathogenesis. Glaucoma development involves processes such as neuroinflammation, impairment of the immune response, disturbance of nerve impulse transmission, and other mechanisms underlying neurodegeneration. Studying them may reveal new possibilities for the prognosis and treatment of glaucoma [5–7].

Synucleins are highly conserved cytosolic proteins involved in the regulation of synaptic transmission in the nervous system [8–10]. The family consists of three highly homologous proteins: α-, β-, and γ-synuclein. α-synuclein is actively studied because of its role in the pathogenesis of neurodegenerative diseases. α-synuclein is the main component of Lewy bodies in Parkinson's disease (PD) and dementia with Lewy bodies (DLB). It is also found in senile plaques in Alzheimer's disease [11–13]. Aggregation of γ-synuclein in the nervous system is detected in amyotrophic lateral sclerosis (ALS) and DLB [14–17]. γ-synuclein is also associated with cancer development [18,19]. All three synucleins are expressed in the retina and the optic nerve, but only γ-synuclein can change the level and distribution of its expression in glaucoma [20]. γ-synuclein could be considered as a specific marker of RGCs because of its high level of expression in these cells. The redistribution of γ-synuclein from one layer of the retina and the optic nerve to another leads to a decrease in its level in the original location. Moreover, the death of RGCs in glaucoma correlates with a decrease in the expression of γ-synuclein [21].

In neurodegenerative diseases, autoantibodies against proteins prone to aggregation are found in the blood [22]. Thus, autoantibodies against α-synuclein are found in patients with PD [23]. We have previously shown that autoantibodies to γ-synuclein are detected in the serum of some patients with ALS and cerebrovascular diseases [24]. The exact role of autoantibodies in the pathogenesis of ocular diseases remains unclear. They can cause autoimmune death of retinal cells. They can also have a protective effect [25,26]. In particular, autoantibodies against γ-synuclein increase the survival of RGCs in RGC-5 cell culture under stress conditions and in primary cultures of porcine retinal explant [27,28]. Therefore, despite the evidence of γ-synuclein involvement in glaucoma development, the particular mechanisms remain unclear.

We found that some patients with primary open-angle glaucoma (POAG) had autoantibodies to γ-synuclein in their blood. Using knockout mice (γ-KO mice) as a model for the loss of function of this protein, we showed that the impact on the neurotransmitter systems of the eye responsible for IOP regulation differently affects the dynamics of IOP changes in γ-KO mice compared to wild-type mice (WT mice). In the tear fluid of γ-KO mice, the activity of chaperone and immune system modulator α2-macroglobulin (α2-MG) was reduced.

## 2. Materials and Methods

### 2.1. Patients

The study involved 25 patients with POAG, including 12 women and 13 men. The control group of apparently healthy individuals without a diagnosis or history of glaucoma included 13 people (7 women and 6 men) (Table 1). Patients were followed at the Helmholtz Moscow Research Institute of Eye Diseases of the Ministry of Health of the Russian Federation. All patients underwent a comprehensive ophthalmological examination: viscometry, pneumotonometry, biomicroscopy, gonioscopy, ophthalmoscopy, and static perimetry. All glaucoma patients received instillation of ocular hypotensive drugs to normalize IOP. The early stage of the disease was diagnosed in 5 (20%) cases; a moderate stage was diagnosed in 7 (28%) cases; an advanced stage was diagnosed in 8 (32%) cases; a severe stage was diagnosed in 4 (16%) cases; and end-stage glaucoma was diagnosed in 1 (4%) case (Table 1). The mean age of patients and volunteers without glaucoma was 73 ± 8 and 66 ± 12 years, respectively. All patients provided written informed consent at admission. The study was conducted according to the guidelines of the Declaration of Helsinki, and approved by the Institutional Ethics Committee of the HMRIED (protocol No. 55/3, 17 June 2021).

**Table 1.** Clinical data of patients and control subjects.

| Patient Number | Gender | Age (Years) | Diagnosis | Autoantibodies to γ-Synuclein | Stage of the Disease | Surgical Treatment of Glaucoma |
|---|---|---|---|---|---|---|
| Control group | | | | | | |
| 1 | M | 44 | No glaucoma symptoms | - | NA | NA |
| 2 | F | 91 | No glaucoma symptoms | - | NA | NA |
| 3 | M | 70 | No glaucoma symptoms | - | NA | NA |
| 4 | M | 85 | No glaucoma symptoms | - | NA | NA |
| 5 | M | 64 | No glaucoma symptoms | - | NA | NA |
| 6 | M | 64 | No glaucoma symptoms | - | NA | NA |
| 7 | F | 80 | No glaucoma symptoms | - | NA | NA |
| 8 | F | 59 | No glaucoma symptoms | - | NA | NA |
| 9 | F | 48 | No glaucoma symptoms | - | NA | NA |
| 10 | F | 67 | No glaucoma symptoms | - | NA | NA |
| 11 | F | 63 | No glaucoma symptoms | - | NA | NA |
| 12 | M | 62 | No glaucoma symptoms | - | NA | NA |
| 13 | F | 67 | No glaucoma symptoms | - | NA | NA |
| Primary open-angle glaucoma | | | | | | |
| 14 | M | 80 | POAG | + | 4 OU | Yes |
| 15 | F | 78 | POAG | - | 3 OU | No |
| 16 | M | 76 | POAG | - | 5 OS; 2 OD | Yes |
| 17 | M | 78 | POAG | - | 2OU | Yes |
| 18 | F | 78 | POAG | - | 1 OU | No |
| 19 | M | 78 | POAG | - | 4 OS; 1 OD | No |
| 20 | F | 86 | POAG | - | 1 OU | No |
| 21 | F | 75 | POAG | - | 3 OU | Yes |
| 22 | M | 69 | POAG | - | 4OS; 3 OD | Yes |
| 23 | F | 59 | POAG | - | 1 OS glaucoma suspect; 1 OD | Yes |
| 24 | M | 76 | POAG | + | 3 OS; 1 OD | Yes |
| 25 | F | 69 | POAG | - | OU glaucoma suspect | No |
| 26 | M | 80 | POAG | + | 2 OS; 1 OD | Yes |
| 27 | M | 69 | POAG | - | 1 OS; 3 OD | Yes |
| 28 | F | 80 | POAG | - | 2 OU | No |
| 29 | M | 53 | POAG | - | 3 OU | Yes |
| 30 | M | 76 | POAG | + | 2 OS; 3 OD | Yes |
| 31 | F | 85 | POAG | - | 2 OS, OD anophthalmia | No |
| 32 | M | 64 | POAG | - | 3 OU | No |
| 33 | F | 68 | POAG | - | OU glaucoma suspect | No |
| 34 | F | 71 | POAG | - | 3 OS; 1 OD | No |
| 35 | M | 79 | POAG | - | 4 OS; 3 OD | Yes |
| 36 | F | 68 | POAG | - | 1 OS; 2 OD | Yes |
| 37 | F | 71 | POAG | - | 2 OU | Yes |
| 38 | M | 65 | POAG | + | 1 OS; 2 OD | No |

"+"—positive for autoantibodies to γ-synuclein; "-"—no autoantibodies to γ-synuclein detected; NA—not applicable.

The exclusion criteria were significant refractive errors (high myopia and hyperopia, astigmatism above 2.0 diopters), non-glaucoma pathology of the optic nerve, severe clouding of the cornea, and a history of acute cerebrovascular accidents. Blood taken for the standard biochemical analysis was used for the study. The serum was obtained and stored at a temperature of −80 °C.

### 2.2. Detection of Autoantibodies to γ-Synuclein

Immunoblotting was used to detect autoantibodies to γ-synuclein in blood serum [24]. Electrophoretic separation of 0.15 μg of purified recombinant human γ-synuclein protein in 14% sodium dodecyl sulfate–denaturing polyacrylamide gel was performed. Then, the samples were transferred to a polyvinylidene fluoride (PVDF) membrane (Hybon-P, Amersham, Sheffield, UK) using a semi-dry method. Membrane block was performed in a solution of 4% non-fat dry milk (NFDM) prepared in saline buffer TBST (Tris-buffered saline, 0.1% Tween 20). Then, it was incubated with the serum samples diluted 200 times in a solution of 4% NFDM in TBST overnight at 4 °C. Further incubation was performed with antibodies conjugatedwith horseradish peroxidase against human IgG (Bio-Rad, Hercules, CA, USA) in the same buffer as the serum samples (1:3000 dilution) for 1.5 h at room temperature. The detection of the resulting complex of primary and secondary autoantibodies was performed using a chemiluminescence detection kit (ECL Plus, Thermo Fisher Scientific, Waltham, MA, USA) according to the manufacturer's instructions. X-ray film (Thermo Scientific, Waltham, MA, USA) was used to detect chemiluminescence. A band corresponding to the molecular weight of γ-synuclein 17 kDa was detected on the X-ray film if the antibodies to γ-synuclein were present in the tested serum.

*2.3. Experimental Animals*

We used γ-synuclein (Sncg) knockout mice described previously [29]. The animals were provided by Professor V. Buchman. The name and catalog number of The Jackson Laboratory are B6.129P2-Sncgtm1Vlb/J and 008843, respectively. Homozygous γ-KO mice and WT control mice were generated by intercrossing heterozygous knockout and C57BL/6J mice. Experimental animals were housed at a 12 h light/12 h dark cycle, with food and water supplied ad libitum. The procedures were carried out in accordance with the "Guidelines for accommodation and care of animals. Species-specific provisions for laboratory rodents and rabbits" (GOST 33216-2014) in compliance with the principles enunciated in Directive 2010/63/EU on the protection of animals used for scientific purposes and approved by the local Institute Ethics Review Committee of the IPAC RAS (protocol No. 48, 15 January 2021). All efforts were made to minimize the number of animals and their suffering. All mice were genotyped using PCR analysis of DNA obtained from an ear biopsy, as described elsewhere [30].

Experimental animals (γ-KO and WT mice) underwent single instillations of 10 μL of 1% phenylephrine hydrochloride (Mesaton, Dalkhimfarm, Khabarovsk, Russia), 0.5% timolol maleate (Timolol, Diapharm, Moscow, Russia), 1% pilocarpine (Pilocarpine, Renewal, Novosibirsk, Russia), 0.1% atropine sulfate (Atropine, Dalkhimfarm, Khabarovsk, Russia), 10% dopamine (Sigma-Aldrich, Burlington, MA, USA) in 0.9% NaCl, and 0.25% haloperidol (Haloperidol, Velpharm, Moscow, Russia) in both eyes. IOP was measured under general anesthesia (Avertin, 200 mg/kg, intraperitoneally) using the automatic electronic tonometer Tonovet (Icare Finland Oy, Vantaa, Finland) in the morning, before and after instillation (1 and 2 h later). In our preliminary experiments, it was established that general anesthesia by Avertin does not affect the level of IOP in either γ-KO or WT mice.

Tear fluid was collected from both eyes with a filter paper strip (2.5 mm width wide), which was placed behind the lower eyelid for 5 min. Then, tear fluid components from both eyes from a mouse were eluted for 20 min with saline (50 μL) in one tube, centrifuged for 10 min at 3000 rpm, and the supernatant was used for testing. In the tear fluid eluate, the protein concentration was determined according to Lowry [31], and the activity of α2-MG was determined by the enzymatic method with the specific substrate N-benzoyl-DL-arginine-p-nitroanilide (BAPN), as described previously [32]. The activity of α2-MG was expressed as nmol/min × mL of tear fluid and nmol/min × mg of protein. The optical density of the samples was determined on a multifunctional photometer for microplates Synergy MX (BioTek, Winooski, VT, USA). Statistical analysis was performed using GraphPad Prism 6 software (GraphPad Software, San Diego, CA, USA).

## 3. Results

*3.1. Autoantibodies to γ-Synuclein Were Found in the Blood Serum of Some Patients with Glaucoma*

γ-synuclein autoantibodies were detected in the blood serum of 20% of the patients (5 out of 25) with POAG (Figure 1a,b and Figure S1 and Table 1). In the control group, γ-synuclein autoantibodies were not detected in the blood serum.

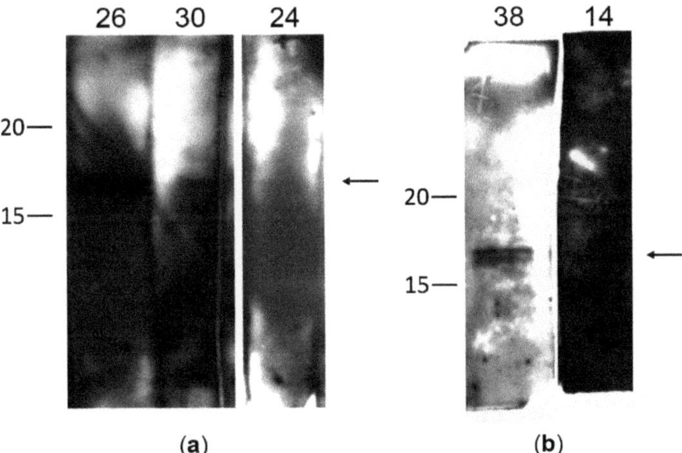



**Figure 1.** Detection of autoantibodies to γ-synuclein in the serum of patients with glaucoma by immunoblotting: (**a**,**b**) correspond to different independent membranes divided into strips. Each strip of membrane was incubated with a serum sample from one patient. On the left, the molecular weight (kDa) is indicated according to the protein ladder. The numbers 14, 24, 26, 30, 38 indicate the numbers of serum samples from patients with glaucoma, in which autoantibodies to γ-synuclein (17 kDa) were detected. The arrow on the right marks the band corresponding to γ-synuclein.

All samples with a positive reaction were reanalyzed to confirm the results. In most cases, immunoblotting showed a single immunoreactive band of 17 kDa, corresponding to γ-synuclein. However, in one case (sample 38), two bands were observed, which may be due to partial degradation of the recombinant protein or contamination of the sample with bacterial proteins. The signal intensity varied for different patients. The high band intensity for some patients (sample 26 in Figure 1a) may indicate a fairly high titer of γ-synuclein autoantibodies in this patient.

Thus, the presence of γ-synuclein autoantibodies in the blood serum was detected in 20% of the patients with POAG. Patients with detected autoantibodies to γ-synuclein were diagnosed with different stages of glaucoma development from 1a to 4a. No correlation was found between the stage of the pathological process, the IOP level, and the presence of γ-synuclein autoantibodies. Subsequent studies on a greater number of patients may be required to obtain more reliable results.

### 3.2. Loss of γ-Synuclein Function Led to Dysregulation of IOP in Mice

We studied IOP changes in γ-KO mice during the instillation of drugs affecting IOP regulation to determine what effect lack of γ-synuclein can have on hydrodynamics of the eye.

We used agonists and antagonists for the three main mediator systems that control IOP: adrenergic agonist phenylephrine, adrenergic antagonist timolol, cholinergic agonist pilocarpine, cholinergic antagonist atropine, dopamine agonist dopamine, and dopamine antagonist haloperidol. There was a difference in the dynamics of changes in IOP between the group of γ-KO mice and WT mice after instillation into the eyes of mice with drugs that affect one of the three mediator systems.

The adrenergic agonist 1% phenylephrine caused IOP reduction in γ-KO mice by an average of 4 mmHg after 1 h ($p = 0.0017$) and 3 mmHg after 2 h ($p = 0.0222$). In WT mice, there was a tendency to the opposite dynamic of changes in IOP. However, there were no significant differences between the first (0 h) and subsequent (1 h and 2 h) time points (Figure 2a). The adrenergic antagonist 0.5% timolol caused IOP reduction in both groups.

However, a significant difference between time points was achieved only for WT mice, with no differences between groups (Figure 2b).

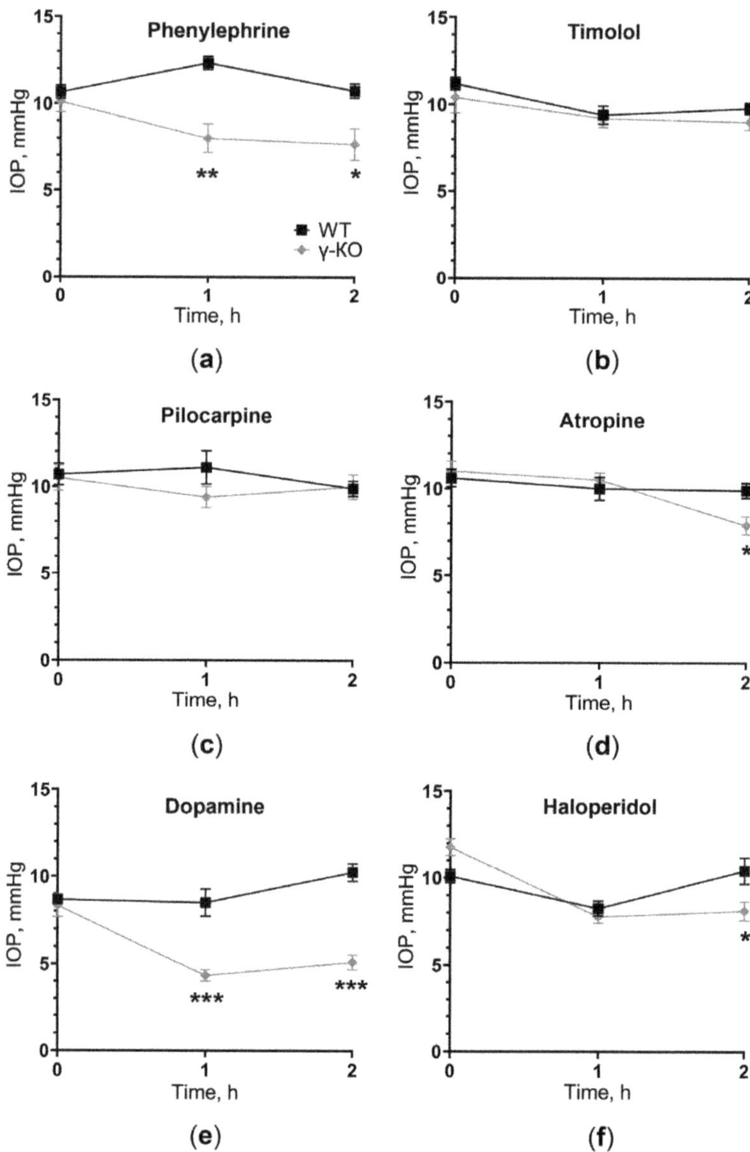

**Figure 2.** IOP change in γ-synuclein knockout (γ-KO) mice compared to wild-type (WT) mice after instillation of drugs affecting different neuromediators: (**a**) adrenergic agonist 1% phenylephrine, (**b**) adrenergic antagonist 0.5% timolol, (**c**) cholinergic agonist 1% pilocarpine, (**d**) cholinergic antagonist 0.1% atropine, (**e**) dopamine agonist 10% dopamine and (**f**) dopamine antagonist 0.25% haloperidol. IOP was measured before (0 h) and 1 and 2 h after drugs installation. Mean $\pm$ SE are presented. Statistical analysis was performed using ANOVA followed by Holm-Sidak's multiple comparisons test between groups (* $p < 0.05$, ** $p < 0.05$, *** $p < 0.001$). For each group, $n = 5$ (10 eyes).

The cholinergic agonist 1% pilocarpine did not cause any changes in IOP dynamics relative to baseline values and between groups (Figure 2c). The cholinergic antagonist 0.1% atropine caused a significant IOP reduction in γ-KO mice after 2 h by an average of 2 mmHg ($p = 0.0271$), compared to that of WT mice, as well as in comparison with the initial value by 3 mmHg (0.0018). In WT mice, atropine did not affect the level of IOP (Figure 2d).

The dopamine agonist 10% dopamine caused IOP reduction in γ-KO mice from 8 to 4 mmHg ($p = 0.0011$) after 1 h and from 8 to 5 mm Hg. ($p = 0.0011$) after 2 h from baseline. IOP in WT mice was twice as high as IOP in γ-KO mice throughout the experiment ($p = 0.0004$ and $p < 0.0001$ for 1 h and 2 h, respectively). IOP in WT mice did not change significantly (Figure 2e). The dopamine antagonist 0.25% haloperidol caused IOP reduction after 1 h in both groups, relative to the initial values. After 2 h, the IOP level in WT mice returned to the original values, while in γ-KO mice it remained lower by 2 mmHg than that of the WT mice ($p = 0.0063$, Figure 2f).

Therefore, the impact on any of the three mediator systems, at least in one direction (stimulation or inhibition), led to changes in the IOP dynamic in γ-KO mice compared to WT mice. The most prominent effect was manifested upon stimulation of dopamine reception, and IOP changes in knockout mice were observed with the use of both an agonist and an antagonist of dopamine receptors.

### 3.3. The Total Protein Concentration and the Activity of the α2-Macroglobulin Changed in the Tear Fluid of γ-KO Mice

In the tear fluid of γ-KO mice, a higher content of total protein was found than in the tear fluid of WT mice (Figure 3a). Measuring α2-MG activity in the tear fluid of γ-KO mice showed a tendency to the activity decrease normalized to the volume of tear fluid (Figure 3b), and a significant decrease by 70% normalized to the total amount of protein (mg) (Figure 3c).

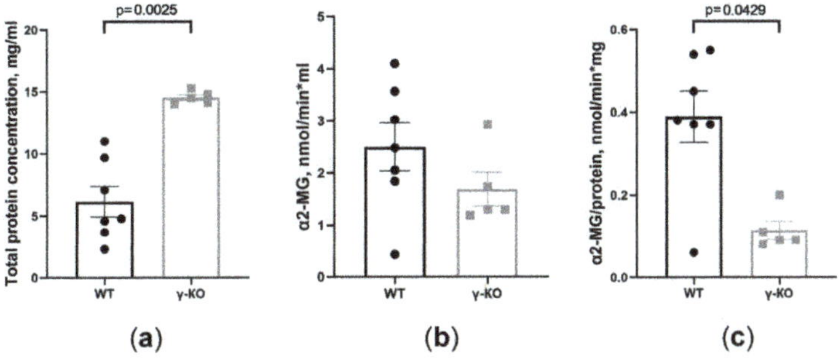

**Figure 3.** Total protein concentration and α2-macroglobulin (α2-MG) activity are different in tear fluid of γ-synuclein knockout (γ-KO) mice than those of wild-type (WT) mice: (**a**) total protein concentration determined by the Lowry assay (mg/mL); (**b**) total activity (nmol/min × mL); (**c**) specific activity (nmol/min × mg) of α2-MG in tear fluid of γ-KO and WT mice. Means ± SE with individual data are presented and p values are given if differences between groups were statistically significant according to the Mann–Whitney test. For the WT mice group, $n = 7$; for the γ-KO group, $n = 5$.

## 4. Discussion

Autoantibodies to γ-synuclein are not detected in individuals without glaucoma. The presence of autoantibodies to γ-synuclein in the blood of 20% of the patients with POAG indicates that this protein, at least in some cases, is involved in the pathogenesis of glaucoma. Based on the multifactorial nature of glaucoma development, we assume that γ-synuclein is involved in this process in varying degrees. Previously, we showed that autoantibodies against γ-synuclein are also detected in patients with chronic cerebral ischemia, with the frequency of detection of one in 7.6 people, and in patients with ALS, with the frequency of detection of one in 20.5 people [24]. The results of this study indicate that this rate is higher in patients with glaucoma (one in five people). We assume that a common mechanism explaining the formation of autoantibodies to γ-synuclein in the above-mentioned diseases is γ-synuclein or its oligomeric-aggregated forms entering the bloodstream in cases of histohematic barriers disruption. Some proteins, including synucleins, change localization and accumulate in certain layers and/or types of retinal cells in neurodegenerative diseases [33,34]. Moreover, in neurodegeneration, the aggregation of protein occurs in the retina, as it does in the brain [35,36]. In glaucoma, changes in the expression and localization of γ-synuclein were noted in the optic nerve. Optic nerve fibers are normally immunopositive for γ-synuclein. However, strong γ-synuclein immunopositive staining of the optic nerve was observed in the area of lamina cribrosa and in the postlamina area only in eyes with glaucoma. A model of glaucoma reproduced by cauterization of episcleral veins shows a decrease in the amount of both γ-synuclein mRNA and the protein in the optic nerve. During the incubation of rat astrocyte culture at elevated hydrostatic pressure, the amount of γ-synuclein decreased [20]. The expression of γ-synuclein at the mRNA and protein levels decreased in the retina of genetic glaucoma (DBA/2J) mice and optic nerve crush (ONC) mice [37,38]. At the same time, accumulation of γ-synuclein was observed only in the optic nerve of ONC mice but not in that of DBA/2J mice [37]. Despite the obtained experimental evidence of changes in the expression and localization of γ-synuclein in the eye during the glaucomatous process, the exact mechanisms of implication of this protein in the pathogenesis of glaucoma remain unclear. The formation of autoantibodies against γ-synuclein and the lack of protein function may contribute to the pathogenesis of glaucoma. Nevertheless, some data indicate that autoantibodies against γ-synuclein can be a protective factor that ensures removing γ-synuclein excess during a critical increase in its intracellular concentration and removing its misfolded and aggregated forms. This is indirectly confirmed by the neuroprotective effect of autoantibodies to γ-synuclein in cultured neuroretinal cells [27,28].

One of the mechanisms of γ-synuclein involvement in glaucoma pathogenesis may be its participation in the neurotransmission of monoamines, which may affect IOP [39,40]. All three members of the synuclein family modulate the transport, expression, and function of monoamine transporters on the cell surface, thereby playing a central role in the regulation of monoamine reuptake [40]. In γ-synuclein knockout mice, behavioral testing revealed a moderate deficiency of the dopaminergic neurotransmitter system [41]. γ-synuclein can regulate the activity of the serotonin transporter that reuptakes serotonin from the synaptic cleft [42]. The level of IOP depends on the rate of outflow and the formation of intraocular fluid, which is influenced by the adrenergic, dopaminergic, and cholinergic systems [39]. In this study, we showed that the IOP dynamic response is different in γ-KO mice compared to that in WT mice when mediator regulators are instilled. IOP decreases in γ-KO mice after instillations of a dopaminergic agonist and antagonist, an adrenergic agonist, and an antagonist of cholinergic receptors. At the same time, IOP does not change or tends to increase in WT mice, as in the case of phenylephrine. This may be due to dysfunction of the neuronal control of IOP in γ-KO mice and disruption of the correct interaction among mediators of the adrenergic, dopaminergic, and cholinergic systems [43,44]. Further investigation is needed with additional selective receptor agonists and antagonists.

One of the markers of a neurodegenerative process, not only in the retina but also in the brain is an increase in the level of α2-MG in the tear fluid [32]. α2-MG is a protein of acute phase inflammation, an inhibitor of a wide range of proteolytic enzymes, and a regulator of the function of many cytokines and growth factors. As an extracellular chaperone, α2-MG prevents the formation of misfolded proteins and their conglomerates, including amyloid-β, which accumulates in the retina in glaucoma [45–49]. We noticed a significant increase in the concentration of total protein in the tear fluid of γ-KO mice. This is possibly related to a decrease in the water flow into the tear fluid from the lacrimal glands and the formation of a more concentrated tear fluid. A significant decrease of α2-MG activity in the tear fluid of γ-KO mice indicates a dysregulation of its protective effect. An increase in α2-MG activity is a response to neuroinflammation and the formation of malformed proteins, aimed at protecting nerve cells from death [46]. Therefore, our data and other data indicate that the lack of γ-synuclein leads to weakening of protective mechanisms that prevent the development of neurodegenerative processes, including such processes in the eyes.

## 5. Conclusions

The detected autoantibodies in 20% of the patients with POAG confirmed the involvement of this factor in the pathogenesis in some forms of glaucoma. The change in IOP response to the instillation of drugs that affect the hydrodynamics of the eye in γ-KO mice indicated that the loss of this protein function can lead to IOP dysregulation. In the tear fluid of γ-KO mice, the activity of α2-MG controlling the processes of neuroinflammation and the formation of misfolded proteins was reduced, whereas the total amount of protein was increased. Therefore, γ-synuclein dysfunction in glaucoma contributes to impaired IOP regulation and development of neurodegenerative processes in the retina—the main factors leading to optic neuropathy and loss of visual functions in glaucoma.

**Supplementary Materials:** The following supporting information can be downloaded at: https://www.mdpi.com/article/10.3390/biomedicines11010060/s1, Figure S1: Original, unmodified images of X-ray films for Western blots. Numbers indicate all positive serum samples. Cntr+ sample—positive serum from our previous work.

**Author Contributions:** Conceptualization, M.S.K., T.A.P. and R.K.O.; methodology, A.Y.R. and N.G.D.; validation, O.A.L., M.S.K. and R.K.O.; formal analysis, M.S.K.; investigation, A.Y.R., N.E.P., I.S.S. and V.O.S.; resources, M.S.K. and N.B.C.; data curation, M.S.K.; writing—original draft preparation, M.S.K. and T.A.P.; writing—review and editing, M.W.E. and N.B.C.; visualization, M.S.K.; supervision, M.S.K. and R.K.O.; funding acquisition, R.K.O. All authors have read and agreed to the published version of the manuscript.

**Funding:** This work was supported by the Russian Science Foundation (grant no. 22-75-00112). The collection of clinical specimens was performed as part of the State Assignment of the HMRIED (Research and Technological Development project No. AAAA-A21-121011190051-2). Animals were provided and supported by Bioresource Collection of IPAC RAS and Centre for Collective Use IPAC RAS facilities and equipment was used to maintain animals in the framework of the State Assignment of IPAC RAS (FFSN-2021-0005).

**Institutional Review Board Statement:** The study was conducted according to the guidelines of the Declaration of Helsinki and approved by the Institutional Ethics Committee of the HMRIED (protocol No. 55/3, 17 June 2021). The animal study protocol was approved by the local Institute Ethics Review Committee of the IPAC RAS (protocol No. 48, 15 January 2021). All animal work was carried out in accordance with the "Guidelines for accommodation and care of animals. Species-specific provisions for laboratory rodents and rabbits" (GOST 33216-2014) in compliance with the principles enunciated in the Directive 2010/63/EU on the protection of animals used for scientific purposes.

**Informed Consent Statement:** Informed consent was obtained from all subjects involved in the study and no private information identifying participants was made public. No additional samples for the purpose of this study were collected. Patients were informed of the research and their nonobjection approval was confirmed.

**Data Availability Statement:** Data are contained within the current article and its supplementary material.

**Conflicts of Interest:** The authors declare no conflict of interest.

## References

1. Tuulonen, A.; Airaksinen, P.J.; Erola, E.; Forsman, E.; Friberg, K.; Kaila, M.; Klemetti, A.; Makela, M.; Oskala, P.; Puska, P.; et al. The Finnish evidence-based guideline for open-angle glaucoma. *Acta Ophthalmol. Scand.* **2003**, *81*, 3–18. [CrossRef] [PubMed]
2. Ophthalmol, B.J. European Glaucoma Society Terminology and Guidelines for Glaucoma, 4th Edition—Chapter 2: Classification and terminologySupported by the EGS Foundation: Part 1: Foreword; Introduction; Glossary; Chapter 2 Classification and Terminology. *Br. J. Ophthalmol.* **2017**, *101*, 73–127. [CrossRef]
3. Quigley, H.A.; Broman, A.T. The number of people with glaucoma worldwide in 2010 and 2020. *Br. J. Ophthalmol.* **2006**, *90*, 262–267. [CrossRef] [PubMed]

4.  Friedman, D.S.; Wilson, M.R.; Liebmann, J.M.; Fechtner, R.D.; Weinreb, R.N. An evidence-based assessment of risk factors for the progression of ocular hypertension and glaucoma. *Am. J. Ophthalmol.* **2004**, *138*, S19-312004. [CrossRef]

5.  Adornetto, A.; Russo, R.; Parisi, V. Neuroinflammation as a target for glaucoma therapy. *Neural. Regen. Res.* **2019**, *14*, 391–394. [CrossRef] [PubMed]

6.  Carelli, V.; La Morgia, C.; Ross-Cisneros, F.N.; Sadun, A.A. Optic neuropathies: The tip of the neurodegeneration iceberg. *Hum. Mol. Genet.* **2017**, *26*, R139-R1502017. [CrossRef]

7.  Jiang, S.; Kametani, M.; Chen, D.F. Adaptive Immunity: New Aspects of Pathogenesis Underlying Neurodegeneration in Glaucoma and Optic Neuropathy. *Front. Immunol.* **2020**, *11*, 65. [CrossRef] [PubMed]

8.  Bras, I.C.; Dominguez-Meijide, A.; Gerhardt, E.; Koss, D.; Lazaro, D.F.; Santos, P.I.; Vasili, E.; Xylaki, M.; Outeiro, T.F. Synucleinopathies: Where we are and where we need to go. *J. Neurochem.* **2020**, *153*, 433–454. [CrossRef]

9.  Surguchov, A. Intracellular Dynamics of Synucleins: "Here, There and Everywhere". *Int. Rev. Cell. Mol. Biol.* **2015**, *320*, 103–169. [CrossRef]

10.  Carnazza, K.E.; Komer, L.E.; Xie, Y.X.; Pineda, A.; Briano, J.A.; Gao, V.; Na, Y.; Ramlall, T.; Buchman, V.L.; Eliezer, D.; et al. Synaptic vesicle binding of alpha-synuclein is modulated by beta- and gamma-synucleins. *Cell. Rep.* **2022**, *39*, 110675. [CrossRef]

11.  Burre, J.; Sharma, M.; Sudhof, T.C. Cell Biology and Pathophysiology of alpha-Synuclein. *Cold Spring Harb. Perspect. Med.* **2018**, *8*, a024091. [CrossRef] [PubMed]

12.  Spillantini, M.G.; Schmidt, M.L.; Lee, V.M.; Trojanowski, J.Q.; Jakes, R.; Goedert, M. Alpha-synuclein in Lewy bodies. *Nature* **1997**, *388*, 839–840. [CrossRef] [PubMed]

13.  Ueda, K.; Fukushima, H.; Masliah, E.; Xia, Y.; Iwai, A.; Yoshimoto, M.; Otero, D.A.; Kondo, J.; Ihara, Y.; Saitoh, T. Molecular cloning of cDNA encoding an unrecognized component of amyloid in Alzheimer disease. *Proc. Natl. Acad. Sci. USA* **1993**, *90*, 11282–11286. [CrossRef] [PubMed]

14.  Ninkina, N.; Peters, O.; Millership, S.; Salem, H.; van der Putten, H.; Buchman, V.L. Gamma-synucleinopathy: Neurodegeneration associated with overexpression of the mouse protein. *Hum. Mol. Genet.* **2009**, *18*, 1779–1794. [CrossRef] [PubMed]

15.  Peters, O.M.; Millership, S.; Shelkovnikova, T.A.; Soto, I.; Keeling, L.; Hann, A.; Marsh-Armstrong, N.; Buchman, V.L.; Ninkina, N. Selective pattern of motor system damage in gamma-synuclein transgenic mice mirrors the respective pathology in amyotrophic lateral sclerosis. *Neurobiol. Dis.* **2012**, *48*, 124–131. [CrossRef]

16.  Peters, O.M.; Shelkovnikova, T.; Highley, J.R.; Cooper-Knock, J.; Hortobagyi, T.; Troakes, C.; Ninkina, N.; Buchman, V.L. Gamma-synuclein pathology in amyotrophic lateral sclerosis. *Ann. Clin. Transl. Neurol.* **2015**, *2*, 29–37. [CrossRef]

17.  Surgucheva, I.; Newell, K.L.; Burns, J.; Surguchov, A. New alpha- and gamma-synuclein immunopathological lesions in human brain. *Acta Neuropathol. Commun.* **2014**, *2*, 132. [CrossRef]

18.  Liu, H.; Liu, W.; Wu, Y.; Zhou, Y.; Xue, R.; Luo, C.; Wang, L.; Zhao, W.; Jiang, J.D.; Liu, J. Loss of epigenetic control of synuclein-gamma gene as a molecular indicator of metastasis in a wide range of human cancers. *Cancer Res.* **2005**, *65*, 7635–7643. [CrossRef]

19.  Surguchov, A. γ-Synuclein as a Cancer Biomarker: Viewpoint and New Approaches. *Oncomedicine* **2016**, *1*, 1–3. [CrossRef]

20.  Surgucheva, I.; McMahan, B.; Ahmed, F.; Tomarev, S.; Wax, M.B.; Surguchov, A. Synucleins in glaucoma: Implication of gamma-synuclein in glaucomatous alterations in the optic nerve. *J. Neurosci. Res.* **2002**, *68*, 97–106. [CrossRef]

21.  Surgucheva, I.; Weisman, A.D.; Goldberg, J.L.; Shnyra, A.; Surguchov, A. Gamma-synuclein as a marker of retinal ganglion cells. *Mol. Vis.* **2008**, *14*, 1540–1548. [PubMed]

22.  Gold, M.; Pul, R.; Bach, J.P.; Stangel, M.; Dodel, R. Pathogenic and physiological autoantibodies in the central nervous system. *Immunol. Rev.* **2012**, *248*, 68–86. [CrossRef] [PubMed]

23.  Papachroni, K.; Ninkina, N.; Papapanagiotou, A.; Hadjigeorgiou, G.; Xiromerisiou, G.; Papadimitriou, A.; Kalofoutis, A.; Buchman, V.L. Autoantibodies to alpha-synuclein in inherited Parkinson's disease. *Neurochem* **2007**, *101*, 749–756. [CrossRef] [PubMed]

24.  Roman, A.Y.; Kovrazhkina, E.A.; Razinskaya, O.D.; Kukharsky, M.S.; Maltsev, A.V.; Ovchinnikov, R.K.; Lytkina, O.A.; Smirnov, A.P.; Moskovtsev, A.A.; Borodina, Y.V.; et al. Detection of autoantibodies to potentially amyloidogenic protein, gamma-synuclein, in the serum of patients with amyotrophic lateral sclerosis and cerebral circulatory disorders. *Dokl. Biochem. Biophys.* **2017**, *472*, 64–67. [CrossRef] [PubMed]

25.  Bell, K.; Und Hohenstein-Blaul, N.V.T.; Teister, J.; Grus, F. Modulation of the Immune System for the Treatment of Glaucoma. *Curr. Neuropharmacol.* **2018**, *16*, 942–958. [CrossRef] [PubMed]

26.  Von Thun Und Hohenstein-Blaul, N.; Kunst, S.; Pfeiffer, N.; Grus, F.H. Biomarkers for glaucoma: From the lab to the clinic. *Eye* **2017**, *31*, 225–231. [CrossRef] [PubMed]

27.  Bell, K.; Wilding, C.; Funke, S.; Perumal, N.; Beck, S.; Wolters, D.; Holz-Muller, J.; Pfeiffer, N.; Grus, F.H. Neuroprotective effects of antibodies on retinal ganglion cells in an adolescent retina organ culture. *J. Neurochem.* **2016**, *139*, 256–269. [CrossRef] [PubMed]

28.  Wilding, C.; Bell, K.; Beck, S.; Funke, S.; Pfeiffer, N.; Grus, F.H. gamma-Synuclein antibodies have neuroprotective potential on neuroretinal cells via proteins of the mitochondrial apoptosis pathway. *PLoS ONE* **2014**, *9*, e90737. [CrossRef]

29.  Ninkina, N.; Papachroni, K.; Robertson, D.C.; Schmidt, O.; Delaney, L.; O'Neill, F.; Court, F.; Rosenthal, A.; Fleetwood-Walker, S.M.; Davies, A.M.; et al. Neurons expressing the highest levels of gamma-synuclein are unaffected by targeted inactivation of the gene. *Mol. Cell. Biol.* **2003**, *23*, 8233–8245. [CrossRef]

30.  Anwar, S.; Peters, O.; Millership, S.; Ninkina, N.; Doig, N.; Connor-Robson, N.; Threlfell, S.; Kooner, G.; Deacon, R.M.; Bannerman, D.M.; et al. Functional alterations to the nigrostriatal system in mice lacking all three members of the synuclein family. *J. Neurosci.* **2011**, *31*, 7264–7274. [CrossRef]

31. Lowry, O.H.; Rosebrough, N.J.; Farr, A.L.; Randall, R.J. Protein measurement with the Folin phenol reagent. *J. Biol. Chem.* **1951**, *193*, 265–275. [CrossRef] [PubMed]

32. Bogdanov, V.; Kim, A.; Nodel, M.; Pavlenko, T.; Pavlova, E.; Blokhin, V.; Chesnokova, N.; Ugrumov, M. A Pilot Study of Changes in the Level of Catecholamines and the Activity of alpha-2-Macroglobulin in the Tear Fluid of Patients with Parkinson's Disease and Parkinsonian Mice. *Int. J. Mol. Sci.* **2021**, *22*, 4736. [CrossRef] [PubMed]

33. Maurage, C.A.; Ruchoux, M.M.; de Vos, R.; Surguchov, A.; Destee, A. Retinal involvement in dementia with Lewy bodies: A clue to hallucinations? *Ann. Neurol.* **2003**, *54*, 542–547. [CrossRef]

34. Surguchev, A.; Surguchov, A. Conformational diseases: Looking into the eyes. *Brain Res. Bull.* **2010**, *81*, 12–24. [CrossRef] [PubMed]

35. Grimaldi, A.; Brighi, C.; Peruzzi, G.; Ragozzino, D.; Bonanni, V.; Limatola, C.; Ruocco, G.; Di Angelantonio, S. Inflammation, neurodegeneration and protein aggregation in the retina as ocular biomarkers for Alzheimer's disease in the 3xTg-AD mouse model. *Cell Death Dis.* **2018**, *9*, 685. [CrossRef] [PubMed]

36. Leger, F.; Fernagut, P.O.; Canron, M.H.; Leoni, S.; Vital, C.; Tison, F.; Bezard, E.; Vital, A. Protein aggregation in the aging retina. *J. Neuropathol. Exp. Neurol.* **2011**, *70*, 63–68. [CrossRef] [PubMed]

37. Liu, Y.; Tapia, M.L.; Yeh, J.; He, R.C.; Pomerleu, D.; Lee, R.K. Differential Gamma-Synuclein Expression in Acute and Chronic Retinal Ganglion Cell Death in the Retina and Optic Nerve. *Mol. Neurobiol.* **2020**, *57*, 698–709. [CrossRef] [PubMed]

38. Soto, I.; Oglesby, E.; Buckingham, B.P.; Son, J.L.; Roberson, E.D.; Steele, M.R.; Inman, D.M.; Vetter, M.L.; Horner, P.J.; Marsh-Armstrong, N. Retinal ganglion cells downregulate gene expression and lose their axons within the optic nerve head in a mouse glaucoma model. *J. Neurosci.* **2008**, *28*, 548–561. [CrossRef] [PubMed]

39. Pescosolido, N.; Parisi, F.; Russo, P.; Buomprisco, G.; Nebbioso, M. Role of dopaminergic receptors in glaucomatous disease modulation. *Biomed. Res. Int.* **2013**, *2013*, 193048. [CrossRef]

40. Oaks, A.W.; Sidhu, A. Synuclein modulation of monoamine transporters. *FEBS Lett.* **2011**, *585*, 1001–1006. [CrossRef]

41. Kokhan, V.S.; Kokhan, T.Y.G.; Samsonova, A.N.; Fisenko, V.P.; Ustyugov, A.A.; Aliev, G. The Dopaminergic Dysfunction and Altered Working Memory Performance of Aging Mice Lacking Gamma-synuclein Gene. *CNS Neurol. Disord. Drug Targets* **2018**, *17*, 604–607. [CrossRef] [PubMed]

42. Wersinger, C.; Sidhu, A. Partial regulation of serotonin transporter function by gamma-synuclein. *Neurosci. Lett.* **2009**, *453*, 157–161. [CrossRef] [PubMed]

43. Lei, S. Cross interaction of dopaminergic and adrenergic systems in neural modulation. *Int. J. Physiol. Pathophysiol. Pharmacol.* **2014**, *6*, 137–142. [PubMed]

44. Schwahn, H.N.; Kaymak, H.; Schaeffel, F. Effects of atropine on refractive development, dopamine release, and slow retinal potentials in the chick. *Vis. Neurosci.* **2000**, *17*, 165–176. [CrossRef]

45. Rehman, A.A.; Ahsan, H.; Khan, F.H. alpha-2-Macroglobulin: A physiological guardian. *J. Cell. Physiol.* **2013**, *228*, 1665–1675. [CrossRef]

46. Cater, J.H.; Wilson, M.R.; Wyatt, A.R. Alpha-2-Macroglobulin, a Hypochlorite-Regulated Chaperone and Immune System Modulator. *Oxid. Med. Cell. Longev.* **2019**, *2019*, 5410657. [CrossRef]

47. Wyatt, A.R.; Constantinescu, P.; Ecroyd, H.; Dobson, C.M.; Wilson, M.R.; Kumita, J.R.; Yerbury, J.J. Protease-activated alpha-2-macroglobulin can inhibit amyloid formation via two distinct mechanisms. *FEBS Lett.* **2013**, *587*, 398–403. [CrossRef]

48. Inyushin, M.; Zayas-Santiago, A.; Rojas, L.; Kucheryavykh, Y.; Kucheryavykh, L. Platelet-generated amyloid beta peptides in Alzheimer's disease and glaucoma. *Histol. Histopathol.* **2019**, *34*, 843–856. [CrossRef]

49. Wang, L.; Mao, X. Role of Retinal Amyloid-beta in Neurodegenerative Diseases: Overlapping Mechanisms and Emerging Clinical Applications. *Int. J. Mol. Sci.* **2021**, *22*, 2360. [CrossRef]

*Article*

# Loss of the Synuclein Family Members Differentially Affects Baseline- and Apomorphine-Associated EEG Determinants in Single-, Double- and Triple-Knockout Mice

Vasily Vorobyov [1,2], Alexander Deev [3], Iuliia Sukhanova [4,5], Olga Morozova [6], Zoya Oganesyan [6], Kirill Chaprov [1,4,5] and Vladimir L. Buchman [1,5,*]

1   School of Biosciences, Sir Martin Evans Building, Cardiff University, Museum Avenue, Cardiff CF10 3AX, UK
2   Institute of Cell Biophysics, Russian Academy of Sciences, 142290 Pushchino, Russia
3   Institute of Theoretical and Experimental Biophysics, Russian Academy of Sciences, 142290 Pushchino, Russia
4   Institute of Physiologically Active Compounds, Russian Academy of Sciences, 142432 Chernogolovka, Russia
5   Center of Pre-Clinical and Clinical Studies, Belgorod State National Research University, 308015 Belgorod, Russia
6   International School "Medicine of the Future", I.M. Sechenov First Moscow State Medical University (Sechenov University), 119991 Moscow, Russia
*   Correspondence: buchmanvl@cf.ac.uk

**Abstract:** Synucleins comprise a family of small proteins highly expressed in the nervous system of vertebrates and involved in various intraneuronal processes. The malfunction of alpha-synuclein is one of the key events in pathogenesis of Parkinson disease and certain other neurodegenerative diseases, and there is a growing body of evidence that malfunction of other two synucleins might be involved in pathological processes in the nervous system. The modulation of various presynaptic mechanisms of neurotransmission is an important function of synucleins, and therefore, it is feasible that their deficiency might affect global electrical activity detected of the brain. However, the effects of the loss of synucleins on the frequency spectra of electroencephalograms (EEGs) have not been systematically studied so far. In the current study, we assessed changes in such spectra in single-, double- and triple-knockout mice lacking alpha-, beta- and gamma-synucleins in all possible combinations. EEGs were recorded from the motor cortex, the putamen, the ventral tegmental area and the substantia nigra of 78 3-month-old male mice from seven knockout groups maintained on the C57BL/6J genetic background, and 10 wild-type C57BL/6J mice for 30 min before and for 60 min after the systemic injection of a DA receptor agonist, apomorphine (APO). We found that almost any variant of synuclein deficiency causes multiple changes in both basal and APO-induced EEG oscillation profiles. Therefore, it is not the absence of any particular synuclein but rather a disbalance of synucleins that causes widespread changes in EEG spectral profiles.

**Keywords:** synucleins; dopamine; electroencephalogram; frequency spectrum

**Citation:** Vorobyov, V.; Deev, A.; Sukhanova, I.; Morozova, O.; Oganesyan, Z.; Chaprov, K.; Buchman, V.L. Loss of the Synuclein Family Members Differentially Affects Baseline- and Apomorphine-Associated EEG Determinants in Single-, Double- and Triple-Knockout Mice. *Biomedicines* **2022**, *10*, 3128. https://doi.org/10.3390/biomedicines10123128

Received: 20 October 2022
Accepted: 24 November 2022
Published: 4 December 2022

**Publisher's Note:** MDPI stays neutral with regard to jurisdictional claims in published maps and institutional affiliations.



## 1. Introduction

The synuclein family comprises three closely related proteins, specifically alpha-, beta- and gamma-synucleins (*alpha*-syn, *beta*-syn and *gamma*-syn), that are widely expressed in the nervous system of vertebrates, where they are involved in multiple intraneuronal processes, particularly those that are associated with synaptic neurotransmission [1,2]. *Alpha*-syn has been originally identified as a protein highly enriched in presynaptic terminals in various brain areas, where it is commonly colocalised with *beta*-syn [3]. The involvement of *alpha*-syn and *beta*-syn in chemical, particularly dopaminergic, neurotransmission has been linked with their role in the molecular pathogenesis of Parkinson disease (PD) [4,5], dementia with Lewy bodies (DLBs) and certain other neurodegenerative diseases (so-called synucleinopathies) [6–8]. *Gamma*-syn is predominantly expressed in sensory neurons; nevertheless, it has been shown to induce cortical astrocyte proliferation with subsequent BDNF expression and release [9]. Furthermore, *gamma*-syn is able to modify *beta*-syn-membrane interaction [10] and control midbrain dopamine (DA) function [11].

Various animal models based on the expression of mutated forms or simply overexpression of these proteins (*alpha*-syn, in particular) have been developed that displayed neurological dysfunction, which supported a notion about the gain of function, i.e., pathological aggregation, as the main cause of neurodegeneration in these diseases [12,13]. However, to better understand pathological mechanisms triggered by malfunction of synucleins, their roles in normal brain physiology need to be clarified. Interpretations of results of such studies are complicated by a potential functional redundancy within the family due to the high similarity of amino acid sequences and overlapping expression patterns [14,15]. In the triple-knockout (TKO) mice, some physiological mechanisms (age dependent, in particular) were found to be affected to a greater extent than in single- or double-synuclein knockouts (KOs) [16–18]. These studies also demonstrated that although synucleins are not essential for basic synaptic functions, they are required for the stabilisation and maintenance of DA level in the mediatory system [19,20]. It has also been revealed that learning abilities in tasks requiring intact spatial and working memory are compromised in *alpha*-syn KO mice, indicating an important role of *alpha*-syn in cognitive processes [21,22]. At the same time, *gamma*-syn KO mice are characterised by increased orientational and exploratory behaviour, reduced state anxiety and enhanced fear memory [23,24].

Synaptic dysfunctions and an imbalance between coordinated activities of different brain structures are hypothesised to be the main cause of abnormal functioning of the diseased brain [25]. It is feasible to suggest that by modifying synaptic transmission the deficiency of synuclein family member(s) would disrupt the neuronal network functioning, accompanied by modifications of electrical oscillations in the affected neuronal circuits [26], thus disrupting their interaction [27]. Superimposed extracellular fields arising from synaptic transmembrane currents of neurons involved in these circuits form the electroencephalogram (EEG) [28]. Changes in EEG patterns have been shown to be associated with PD pathology both in patients [29,30] and in animal models of PD [31,32]. In several recent EEG studies of PD, substantial attention has been paid to neuronal networks within/between the cortical and subcortical brain areas [33,34] and to the role of DA transmission in the functioning of these networks [20,35,36]. However, we are still lacking detailed information on how changes in the composition of synuclein family members affect the DA system. This needs to be clarified given data demonstrating a potential role of neuronal/synaptic plasticity in the DA system associated with synucleins [19]. In a few EEG studies, associations between modifications of *alpha*-syn and changes in the brain electrical activity have been shown [37,38]. Thus, a deeper insight into the activities of different neuronal networks and their changes in different types of synuclein KO mice would be beneficial for obtaining a better understanding of the role of synuclein family members in healthy and neurodegeneration-affected nervous systems.

In this study, we recorded EEGs from the motor cortex (MC), the putamen (Pt) and the DA-producing brain regions (ventral tegmental area (VTA) and substantia nigra (SN)) of adult mice with all possible combinations of *alpha*-, *beta*-, and *gamma*-syn KOs, before and after the systemic injection of a DA receptor agonist, apomorphine (APO). Genotype-specific and brain area–specific differences between various synuclein KO and control wild-type (WT) mice were revealed in the frequency spectra of both baseline and APO-evoked EEG from these brain areas.

## 2. Materials and Methods

### 2.1. Experimental Animals

In this study, 3-month-old male mice with different combinations of *alpha* B6(Cg)-Snca $^{tm1.2Vlb}$, *beta* B6(Cg)-Sncb $^{tm1Sud}$ and *gamma* B6(Cg)-Sncg $^{tm1Vlb}$ synuclein knockouts, all maintained on the C57BL/6J genetic background [39,40] and where WT mice originated from the same breeding programme, were used. Overall, eight cohorts of mice were compared: WT (A+B+G+, n = 10), ABG-KO (A-B-G-, n = 8), AG-KO (A-B+G-, n = 10), B-KO

(A+B-G+, n = 11), A-KO (A-B+G+, n = 12), BG-KO (A+B-G-, n = 13), G-KO (A+B+G-, n = 11) and AB-KO (A-B-G+, n = 13).

Up to the age of 2 months, animals were housed in groups of five per cage, and thereafter, each of them was kept for 1 month in an individual cage. Mice were housed in a standard environment (12 h light/dark cycle, 22–25 °C RT, 50%–55% relative humidity) with food and water ad libitum. The procedures were carried out in accordance with the "Guidelines for accommodation and care of animals. Species-specific provisions for laboratory rodents and rabbits" (GOST 33216-2014), in compliance with the principles enunciated in the Directive 2010/63/EU on the protection of animals used for scientific purposes and approved by the local Institute Ethics Review Committee (protocol № 48, 15.01.2021). All efforts were made to minimise the number of the animals and their suffering. All mice were genotyped using a PCR analysis of DNA obtained from the ear biopsy, as described elsewhere [20].

### 2.2. Implantation of Electrodes and EEG Recording

After 1 month of adaptation to the individual cage, each mouse was anesthetised with the subcutaneous (s.c.) injection of a combination of dissolved tiletamine/zolazepam (Zoletil®, Virbac, Carros, France) and xylazine solution (Rometar®, Bioveta, Ivanovice na Hané, Czech Republic) at doses of 25 mg/kg and 2.5 mg/kg, respectively. Four recording electrodes were implanted into the left MC and Pt (MC and Pt; AP: +1.1 mm anterior to bregma; ML: ±1.5 mm lateral to midline; DV: −0.75 and −2.75 mm depths from skull surface, respectively), into the left VTA (AP: −3.1, ML: −0.4, DV: −4.5) and into the right SN (AP: −3.2, ML: +1.3, DV: −4.3) [41] (DV was measured from the skull surface). Within brain areas analysed in this study, the opposite hemisphere for SN was chosen, first because of its proximity to VTA, which meant we could not exclude possible mutual damage during electrode implantation in the same hemisphere. Second, it is well known that the contralateral SN is the dominant source of DA in the opposite hemisphere. Custom-made electrodes were constructed from two varnish-insulated nichrome wires (100 μm diameter) glued together (3M Vetbond™ Tissue Adhesive, St. Paul, MN, USA) with 100 μm tips, free from insulation. Thus, the electrodes were sufficiently inflexible and had higher effective surface–volume ratio than a monowire electrode of a 200 μm diameter. The reference and ground electrodes (stainless steel wire, 0.4 mm in diameter) were placed symmetrically into the caudal cavities behind the cerebrum (AP: −5.3, ML: ±1.8, DV: −0.5). All electrodes were positioned using a computerised 3D stereotaxic StereoDrive (Neurostar, Tübingen, Germany), fixed to the skull with dental cement and soldered to a dual row socket connector (Sullins Connector Solutions, San Marcos, CA, USA). Each of nichrome wires was soldered to one of the connector's pins. After electrode implantation, animals were housed individually for the recovery, followed by the experimental sessions. The postmortem verification of the electrode tip location included a preliminary anodal current (80–100 μA, 1 s) coagulation of the adjacent tissue and extirpation of the brain. The brains were fixed in Carnoy's (60% ethanol, 30% chloroform, 10% glacial acetic acid) at 4 °C overnight following dehydration in alcohol series and embedding in paraffin blocks (see details in [14,42]). Furthermore, 8 μm thick coronal brain sections were cut using Leica Biosystems (Deer Park, IL, USA) microtome and mounted onto poly-L-lysine–coated slides as described previously [18]. One of five slides from each of eight series was stained with hematoxylin and eosin. Electrode positions in Pt and MC were visualised without additional immunostaining, whereas for those in SN and VTA, an adjacent slide was stained with antibodies against tyrosine hydroxylase (TH, mouse monoclonal antibody, clone TH-2, Sigma, diluted 1:1000) and secondary Goat anti-mouse IgG (H+L) highly cross-adsorbed second antibodies (Alexa Fluor 488, Thermo A11029 diluted 1:1000) as described previously [43]. The borders of SN and VTA on histological sections were outlined using the atlas of TH-positive cells distribution [44]. Representative images demonstrating coagulated tissues at the position of electrode tip either in the SN region or in the VTA region are shown in the Supplementary Figure S1.

Effective electrode targeting of the chosen brain areas was based on a precise measurement of the bregma and lambda coordinates and the providing of coordinates corrections for individual brain areas given that the value that was used for the preparation of the stereotaxic atlas [41] was equal to $4.2 \pm 0.25$ mm. In several cases, when the electrolytic marker was relatively enlarged, the electrode tip position was assigned to the point where the effect of electrolysis within the coagulated area was maximal.

Three days after electrode implantation, each mouse was adapted for four days (1 h/day) to both an experimental cage (Perspex, 15 cm $\times$ 17 cm $\times$ 20 cm) in an electrically shielded chamber and to a cable (five 36-gauge wires, Plexon Inc, Dallas, TX, USA) plugged into a digital Neuro-MEP amplifier (Neurosoft Ltd., Ivanovo, Russian Federation). On day 8, a baseline EEG was recorded for 30 min, starting 20 min after placing the animal into the box. EEG recordings were continued for 60 min after the s.c. injection of either saline (control) or, on the next day, apomorphine (APO, Sigma, Milan, Italy), at a dose of 1.0 mg/kg. To minimise the effect of oxidation, only freshly dissolved APO was used. All experiments were performed from 9:00 a.m. to 6:00 p.m. in daylight combined with an artificial light source, keeping illumination at a relatively stable level.

### 2.3. Computation of EEG Frequency Spectra

Monopolar EEG signals measured between the active and reference electrodes were amplified, filtered (0.1–35 Hz) and sampled (1 kHz) online by using the amplifier and kept in an operational computer for further analysis. The frequency spectra of successive 12 s EEG epochs were studied using a modified version of period-amplitude analysis [45], which, in contrary to the Fourier transform, was not affected by the well-known nonstationary nature of the EEG signals. The absolute values of the half-wave amplitudes with periods/frequencies in each of selected narrow EEG frequency sub-bands were summed, followed by their normalisation to the summarised values. The programme allowed both the automatic and the manual rejection of EEG fragments containing artefacts and electrographic seizures. However, artefacts appeared seldom, because of tight connections in the recording cable sockets and insertion of the cable into a thin, flexible, grounded silvered shield, to protect EEGs against so-called capacity artefacts. In this study, 25 sub-bands in the 0.48–31.5 Hz range were analysed: 0.48–0.53 (0.5), 0.83–0.92 (0.9), 1.20–1.33 (1.3), 1.59–1.76 (1.7), 1.99–2.20 (2.1), 2.42–2.67 (2.5), 2.86–3.17 (3.0), 3.34–3.69 (3.5), 3.83–4.24 (4.0), 4.36–4.82 (4.6), 4.92–5.44 (5.2), 5.52–6.10 (5.8), 6.17–6.82 (6.5), 6.87–7.59 (7.2), 7.62–8.43 (8.0), 8.45–9.34 (8.9), 9.37–10.36 (9.9), 10.40–11.49 (10.9), 11.56–12.77 (12.2), 12.90–14.26 (13.6), 14.49–16.01 (15.3), 16.43–18.16 (17.3), 18.93–20.93 (19.9), 22.47–24.83 (23.6) and 28.50–31.50 (30.0). The sub-bands are marked in figures by their centre (mean) frequency values (see in brackets above). The final analyse was performed for "classical" EEG bands: *delta 1* (0.5–1.7 Hz), *delta 2* (2.1–3.5 Hz), *theta* (4.0–8.0 Hz), *alpha* (8.9–12.2 Hz), *beta 1* (13.6–17.3) and *beta 2* (19.9–30.0).

The frequency spectra of 12 s EEG epochs were individually averaged for every successive 10 min interval for each mouse, followed by the separate averaging of the individual values in WT and in each of the KO groups. The differences in the averaged spectra of baseline EEG in different groups allow the evaluation of a role of the synucleins (and/or their lack) in the modification of the EEG frequency spectra, whereas the differences in EEG effects of APO are expected to be associated with specific modifications of the DA system by synucleins.

### 2.4. Statistics

Differences in the "classical" frequency ranges of the averaged EEG spectra from each brain area were evaluated by two-way ANOVA for repeated measures between different cohorts of animals in the baseline 30 min interval and for 60 min after APO (vs. saline) injection. For multiple comparisons, a Bonferroni post hoc test was employed. The group data were expressed as the means $\pm$ SEM; differences were considered significant at $p < 0.05$. For two-way ANOVA, STATISTICA 10 (StatSoft, Inc., Tulsa, OK, USA) was used.

## 3. Results

### 3.1. Baseline EEG

During baseline EEG recordings, mice with knockouts of synuclein genes behaved similarly to WT control mice: displayed intensive exploration of the experimental box with stochastically scattered sleep-like bouts.

Baseline EEGs in WT (A+B+G+) and TKO (A-B-G-) mice (Figure 1A and B, respectively) were characterised by patterns of relatively slow oscillations of 6–12 Hz, more powerfully expressed in WT mouse. In contrast, the fastest EEG activity of 19.9–30 Hz predominated in Pt, VTA, and SN in a TKO mouse. These EEG patterns were represented in their frequency spectra by higher peaks in the *upper theta-alpha* range and *beta* 2 band in WT and TKO mice, respectively (Figure 1C–F). These differences between the groups were stable in EEG spectra averaged over consecutive 10 min intervals and, thus, were evidently observed in the spectral profiles that characterised the whole (30-min) baseline period (Figure 2).

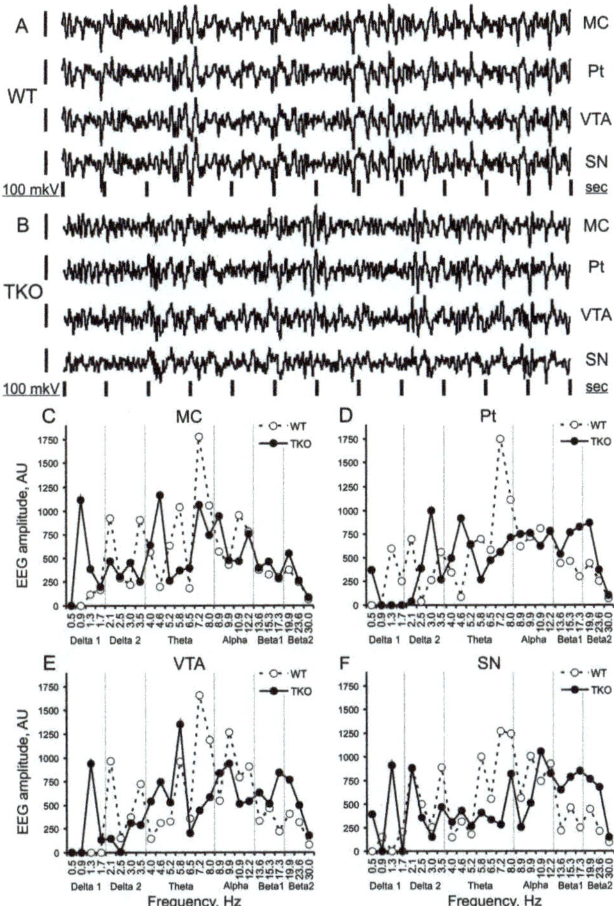

**Figure 1.** Baseline EEG fragments and their frequency spectra in a 3-month-old triple-knockout mouse vs. a wild-type littermate. Typical patterns in 12 s fragments of baseline EEG in wakeful and

behaviourally active wild-type (WT) and triple (*alpha-, beta-, and gamma*-synucleins) knockout (TKO) mice ((**A**) and (**B**), respectively) and their frequency spectra (**C–F**) in the motor cortex (MC), putamen (Pt), ventral tegmental area (VTA) and substantia nigra (SN). On A and B, time calibration is 1 s and amplitude calibration is 100 µV. On (**C–F**), abscissa is a frequency sub-band marked with its mean value, in hertz ordinate is the summed amplitudes of EEG in each of the 25 sub-bands, normalised to a sum of all amplitude values, in arbitrary units. Vertical grey lines separate "classical" EEG frequency bands: *delta 1, delta 2, theta, alpha, beta 1, and beta 2.*

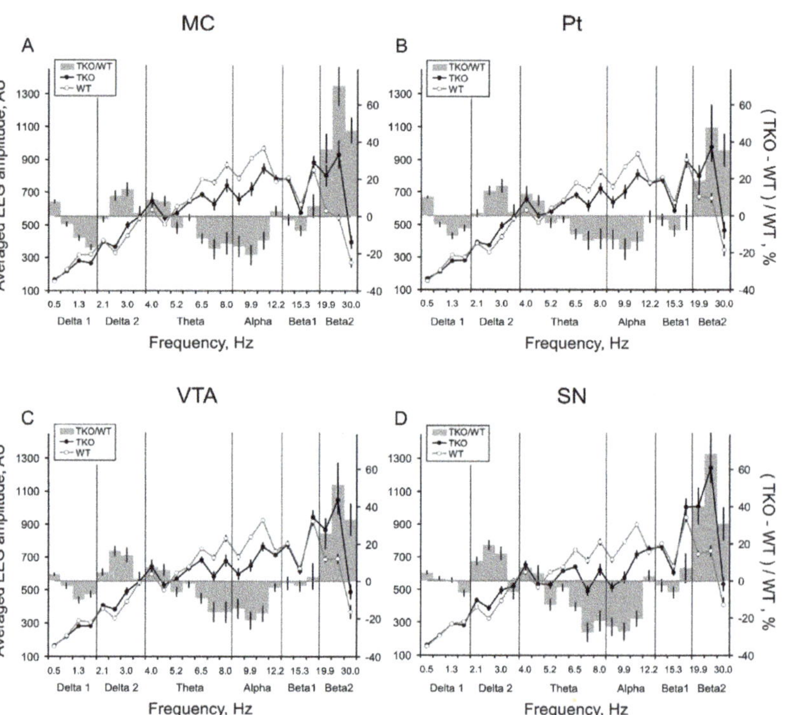

**Figure 2.** Averaged baseline EEG spectra in triple-knockout mice vs. wild-type littermates. Averaged amplitude-frequency spectra of 12 s baseline EEG fragments recorded from the motor cortex (**A**), putamen (**B**), VTA (**C**) and SN (**D**) for 30 min in wild-type (WT, n = 10) and triple (*alpha-, beta-, and gamma*-synucleins) knockout (TKO, n = 8) mice (dashed and solid lines, respectively) and spectral ratios (grey bars) between the groups (TKO/WT) in %. Abscissa is a frequency sub-band marked with its mean value, in hertz; the left ordinate is summed absolute values of EEG amplitudes in each of 20 sub-bands, normalised to sum of all amplitude values, in arbitrary units; and the right ordinate is a ratio of the EEG amplitudes, calculated as (TKO-WT) / WT, in %. Vertical lines are ± 1 SEM. Vertical grey lines separate "classical" EEG frequency bands: *delta 1, delta 2, theta, alpha, beta 1, and beta 2.*

During this period, baseline EEG activity in MC in TKO vs. WT mice (Figure 3A–F) was significantly suppressed in both *theta* and *alpha* bands (Figure 3C,I; two-way ANOVA: $F_{1,48}$ = 9.2 and 8.2, respectively, $p < 0.01$ for both) and enhanced in *beta 2* band (Figure 3F; two-way ANOVA: $F_{1,48}$ = 14.9, $p < 0.001$). In Pt (Figure 3G–L), the EEG differences in the *theta, alpha* and *beta 2* bands were similar to those in MC (two-way ANOVA: $F_{1,48}$ = 5.9, 6.1 and 8.1, $p < 0.05$ for both, and $< 0.01$, respectively). In VTA (Figure 4A–F) and SN (Figure 4G–F), significant differences between TKO and WT mice were observed in the same frequency ranges ($F_{1,48}$ = 4.8, 14.3 and 11.5, $p < 0.05$, 0.001 and 0.01, respectively, for

VTA, and $F_{1,48}$ = 16.3, 19.8 and 18.3, respectively, $p < 0.001$ for SN). All two-way ANOVA evaluations vs. WT group are presented in Appendix A Figure A1.

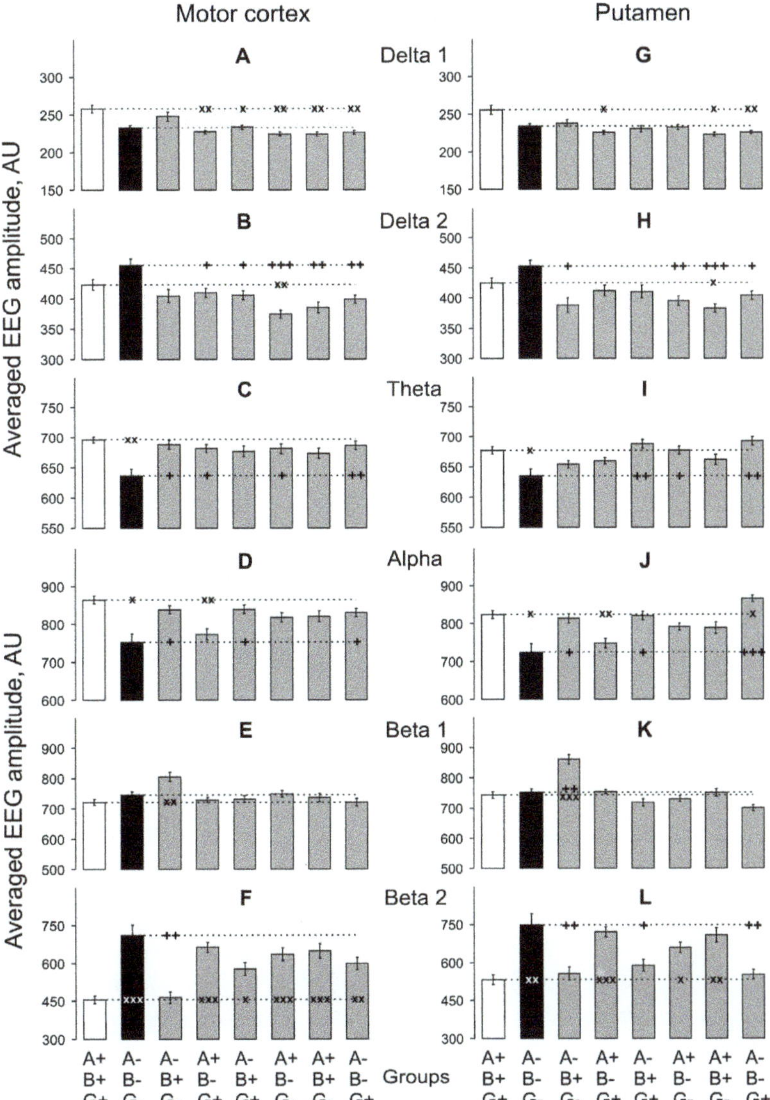



**Figure 3.** Relations between averaged amplitudes in "classical" frequency bands of 12 s baseline EEG fragments recorded from the motor cortex (**A–F**) and putamen (**G–L**) for 30 min in knockout mice of different types denoted on the horizontal axes. Ordinate is the averaged absolute values of EEG amplitudes in each of the "classical" bands, in arbitrary units (vertical lines are ±1 SEM). Horizontal dashed lines denote the values in wild-type and triple-knockout groups: x and + symbols denote significant two-way ANOVA differences from the wild-type and triple-knockout mice, respectively (one, two and three symbols denote $p < 0.05$, <0.01 and 0.001, respectively).

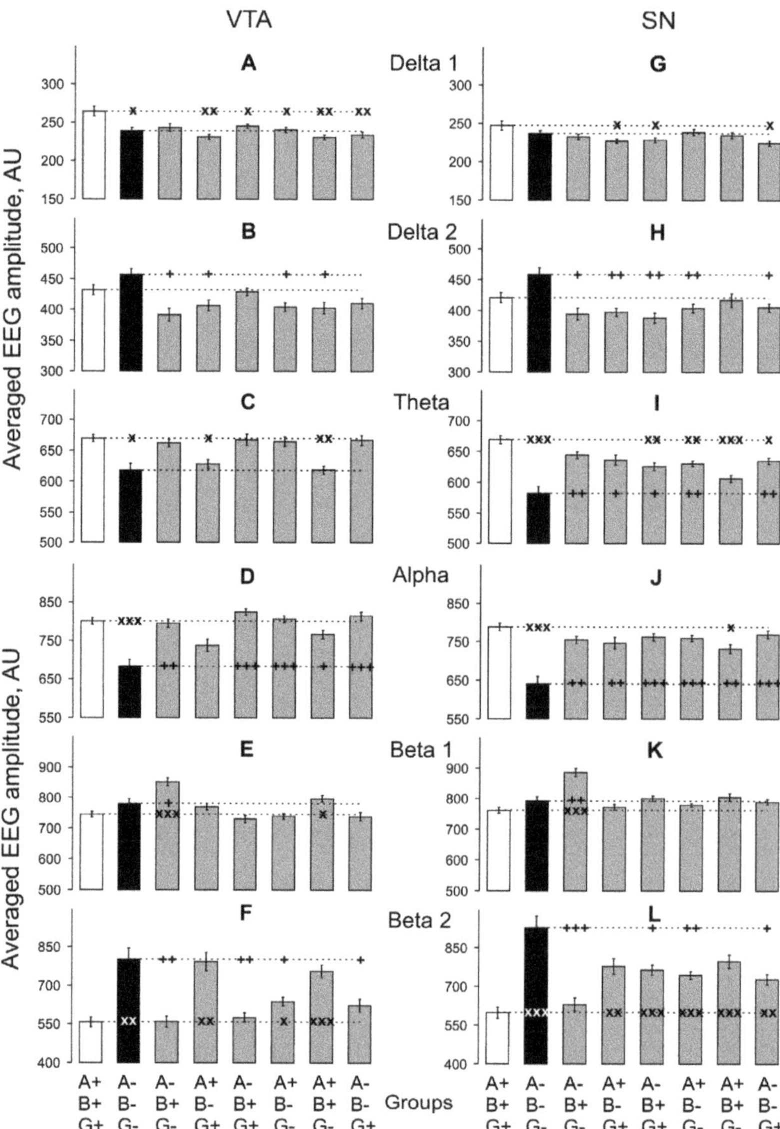



**Figure 4.** Relations between averaged amplitudes in "classical" frequency bands of 12 s baseline EEG fragments recorded from VTA (**A–F**) and SN (**G–L**) for 30 min in knockout mice of different types denoted on the horizontal axes. Ordinate is the averaged absolute values of EEG amplitudes in each of the "classical" bands, in arbitrary units (vertical lines are ±1 SEM). Horizontal dashed lines denote the values in wild-type and triple-knockout groups: x and + symbols denote significant two-way ANOVA differences from the wild-type and triple-knockout mice, respectively (one, two and three symbols denote $p < 0.05$, $< 0.01$ and 0.001, respectively).

To assess whether the changes in the EEG spectra observed in the brain areas of TKO mice developed only as the result of the absence of all three synucleins or whether the depletion of certain family member(s) could be sufficient for the development of the same changes, we measured these spectra in the brain areas of mice lacking one or two synucleins in all six possible combinations.

Changes in *delta 1* activity were found to be quite diverse between genotypes and brain regions when compared with WT mice, although this activity was lower in all cases when the difference was significant (Figures 3A,G and 4A,G). Conversely, *delta 2* activity that was not affected in any of studied brain areas of TKO mice showed no significant changes in the SN and VTA of mice of all single- and double-KO genotypes, whereas a decrease was observed in the Pt of A + B+G- and MC of A + B-G- mice when compared with WT mice. The *theta* activity that was depressed in the TKO vs. WT group in all brain areas (Figures 3C,I and 4C,I) appeared to be unaffected in the MC and Pt of all other KO-genotype mouse groups (Figure 3C,I) but was depressed in the VTA of mice lacking either only *beta*-syn (A+B-G+) or only *gamma*-syn (A+B+G-) (Figure 4C,I). In the SN, an even-more-diverse pattern of the *theta* activity was observed with its depression in mice either lacking *alpha*-syn (A-B+G+) or expressing this protein in the absence of the other two synucleins (A+B-G-)— and either lacking *gamma*-syn (A+B+G-) or expressing this protein in the absence of the other two synucleins (A-B-G+) (Figure 4C,I). The *alpha* activity that was also depressed in the TKO vs. WT group in all brain areas (Figures 3D,J and 4D,J) showed no such changes in most of the KO genotypes and brain areas, except for the MC and Pt of mice lacking *beta*-syn (A+B-G+) (Figure 3D,J) and the SN of mice lacking *gamma*-syn (A+B+G-) (Figure 4J). The *beta 1* activity was characterised by the most synuclein-independent pattern of activity throughout brain areas and genotypes, but it was consistently enhanced in all brain areas of mice expressing only *beta*-syn (Figures 3E,K and 4E,K). In contrast, these mice showed no changes in *beta 2* activity in all four studied brain areas when compared with WT mice (Figures 3F,L and 4F,L). Similarly, no changes in *beta 2* activity were found in the MC and VTA of mice either lacking *alpha*-syn (A-B+G+) or expressing this protein in the absence of the other two synucleins (A+B-G-), whereas all other combinations of KO genotypes, including TKO, and brain area enhanced *beta 2* activity (Figures 3F,L and 4F,L).

### 3.2. Apomorphine Effects

In both WT and synuclein KO mice, APO initiated stereotyped behaviour, i.e., short-lasting freezing followed by uninterrupted licking of the floor and raising of the erected tail for about 30 min after injection. Also, occasional short sleep-like bouts were observed during this period in all APO-treated groups of mice.

### 3.2.1. Apomorphine vs. Saline

In MC (Figure 5), APO significantly suppressed *delta 2* and *alpha* activities and enhanced the *theta* in WT (A+B+C+) mice (two-way ANOVA: $F_{1,108} = 4.3, 9.8$ and $6.7, p = 0.04$, 0.002 and 0.009, respectively). In TKO (A-B-G-) mice, APO produced *delta 2* suppression in MC (two-way ANOVA: $F_{1,84} = 9.9, p = 0.002$), whereas the A-B+G- group was characterised by *alpha* suppression and *beta 2* enhancement (two-way ANOVA: $F_{1,108} = 27.5$ and $14.7$, respectively, $p < 0.001$ for both). In A+B-G+ mice, APO suppressed *delta 2* activity (two-way ANOVA: $F_{1,120} = 9.8, p = 0.002$), whereas in A-B+G+ mice, it produced significant attenuation of *alpha* activity (two-way ANOVA: $F_{1,132} = 14.9, p < 0.001$). In the A+B-G- group, *alpha* suppression and *beta 2* enhancement were observed (two-way ANOVA: $F_{1,144} = 11.4$ and $5.8, p < 0.001$ and $= 0.02$, respectively). The mice most sensitive to APO were in the A+B+G- group, for which significant suppression of *delta 2* and *alpha* activities (two-way ANOVA: $F_{1,120} = 9.2$ and $31.2, p = 0.003$ and $< 0.001$, respectively) and enhancement of *beta 1* and *beta 2* ones (two-way ANOVA: $F_{1,120} = 9.2$ and $25.0$ $p = 0.003$ and $< 0.001$, respectively) were revealed.

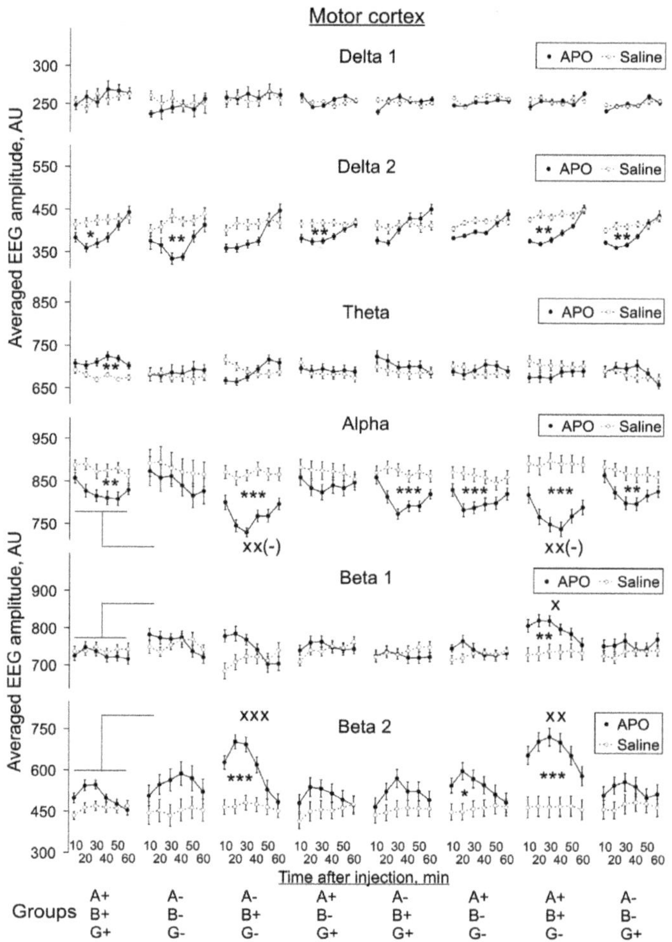


**Figure 5.** The evolution of apomorphine (APO, 1.0 mg/kg, s.c.) vs. saline effects in "classical" frequency bands of 12 s fragments of EEG from the motor cortex averaged for consecutive 10 min intervals (abscissa) in knockout mice of different types denoted on the horizontal axes. Ordinate is the averaged absolute values of EEG amplitudes in each of the "classical" bands, in arbitrary units (vertical lines are ±1 SEM), obtained in experiments with saline and APO (grey and black lines, respectively) in each group. Differences between baseline EEGs in knockout and wild-type mice were used to normalise APO effects in knockout groups to those in control. Symbols * and x denote significant differences in APO vs. saline and knockout mice vs. wild-type control, respectively (one, two and three symbols denote $p < 0.05$, $<0.01$ and $0.001$, respectively). Symbol x(-) denotes the significant enhancement of the APO suppressive effect, for clarity.

In Pt (Figure 6), the distribution of significant differences after APO vs. saline injections was similar to that observed in MC, but in WT (A+B+G+) mice, *delta 2* suppression did not reach significant values (two-way ANOVA: $F_{1,108} = 3.7$, $p = 0.056$), whereas *beta 2* activity was significantly enhanced in the A-B+G+ group (two-way ANOVA: $F_{1,132} = 7.5$, $p = 0.007$), and *theta* activity was significantly supressed in the A+B+G- group (two-way ANOVA: $F_{1,120} = 8.3$, $p = 0.005$).

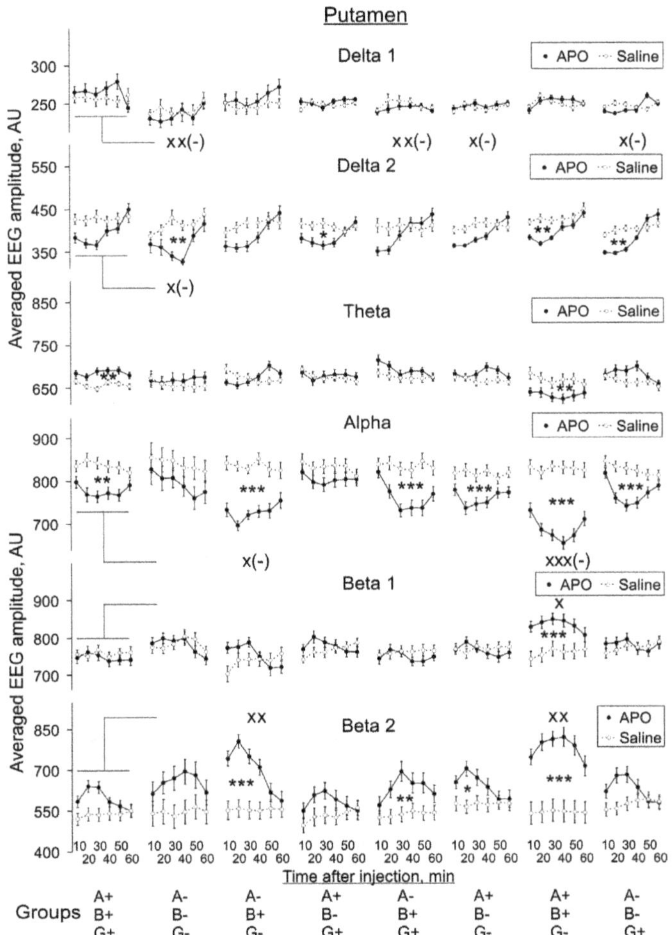



**Figure 6.** Evolution of apomorphine (APO, 1.0 mg/kg, s.c.) vs. saline effects in "classical" frequency bands of 12 s fragments of EEG from the putamen averaged for consecutive 10 min intervals (abscissa) in knockout mice of different types denoted on the horizontal axes. Ordinate is the averaged absolute values of EEG amplitudes in each of the "classical" bands, in arbitrary units (vertical lines are ±1 SEM), obtained in experiments with saline and APO (grey and black lines, respectively) in each group. Differences between baseline EEGs in knockout and wild-type mice were used to normalise the APO effects in knockout groups to those in control. Symbols * and x denote the significant differences in APO vs. saline and knockout mice vs. wild-type control, respectively (one, two and three symbols denote $p < 0.05$, $<0.01$ and $0.001$, respectively). Symbol x(-) denotes the significant enhancement of the APO suppressive effect, for clarity.

All changes in EEG following APO administration observed in Pt and MC were also observed in the VTA of mice of corresponding genotypes (Figure 7), except for the lack of suppression of *alpha* activity in the A-B-G+ group (two-way ANOVA: $F_{1,144} = 3.6$, $p = 0.059$). In addition, in the A-B+G- and A+B-G- groups, *delta* 2 activity was suppressed (two-way ANOVA: $F_{1,108} = 4.1$, $p = 0.045$, and $F_{1,144} = 11.1$, $p = 0.001$, respectively): in the A+B-G+ group, *theta* activity was increased (two-way ANOVA: $F_{1,120} = 6.3$, $p = 0.014$), as was *beta* 2 activity in the WT and A-B-G- groups (two-way ANOVA: $F_{1,108} = 5.1$, $p = 0.026$, and $F_{1,84} = 4.5$, $p = 0.038$, respectively).

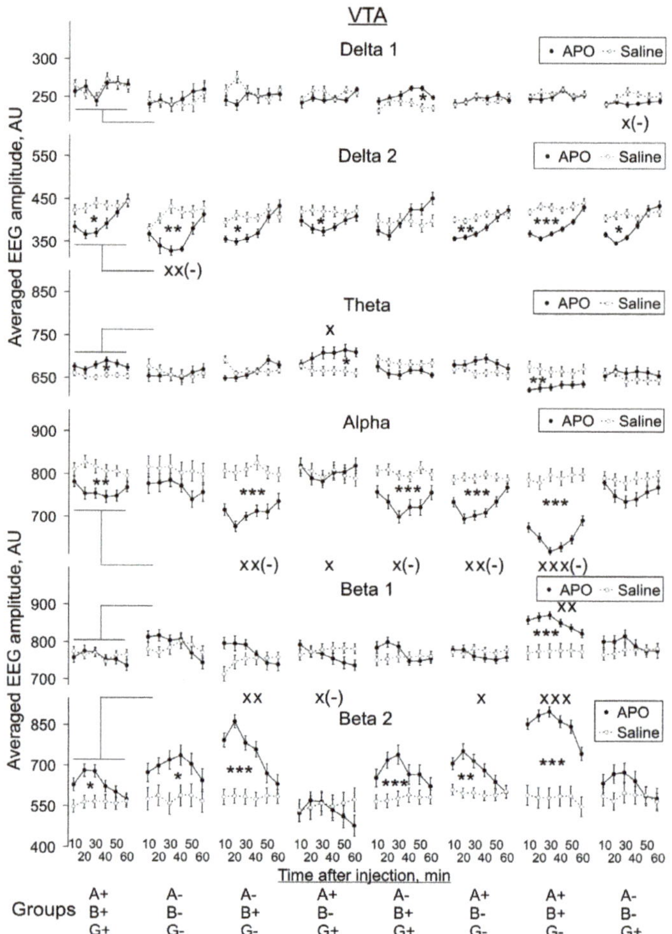



**Figure 7.** Evolution of apomorphine (APO, 1.0 mg/kg, s.c.) vs. saline effects in "classical" frequency bands of 12 s fragments of EEG from VTA averaged for consecutive 10 min intervals (abscissa) in knockout mice of different types. Ordinate is the averaged absolute values of EEG amplitudes in each of the "classical" bands, in arbitrary units (vertical lines are ±1 SEM), obtained in experiments with saline and APO (grey and black lines, respectively) in each group. Differences between baseline EEGs in knockout and wild-type mice were used to normalise the APO effects in knockout groups to those in control. Symbols * and x denote the significant differences of APO vs. saline and knockout mice vs. wild-type control, respectively (one, two and three symbols denote $p < 0.05$, <0.01 and 0.001, respectively). Symbol x(-) denotes the significant enhancement of the APO suppressive effect, for clarity.

In SN the patterns of *alpha*, *beta 1* and *beta 2* activities following APO treatment were the same as those in Pt (Figures 7 and 8), except for *alpha* activity in the A-B-G+ group, in which no significant difference was detected (two-way ANOVA: $F_{1,144} = 3.0$, $p = 0.084$). An increase of *theta* activity in SN was observed only in the A+B-G+ group (two-way ANOVA: $F_{1,120} = 5.8$, $p = 0.018$) and decreased *delta 2* activity in the same five groups as in VTA but not in the WT and A-B-G- groups (two-way ANOVA: $F_{1,108} = 1.7$, $p = 0.197$, and $F_{1,84} = 1.6$, $p = 0.802$, respectively). All two-way ANOVA evaluations vs. saline are presented in Appendix A Figure A2.

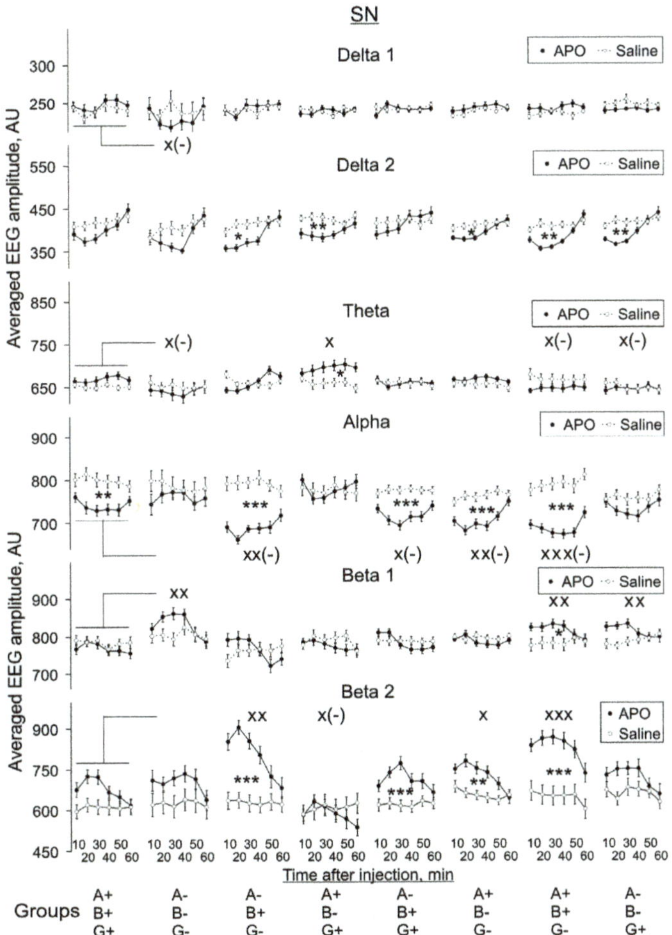



**Figure 8.** Evolution of apomorphine (APO, 1.0 mg/kg, s.c.) vs. saline effects in "classical" frequency bands of 12 s fragments of EEG from SN averaged for consecutive 10 min intervals (abscissa) in knockout mice of different types. Ordinate is the averaged absolute values of EEG amplitudes in each of the "classical" bands, in arbitrary units (vertical lines are $\pm 1$ SEM), obtained in experiments with saline and APO (grey and black lines, respectively) in each group. Differences between baseline EEGs in knockout and wild-type mice were used to normalise the APO effects in knockout groups to those in control. Symbols * and x denote the significant differences of APO vs. saline and knockout mice vs. wild-type control, respectively (one, two and three symbols denote $p < 0.05$, $<0.01$ and 0.001, respectively). Symbol x(-) denotes the significant enhancement of the APO suppressive effect, for clarity.

### 3.2.2. APO Effects in Different Groups vs. Those in WT Mice

We also compared APO-induced changes in EEG spectra in KO groups with those in the WT group, and these changes are also indicated in Figures 5–8. The two-way ANOVA evaluations shown in Appendix A Figure A3 demonstrate the statistical significance of the differences in the degree of these changes, independently of whether APO injection induced a decrease or whether it induced an increase of a particular EEG band activity compared with saline injection.

The effect of animal genotype on APO-induced changes in EEG spectra was most profound in *gamma*-syn KO (A+B+G−) mice: a statistically significant decrease in the degree of changes was revealed for *alpha* activity and a statistically significant increase for *beta 1* and *beta 2* activities in all four studied brain areas. The effect on *alpha* and *beta 2* activity remained when both *gamma*-syn and *alpha*-syn synuclein were absent in double-KO animals (A-B+G-), but the effect on *beta 1* could not be seen anymore. The further removal of *beta*-syn abolished the effects of APO treatment on *alpha, beta 1* and *beta 2* activities in all four brain areas of TKO (A-B-G-) mice, with the exception of *beta 1* in SN (Figures 5–8, Appendix A Figure A3).

In MC, the only detected APO-induced changes across synuclein KO genotypes were described above changes in *alpha* and *beta* activities, but all the other three studied brain areas were characterised by a more profound role of dopamine neurotransmission. Such changes were also observed for *delta 1* activity, particularly its decrease in Pt, and for *theta* activity, particularly in SN. A statistically significant decrease in *delta 2* activity was noticed only in the Pt and VTA of TKO (A-B-G-) mice (Figures 5–8, Appendix A Figure A3).

## 4. Discussion

In this study, we have revealed the effects of all possible combinations of synuclein family members' depletion (three single-, three double- and the triple-synuclein gene KOs) on *baseline* and *apomorphine-modified EEGs* recorded from the different brain areas of mice: motor cortex (MC), putamen (Pt), ventral tegmental area (VTA) and substantia nigra (SN).

Not surprisingly, across the studied brain area, the most frequent changes in baseline EEG spectral profiles, when compared with those in WT mice, were observed in the absence of all three synucleins, i.e., in TKO mice (schematically illustrated in Supplementary Figure S2), although no changes in *delta 2* or *beta 1* spectra were detected in these animals. This is consistent with substantial changes in synaptic morphology (e.g., decreased presynaptic terminal area of CA3 excitatory synapses) and activity (e.g., changes in the amplitude of the field excitatory postsynaptic potentials in the hippocampus) previously observed in TKO mice [16]. Interestingly, among single- and double-synuclein KO genotypes, the pattern of changes seen in mice lacking only *gamma*-syn was the most similar to the pattern seen in TKO mice, particularly in basal ganglia, SN and VTA (Supplementary Figure S2). This is in line with the recently obtained evidence that *gamma*-syn transcription in DA neurons modifies DA mediation in the brain [11]. However, this similarity of patterns gets lost in the absent of an additional member of the synuclein family, i.e., in *beta*-syn/*gamma*-syn and particularly in *alpha*-syn/*gamma*-syn double-KO mice (Supplementary Figure S2). Moreover, *gamma*-syn on its own, i.e., in *alpha*-syn/*beta*-syn double-KO mice, is not able to restore the WT pattern. It is *beta*-syn that singularly can normalise average EEG amplitudes at all but *beta 1* frequencies in *alpha*-syn/*gamma*-syn double-KO mice (Supplementary Figure S2), which is consistent the ability of *beta*-syn to potentiate neurotransmitter uptake by synaptic vesicles in the absence of other synucleins [6] and would suggest a key role of this protein in the regulation of EEG oscillations. Yet the absence of *beta*-syn either singularly or in combination with another synuclein (i.e., in *beta*-syn/*alpha*-syn or *beta*-syn/*gamma*-syn double-KO mice) causes changes in only some oscillation frequencies in certain brain areas compared with corresponding areas in the brain of WT mice. Thus, it is not the absence of any particular synuclein but rather a disbalance of synucleins that causes widespread changes in EEG spectral profiles.

Another observation that would need further investigation is that independently of the genotype and the brain area, the disbalance of synucleins always alters the vector of *beta 1* and *beta 2* level changes towards their enhancement but for other, higher frequencies, towards their suppression. A link between alterations in EEG oscillations, particularly the elevation of low-frequency *beta* oscillations, with motor impairment and neurodegeneration in PD patients and animal models of the disease has been reported in multiple studies [46–48]. Moreover, the suppression of *beta* oscillations correlates with the positive effects of symptomatic treatments of PD by levodopa or deep brain stimulation (DBS) [49–51]. We found that oscillations recorded from SN and VTA areas appeared to be substantially affected in all synuclein KO mice. Together with the aforementioned enhancement of *beta* oscillations in these mice, which resembles changes in *beta* frequencies EEG recordings in PD patients, these observations suggested that the modulation of dopaminergic neurotransmission in mice lacking certain synucleins might have specific effects on EEG recordings, particularly from these two brain areas. To test this, we treated WT and synuclein KO mice with APO, a DA agonist of several DA receptors and thus an activator of DA signalling.

APO treatment causes fewer changes in EEG oscillations in TKO mice than in all other studied genotypes, whereas the most changes were observed in mice lacking *gamma*-syn in the presence of one or two other members of the family (Supplementary Figure S3). This suggests that compensation for the loss of *gamma*-syn function by other synuclein(s) exerts much-more-profound effects on EEG spectral profiles, i.e., elevation of *beta* and suppression of *delta 2*, *theta* and *alpha* bands, when DA signalling has been activated (Supplementary Figure S3). This effect was also obvious when APO-induced changes in EEG spectral profiles observed in synuclein KO mice were compared with changes observed in WT mice, i.e., elevation of *beta*, particularly *beta 2*, and the suppression of higher frequency bands, particularly *alpha* (Supplementary Figure S4). Taken together, these observations again point to an importance of a balance of synucleins for neuronal function.

## 5. Conclusions

We found that changes in the composition of synucleins significantly affect EEG oscillation profiles in all studied areas of the nervous system and that the activation of DA signalling by APO treatment causes further genotype- and brain area–specific alterations in these profiles. Further studies should unveil molecular and cellular mechanisms linking a disbalance of synucleins and changes in the electrical activity of the brain, as well as whether and how EEG spectral analyses can be applied for the early differential diagnostics of synucleinopathies.

**Supplementary Materials:** The following supporting information can be downloaded at https://www.mdpi.com/article/10.3390/biomedicines10123128/s1, Figure S1—verification of the position of the electrode tip (arrows) following electrocoagulation of surrounding tissues; Figure S2—baseline changes of EEG recordings from four brain areas of synuclein KO mice compared with WT mice.; Figure S3—APO-induced changes of EEG recordings from four brain areas of synuclein KO and WT mice; Figure S4—APO-induced changes in EEG recordings from four brain areas of synuclein KO compared with APO-induced changes in WT mice.

**Author Contributions:** Conceptualisation, V.V. and V.L.B.; methodology, V.V.; validation, V.V., K.C. and A.D.; formal analysis, A.D., Z.O., V.V., K.C. and V.L.B.; investigation, I.S., K.C., Z.O. and O.M.; resources, O.M. and I.S.; data curation, V.V.; writing—original draft preparation, V.V. and K.C.; writing—review and editing, V.L.B. All authors have read and agreed to the published version of the manuscript.

**Funding:** This research was funded by Ministry of Science and Higher Education of the Russian Federation (grant agreement No. 075-15-2020-795, state contract No. 13.1902.21.0027 of 29 September 2020 unique project ID: RF-190220X0027).

**Institutional Review Board Statement:** The animal study protocol was approved by the local Institute Ethics Review Committee (protocol No. 48, 15.01.2021). All animal work was carried out in accordance with the "Guidelines for accommodation and care of animals. Species-specific provisions for laboratory rodents and rabbits" (GOST 33216-2014) in compliance with the principles enunciated in the Directive 2010/63/EU on the protection of animals used for scientific purposes.

**Informed Consent Statement:** Not applicable.

**Data Availability Statement:** Data are contained in the current article and its supplementary material.

**Conflicts of Interest:** The authors declare no conflict of interest.

## Appendix A

| Two-way ANOVA for baseline EEG vs. A+B+G+ | | | | | | | | | | | | | |
|---|---|---|---|---|---|---|---|---|---|---|---|---|---|
| **Groups** | A-B-G- | | A-B+G- | | A+B-G+ | | A-B+G+ | | A+B-G- | | A+B+G- | | A-B-G+ | |
| **MC** | $F_{48}$ | P | $F_{54}$ | P | $F_{57}$ | P | $F_{60}$ | P | $F_{63}$ | P | $F_{57}$ | P | $F_{63}$ | P |
| Delta 1 | 2.81 | 0.099 | 0.38 | 0.539 | 7.17 | 0.009 | 4.70 | 0.034 | 10.3 | 0.002 | 9.40 | 0.003 | 9.15 | 0.004 |
| Delta 2 | 2.28 | 0.138 | 0.47 | 0.497 | 0.77 | 0.497 | 0.84 | 0.362 | 7.36 | 0.009 | 3.57 | 0.064 | 2.09 | 0.154 |
| Theta | 9.21 | 0.004 | 0.15 | 0.701 | 0.93 | 0.339 | 0.68 | 0.412 | 0.35 | 0.557 | 1.49 | 0.227 | 0.10 | 0.757 |
| Alpha | 8.25 | 0.006 | 1.34 | 0.251 | 10.1 | 0.002 | 1.15 | 0.287 | 2.65 | 0.108 | 2.52 | 0.118 | 2.01 | 0.161 |
| Beta 1 | 0.68 | 0.414 | 8.14 | 0.006 | 0.07 | 0.797 | 0.06 | 0.810 | 1.16 | 0.285 | 0.28 | 0.598 | 0.01 | 0.921 |
| Beta 2 | 14.9 | 0.000 | 0.07 | 0.797 | 25.1 | 0.000 | 6.48 | 0.014 | 12.1 | 0.0006 | 12.9 | 0.000 | 8.63 | 0.005 |
| **Putamen** | $F_{48}$ | P | $F_{54}$ | P | $F_{57}$ | P | $F_{60}$ | P | $F_{63}$ | P | $F_{57}$ | P | $F_{63}$ | P |
| Delta 1 | 1.98 | 0.166 | 0.85 | 0.360 | 6.21 | 0.016 | 3.42 | 0.069 | 3.46 | 0.068 | 6.58 | 0.013 | 7.36 | 0.009 |
| Delta 2 | 2.33 | 0.133 | 1.62 | 0.208 | 0.13 | 0.723 | 0.17 | 0.678 | 1.90 | 0.173 | 4.75 | 0.033 | 0.72 | 0.400 |
| Theta | 5.90 | 0.019 | 3.12 | 0.083 | 3.57 | 0.064 | 0.41 | 0.524 | 0.01 | 0.911 | 1.46 | 0.232 | 0.67 | 0.417 |
| Alpha | 6.07 | 0.017 | 0.08 | 0.784 | 7.27 | 0.009 | 0.01 | 0.918 | 1.59 | 0.212 | 1.07 | 0.305 | 4.55 | 0.037 |
| Beta 1 | 0.08 | 0.780 | 14.1 | 0.000 | 0.28 | 0.600 | 0.58 | 0.448 | 0.38 | 0.540 | 0.08 | 0.780 | 3.11 | 0.082 |
| Beta 2 | 8.09 | 0.006 | 0.23 | 0.633 | 16.1 | 0.000 | 1.38 | 0.245 | 6.94 | 0.011 | 9.33 | 0.003 | 0.20 | 0.659 |
| **VTA** | $F_{48}$ | P | $F_{54}$ | P | $F_{57}$ | P | $F_{60}$ | P | $F_{63}$ | P | $F_{57}$ | P | $F_{63}$ | P |
| Delta 1 | 5.13 | 0.028 | 3.10 | 0.084 | 9.08 | 0.004 | 4.59 | 0.036 | 6.63 | 0.012 | 10.4 | 0.002 | 9.23 | 0.004 |
| Delta 2 | 1.26 | 0.268 | 2.67 | 0.108 | 1.48 | 0.229 | 0.00 | 0.982 | 2.07 | 0.155 | 1.61 | 0.210 | 0.89 | 0.350 |
| Theta | 4.77 | 0.034 | 0.25 | 0.618 | 6.14 | 0.016 | 0.02 | 0.894 | 0.01 | 0.918 | 12.1 | 0.000 | 0.00 | 0.996 |
| Alpha | 14.3 | 0.000 | 0.29 | 0.591 | 5.15 | 0.027 | 0.77 | 0.383 | 0.00 | 0.947 | 2.73 | 0.104 | 0.12 | 0.732 |
| Beta 1 | 2.09 | 0.155 | 17.2 | 0.000 | 1.51 | 0.224 | 0.23 | 0.636 | 0.00 | 0.973 | 5.79 | 0.019 | 0.00 | 0.995 |
| Beta 2 | 11.5 | 0.001 | 0.01 | 0.985 | 11.8 | 0.001 | 0.21 | 0.646 | 4.12 | 0.047 | 15.4 | 0.000 | 1.58 | 0.213 |
| **SN** | $F_{48}$ | P | $F_{54}$ | P | $F_{57}$ | P | $F_{60}$ | P | $F_{63}$ | P | $F_{57}$ | P | $F_{63}$ | P |
| Delta 1 | 1.38 | 0.245 | 2.96 | 0.091 | 5.10 | 0.028 | 4.24 | 0.044 | 0.62 | 0.432 | 1.26 | 0.267 | 6.31 | 0.015 |
| Delta 2 | 2.47 | 0.123 | 1.12 | 0.295 | 1.72 | 0.195 | 2.79 | 1.000 | 1.29 | 0.260 | 0.03 | 0.853 | 0.80 | 0.373 |
| Theta | 16.3 | 0.000 | 2.06 | 0.157 | 2.81 | 0.099 | 7.17 | 0.009 | 7.40 | 0.008 | 16.3 | 0.000 | 4.75 | 0.033 |
| Alpha | 19.8 | 0.000 | 2.18 | 0.146 | 1.78 | 0.188 | 1.23 | 0.272 | 1.88 | 0.175 | 4.92 | 0.031 | 0.52 | 0.471 |
| Beta 1 | 1.24 | 0.27 | 20.6 | 0.000 | 0.201 | 0.655 | 2.76 | 0.102 | 0.76 | 0.387 | 2.11 | 0.152 | 1.04 | 0.311 |
| Beta 2 | 18.3 | 0.000 | 0.43 | 0.514 | 8.09 | 0.006 | 12.3 | 0.000 | 12.4 | 0.000 | 12.5 | 0.000 | 7.7 | 0.007 |

**Figure A1.** Results of a two-way ANOVA analysis of baseline changes in EEG recordings from four brain areas of synuclein KO mice vs. WT (A+B+G+) mice. Data with $p < 0.05$ are highlighted blue, $p < 0.01$ highlighted pink and $p < 0.001$ highlighted red.

Two-way ANOVA for EEG effects of apomorphine vs. Saline

| Groups | A+B+G+ | | A-B-G- | | A-B+G- | | A+B-G+ | | A-B+G+ | | A+B-G- | | A+B+G- | | A-B-G+ | |
|---|---|---|---|---|---|---|---|---|---|---|---|---|---|---|---|---|
| **MC** | $F_{108}$ | P | $F_{84}$ | P | $F_{108}$ | P | $F_{120}$ | P | $F_{132}$ | P | $F_{144}$ | P | $F_{120}$ | P | $F_{144}$ | P |
| Delta 1 | 0.17 | 0.678 | 2.75 | 0.101 | 0.17 | 0.683 | 0.85 | 0.359 | 0.66 | 0.419 | 1.84 | 0.177 | 0.25 | 0.617 | 1.76 | 0.186 |
| Delta 2 | 4.34 | 0.040 | 9.88 | 0.002 | 2.29 | 0.134 | 9.84 | 0.002 | 0.09 | 0.760 | 3.06 | 0.083 | 9.25 | 0.003 | 8.01 | 0.005 |
| Theta | 9.82 | 0.002 | 0.19 | 0.665 | 0.40 | 0.531 | 0.20 | 0.650 | 0.88 | 0.366 | 0.09 | 0.76 | 2.63 | 0.108 | 0.40 | 0.526 |
| Alpha | 6.71 | 0.0098 | 0.82 | 0.366 | 27.5 | 0.000 | 1.69 | 0.196 | 14.9 | 0.000 | 11.4 | 0.000 | 31.2 | 0.000 | 8.18 | 0.005 |
| Beta 1 | 0.60 | 0.439 | 0.10 | 0.756 | 1.48 | 0.227 | 0.53 | 0.468 | 0.52 | 0.474 | 0.90 | 0.344 | 9.25 | 0.003 | 1.10 | 0.295 |
| Beta 2 | 2.96 | 0.088 | 2.92 | 0.091 | 14.7 | 0.000 | 2.22 | 0.138 | 2.94 | 0.089 | 5.85 | 0.017 | 25.0 | 0.000 | 2.35 | 0.127 |
| **Putamen** | $F_{108}$ | P | $F_{84}$ | P | $F_{108}$ | P | $F_{120}$ | P | $F_{132}$ | P | $F_{144}$ | P | $F_{120}$ | P | $F_{144}$ | P |
| Delta 1 | 2.53 | 0.115 | 2.53 | 0.115 | 0.43 | 0.514 | 0.02 | 0.904 | 0.44 | 0.510 | 0.17 | 0.682 | 0.32 | 0.571 | 0.01 | 0.909 |
| Delta 2 | 3.73 | 0.056 | 8.32 | 0.005 | 2.11 | 0.149 | 4.95 | 0.028 | 0.81 | 0.371 | 3.21 | 0.075 | 11.3 | 0.001 | 7.72 | 0.006 |
| Theta | 7.55 | 0.007 | 0.51 | 0.478 | 0.10 | 0.75 | 0.01 | 0.900 | 1.89 | 0.171 | 2.75 | 0.099 | 8.27 | 0.005 | 2.74 | 0.100 |
| Alpha | 7.91 | 0.006 | 1.65 | 0.202 | 24.3 | 0.000 | 2.39 | 0.125 | 19.0 | 0.000 | 16 | 0.000 | 56.1 | 0.000 | 15.5 | 0.000 |
| Beta 1 | 0.42 | 0.518 | 0.02 | 0.953 | 0.40 | 0.528 | 0.27 | 0.602 | 0.41 | 0.521 | 0.29 | 0.593 | 12.0 | 0.000 | 0.07 | 0.797 |
| Beta 2 | 3.54 | 0.059 | 2.95 | 0.09 | 13.2 | 0.000 | 1.98 | 0.162 | 7.52 | 0.007 | 4.97 | 0.027 | 31.6 | 0.000 | 3.72 | 0.056 |
| **VTA** | $F_{108}$ | P | $F_{84}$ | P | $F_{108}$ | P | $F_{120}$ | P | $F_{132}$ | P | $F_{144}$ | P | $F_{120}$ | P | $F_{144}$ | P |
| Delta 1 | 0.13 | 0.722 | 0.81 | 0.370 | 0.37 | 0.546 | 0.72 | 0.398 | 5.81 | 0.017 | 0.16 | 0.69 | 0.09 | 0.766 | 2.55 | 0.113 |
| Delta 2 | 5.5 | 0.021 | 10.6 | 0.002 | 4.12 | 0.045 | 5.71 | 0.019 | 0.06 | 0.812 | 6.86 | 0.009 | 31.0 | 0.000 | 6.44 | 0.012 |
| Theta | 6.19 | 0.014 | 0.17 | 0.679 | 0.39 | 0.532 | 6.28 | 0.014 | 0.01 | 0.900 | 3.28 | 0.072 | 10.4 | 0.002 | 0.67 | 0.415 |
| Alpha | 7.59 | 0.007 | 2.58 | 0.112 | 21.9 | 0.000 | 0.00 | 0.964 | 39.4 | 0.000 | 31.2 | 0.000 | 83.0 | 0.000 | 3.63 | 0.059 |
| Beta 1 | 0.40 | 0.526 | 0.15 | 0.6990 | 0.81 | 0.37 | 1.01 | 0.316 | 0.04 | 0.839 | 0.73 | 0.394 | 19.1 | 0.000 | 1.05 | 0.306 |
| Beta 2 | 5.08 | 0.026 | 4.46 | 0.038 | 18.8 | 0.000 | 0.26 | 0.614 | 13.8 | 0.000 | 11.1 | 0.001 | 57.2 | 0.000 | 1.44 | 0.231 |
| **SN** | $F_{108}$ | P | $F_{84}$ | P | $F_{108}$ | P | $F_{120}$ | P | $F_{132}$ | P | $F_{144}$ | P | $F_{120}$ | P | $F_{144}$ | P |
| Delta 1 | 0.72 | 0.397 | 2.14 | 0.148 | 0.05 | 0.945 | 1.87 | 1.000 | 0.00 | 0.926 | 0.57 | 0.453 | 2.14 | 0.146 | 4.12 | 0.054 |
| Delta 2 | 1.69 | 0.197 | 1.61 | 0.208 | 4.99 | 0.028 | 10.9 | 0.001 | 0.12 | 0.732 | 4.67 | 0.032 | 10.9 | 0.001 | 7.60 | 0.007 |
| Theta | 3.78 | 0.054 | 0.61 | 0.437 | 0.00 | 0.900 | 5.75 | 0.018 | 0.03 | 0.86 | 3.23 | 0.074 | 4.02 | 0.051 | 0.33 | 0.564 |
| Alpha | 10.5 | 0.002 | 1.36 | 0.247 | 32.2 | 0.000 | 0.00 | 0.961 | 24.9 | 0.000 | 22.6 | 0.0001 | 46.9 | 0.000 | 3.03 | 0.084 |
| Beta 1 | 0.46 | 0.499 | 1.65 | 0.203 | 0.14 | 0.71 | 0.59 | 0.445 | 0.21 | 0.644 | 1.64 | 0.229 | 4.5 | 0.035 | 3.23 | 0.074 |
| Beta 2 | 3.64 | 0.591 | 1.85 | 0.177 | 16.7 | 0.000 | 0.32 | 0.575 | 12.6 | 0.000 | 8.88 | 0.003 | 23.8 | 0.0001 | 3.40 | 0.067 |

**Figure A2.** Results of a two-way ANOVA analysis of APO-induced changes in EEG recordings from four brain areas of synuclein KO mice and WT (A+B+G+) mice. Recoding from the same brain areas of mice injected with saline were used as controls. Data with $p < 0.05$ are highlighted blue, $p < 0.01$ highlighted pink and $p < 0.001$ highlighted red.

| Two-way ANOVA for EEG effects of apomorphine vs. A+B+G+ | | | | | | | | | | | | | |
|---|---|---|---|---|---|---|---|---|---|---|---|---|---|
| **Groups** | A-B-G- | | A-B+G- | | A+B-G+ | | A-B+G+ | | A+B-G- | | A+B+G- | | A-B-G+ | |
| **MC** | $F_{96}$ | P | $F_{108}$ | P | $F_{114}$ | P | $F_{120}$ | P | $F_{126}$ | P | $F_{114}$ | P | $F_{126}$ | P |
| Delta 1 | 3.23 | 0.076 | 0.26 | 0.614 | 0.36 | 0.552 | 5.56 | 0.020 | 0.46 | 0.498 | 0.04 | 0.837 | 0.73 | 0.393 |
| Delta 2 | 3.1 | 0.082 | 0.00 | 0.97 | 0.06 | 0.884 | 1.69 | 0.196 | 0.87 | 0.353 | 0.02 | 0.897 | 0.13 | 0.721 |
| Theta | 1.55 | 0.217 | 3.57 | 0.061 | 0.7 | 0.405 | 0.03 | 0.87 | 0.7 | 0.404 | 3.65 | 0.059 | 1.21 | 0.273 |
| Alpha | 0.43 | 0.513 | 8.77 | 0.004 | 0.15 | 0.702 | 1.94 | 0.166 | 2.20 | 0.141 | 9.77 | 0.002 | 0.13 | 0.718 |
| Beta 1 | 2.76 | 0.099 | 0.78 | 0.38 | 1.14 | 0.288 | 0.59 | 0.808 | 0.49 | 0.485 | 4.39 | 0.038 | 1.96 | 0.164 |
| Beta 2 | 1.87 | 0.175 | 12.6 | 0.000 | 0.28 | 0.6 | 0.57 | 0.452 | 2.44 | 0.121 | 10.1 | 0.002 | 1.14 | 0.288 |
| **Putamen** | $F_{96}$ | P | $F_{108}$ | P | $F_{114}$ | P | $F_{120}$ | P | $F_{126}$ | P | $F_{114}$ | P | $F_{126}$ | P |
| Delta 1 | 8.82 | 0.004 | 0.51 | 0.477 | 1.7 | 0.863 | 6.98 | 0.009 | 4.54 | 0.035 | 1.91 | 0.17 | 6.15 | 0.014 |
| Delta 2 | 4.54 | 0.036 | 0.52 | 0.477 | 1.34 | 0.25 | 0.13 | 0.721 | 0.64 | 0.425 | 0.06 | 0.807 | 3.1 | 0.081 |
| Theta | 0.90 | 0.345 | 0.80 | 0.374 | 0.14 | 0.711 | 0.66 | 0.417 | 0.03 | 0.866 | 3.48 | 0.065 | 0.00 | 0.960 |
| Alpha | 0.18 | 0.673 | 6.63 | 0.011 | 1.03 | 0.313 | 1.81 | 0.181 | 2.61 | 0.109 | 12.3 | 0.000 | 1.00 | 0.322 |
| Beta 1 | 2.67 | 0.106 | 0.22 | 0.639 | 2.42 | 0.123 | 0.08 | 0.775 | 1.08 | 0.30 | 5.36 | 0.022 | 2.25 | 0.136 |
| Beta 2 | 2.27 | 0.135 | 10.3 | 0.002 | 0.01 | 0.945 | 1.85 | 0.176 | 2.91 | 0.09 | 9.40 | 0.003 | 1.72 | 0.192 |
| **VTA** | $F_{96}$ | P | $F_{108}$ | P | $F_{114}$ | P | $F_{120}$ | P | $F_{126}$ | P | $F_{114}$ | P | $F_{126}$ | P |
| Delta 1 | 1.68 | 0.198 | 0.56 | 0.457 | 1.50 | 0.224 | 1.04 | 0.311 | 2.39 | 0.125 | 1.00 | 0.319 | 5.80 | 0.018 |
| Delta 2 | 7.30 | 0.008 | 1.01 | 0.316 | 0.41 | 0.524 | 0.34 | 0.558 | 2.19 | 0.142 | 1.95 | 0.165 | 0.85 | 0.357 |
| Theta | 2.56 | 0.113 | 1.35 | 0.248 | 5.11 | 0.026 | 0.66 | 0.419 | 0.57 | 0.451 | 3.08 | 0.082 | 0.86 | 0.354 |
| Alpha | 0.01 | 0.931 | 7.02 | 0.009 | 4.38 | 0.039 | 4.93 | 0.028 | 7.93 | 0.006 | 28.6 | 0.000 | 0.21 | 0.646 |
| Beta 1 | 3.83 | 0.053 | 0.62 | 0.413 | 0.05 | 0.831 | 0.57 | 0.451 | 0.12 | 0.727 | 8.88 | 0.004 | 3.62 | 0.059 |
| Beta 2 | 3.43 | 0.067 | 11.3 | 0.001 | 5.85 | 0.017 | 3.64 | 0.059 | 4.93 | 0.028 | 22.7 | 0.000 | 1.36 | 0.245 |
| **SN** | $F_{96}$ | P | $F_{108}$ | P | $F_{114}$ | P | $F_{120}$ | P | $F_{126}$ | P | $F_{114}$ | P | $F_{126}$ | P |
| Delta 1 | 4.27 | 0.041 | 0.32 | 0.574 | 1.83 | 0.179 | 0.64 | 0.426 | 0.42 | 0.515 | 0.31 | 0.579 | 0.83 | 0.364 |
| Delta 2 | 1.65 | 0.202 | 1.02 | 0.316 | 0.73 | 0.396 | 1.13 | 0.290 | 0.62 | 0.432 | 3.89 | 0.051 | 0.50 | 0.479 |
| Theta | 4.96 | 0.028 | 1.04 | 0.31 | 4.70 | 0.032 | 1.23 | 0.270 | 0.01 | 0.928 | 6.59 | 0.012 | 5.41 | 0.022 |
| Alpha | 0.60 | 0.440 | 10.3 | 0.002 | 3.56 | 0.062 | 4.84 | 0.030 | 9.25 | 0.003 | 19.4 | 0.000 | 0.94 | 0.335 |
| Beta 1 | 11.6 | 0.000 | 0.02 | 0.88 | 0.00 | 0.960 | 0.45 | 0.500 | 0.85 | 0.359 | 7.69 | 0.006 | 6.86 | 0.010 |
| Beta 2 | 0.74 | 0.393 | 10.2 | 0.002 | 6.30 | 0.013 | 2.06 | 0.153 | 4.39 | 0.038 | 14.6 | 0.000 | 2.28 | 0.134 |

**Figure A3.** Results of a two-way ANOVA analysis of APO-induced changes in EEG recordings from four brain areas of synuclein KO mice vs. WT (A+B+G+) mice. Data with $p < 0.05$ are highlighted blue, $p < 0.01$ highlighted pink and $p < 0.001$ highlighted red.

## References


1. George, J.M. The synucleins. *Gen. Biol.* **2001**, *3*, 3002.1–3002.6. [CrossRef] [PubMed]
2. Burré, J. The Synaptic Function of α-Synuclein. *J. Park. Dis.* **2015**, *5*, 699–713. [CrossRef] [PubMed]
3. Brown, J.W.P.; Buell, A.K.; Michaels, T.C.T.; Meisl, G.; Carozza, J.; Flagmeier, P.; Vendruscolo, M.; Knowles, T.P.J.; Dobson, C.M.; Galvagnion, C. β-Synuclein suppresses both the initiation and amplification steps of α-synuclein aggregation via competitive binding to surfaces. *Sci. Rep.* **2016**, *6*, 36010. [CrossRef] [PubMed]
4. Venda, L.L.; Cragg, S.J.; Buchman, V.L.; Wade-Martins, R. α-Synuclein and dopamine at the crossroads of Parkinson's disease. *Trends Neurosci.* **2010**, *33*, 559–568. [CrossRef]
5. Raina, A.; Leite, K.; Chakrabarti, K.S.; Guerin, S.; Mahajani, S.U.; Voll, D.; Becker, S.; Griesinger, C.; Bähr, M.; Kügler, S. Dopamine promotes the neurodegenerative potential of β-synuclein. *J. Neurochem.* **2021**, *156*, 674–691. [CrossRef]
6. Ninkina, N.; Millership, S.J.; Peters, O.M.; Connor-Robson, N.; Chaprov, K.; Kopylov, A.T.; Montoya, A.; Kramer, H.; Withers, D.J.; Buchman, V.L. β-synuclein potentiates synaptic vesicle dopamine uptake and rescues dopaminergic neurons from MPTP-induced death in the absence of other synucleins. *J. Biol. Chem.* **2021**, *297*, 101375. [CrossRef]
7. Dev, K.K.; Hofele, K.; Barbieri, S.; Buchman, V.L.; Van der Putten, H. Alpha-synuclein and its molecular pathophysiological role in neurodegenerative disease. *Neuropharmacology* **2003**, *45*, 14–44. [CrossRef]
8. Carnazza, K.E.; Komer, L.E.; Xie, Y.X.; Pineda, A.; Briano, J.A.; Gao, V.; Na, Y.; Ramlall, T.; Buchman, V.L.; Eliezer, D.; et al. Synaptic vesicle binding of α-synuclein is modulated by β- and γ-synucleins. *Cell Rep.* **2022**, *39*, 110675. [CrossRef]
9. Winham, C.L.; Le, T.; Jellison, E.R.; Silver, A.C.; Levesque, A.A.; Koob, A.O. γ-Synuclein induces human cortical astrocyte proliferation and subsequent BDNF expression and release. *Neuroscience* **2019**, *410*, 41–54. [CrossRef]
10. Ducas, V.C.; Rhoades, E. Quantifying interactions of β-synuclein and γ-synuclein with model membranes. *J. Mol. Biol.* **2012**, *423*, 528–539. [CrossRef]
11. Pavia-Collado, R.; Rodríguez-Aller, R.; Alarcón-Arís, D.; Miquel-Rio, L.; Ruiz-Bronchal, E.; Paz, V.; Campa, L.; Galofré, M.; Sgambato, V.; Bortolozzi, A. Up and Down γ-Synuclein Transcription in Dopamine Neurons Translates into Changes in Dopamine Neurotransmission and Behavioral Performance in Mice. *Int. J. Mol. Sci.* **2022**, *23*, 1807. [CrossRef] [PubMed]
12. Visanji, N.P.; Brotchie, J.M.; Kalia, L.V.; Koprich, J.B.; Tandon, A.; Watts, J.C.; Lang, A.E. α-Synuclein-Based Animal Models of Parkinson's Disease: Challenges and Opportunities in a New Era. *Trends Neurosci.* **2016**, *39*, 750–762. [CrossRef]
13. Buchman, V.L.; Ninkina, N. Modulation of α-synuclein expression in transgenic animals for modelling synucleinopathies—Is the juice worth the squeeze? *Neurotox. Res.* **2008**, *14*, 329–341. [CrossRef] [PubMed]
14. Robertson, D.C.; Schmidt, O.; Ninkina, N.; Jones, P.A.; Sharkey, J.; Buchman, V.L. Developmental loss and resistance to MPTP toxicity of dopaminergic neurones in substantia nigra pars compacta of *gamma*-synuclein, *alpha*-synuclein and double *alpha/gamma*-synuclein null mutant mice. *J. Neurochem.* **2004**, *89*, 1126–1136. [CrossRef] [PubMed]
15. Maroteaux, L.; Campanelli, J.T.; Scheller, R.H. Synuclein: A neuron-specific protein localized to the nucleus and presynaptic nerve terminal. *J. Neurosci.* **1988**, *8*, 2804–2815. [CrossRef]
16. Greten-Harrison, B.; Polydoro, M.; Morimoto-Tomita, M.; Diao, L.; Williams, A.M.; Nie, E.H.; Makani, S.; Tian, N.; Castillo, P.E.; Buchman, V.L.; et al. αβγ-Synuclein triple knockout mice reveal age-dependent neuronal dysfunction. *Proc. Natl. Acad. Sci. USA* **2010**, *107*, 19573–19578. [CrossRef] [PubMed]
17. Ninkina, N.; Tarasova, T.V.; Chaprov, K.D.; Roman, A.Y.; Kukharsky, M.S.; Kolik, L.G.; Ovchinnikov, R.; Ustyugov, A.A.; Durnev, A.D.; Buchman, V.L. Alterations in the nigrostriatal system following conditional inactivation of α-synuclein in neurons of adult and aging mice. *Neurobiol. Aging* **2020**, *91*, 76–87. [CrossRef] [PubMed]
18. Connor-Robson, N.; Peters, O.M.; Millership, S.; Ninkina, N.; Buchman, V.L. Combinational losses of synucleins reveal their differential requirements for compensating age-dependent alterations in motor behavior and dopamine metabolism. *Neurobiol. Aging* **2016**, *46*, 107–112. [CrossRef]
19. Chandra, S.; Fornai, F.; Kwon, H.-B.; Yazdani, U.; Atasoy, D.; Liu, X.; Hammer, R.E.; Battaglia, G.; German, D.C.; Castillo, P.E.; et al. Double-knockout mice for *alpha*- and *beta*-synucleins: Effect on synaptic functions. *Proc. Natl. Acad. Sci. USA* **2004**, *101*, 14966–14971. [CrossRef]
20. Anwar, S.; Peters, O.; Millership, S.; Ninkina, N.; Doig, N.; Connor-Robson, N.; Threlfell, S.; Kooner, G.; Deacon, R.M.; Bannerman, D.M.; et al. Functional alterations to the nigrostriatal system in mice lacking all three members of the synuclein family. *J. Neurosci.* **2011**, *31*, 7264–7274. [CrossRef]
21. Saleh, H.; Saleh, A.; Yao, H.; Cui, J.; Shen, Y.; Li, R. Mini review: Linkage between alpha-Synuclein protein and cognition. *Transl. Neurodegener.* **2015**, *4*, 5. [CrossRef] [PubMed]
22. Kokhan, V.S.; Afanasyeva, M.A.; Van'kin, G.I. α-Synuclein knockout mice have cognitive impairments. *Behav. Brain Res.* **2012**, *231*, 226–230. [CrossRef] [PubMed]
23. Kokhan, V.S.; Bolkunov, A.V.; Ustiugov, A.A.; Van'kin, G.I.; Shelkovnikova, T.A.; Redkozubova, O.M.; Strekalova, T.V.; Bukhman, V.L.; Ninkina, N.N.; Bachurin, S.O. Targeted inactivation of gamma-synuclein gene affects anxiety and exploratory behaviour of mice. *Zhurnal Vyss. Nervn. Deiatelnosti Im. IP Pavlov.* **2011**, *61*, 85–93.
24. Kokhan, V.S.; Van'kin, G.I.; Bachurin, S.O.; Shamakina, I.Y. Differential involvement of the gamma-synuclein in cognitive abilities on the model of knockout mice. *BMC Neurosci.* **2013**, *14*, 53. [CrossRef] [PubMed]
25. Oswal, A.; Brown, P.; Litvak, V. Synchronized neural oscillations and the pathophysiology of Parkinson's disease. *Curr. Opin. Neurol.* **2013**, *26*, 662–670. [CrossRef]

26. Nimmrich, V.; Draguhn, A.; Axmacher, N. Neuronal Network Oscillations in Neurodegenerative Diseases. *Neuromolecular Med.* **2015**, *17*, 270–284. [CrossRef]

27. Womelsdorf, T.; Schoffelen, J.M.; Oostenveld, R.; Singer, W.; Desimone, R.; Engel, A.K.; Fries, P. Modulation of neuronal interactions through neuronal synchronization. *Science* **2007**, *316*, 1609–1612. [CrossRef]

28. Buzsáki, G.; Anastassiou, C.A.; Koch, C. The origin of extracellular fields and currents–EEG, ECoG, LFP and spikes. *Nat. Rev. Neurosci.* **2012**, *13*, 407–420. [CrossRef]

29. Brown, P. Oscillatory nature of human basal ganglia activity: Relationship to the pathophysiology of Parkinson's disease. *Mov. Disord.* **2003**, *18*, 357–363. [CrossRef]

30. Bousleiman, H.; Zimmermann, R.; Ahmed, S.; Hardmeier, M.; Hatz, F.; Schindler, C.; Roth, V.; Gschwandtner, U.; Fuhr, P. Power spectra for screening parkinsonian patients for mild cognitive impairment. *Ann. Clin. Transl. Neurol.* **2014**, *1*, 884–890. [CrossRef]

31. Vorobyov, V.; Sengpiel, F. Apomorphine-induced differences in cortical and striatal EEG and their glutamatergic mediation in 6-hydroxydopamine-treated rats. *Exp. Brain Res.* **2008**, *191*, 277–287. [CrossRef]

32. Wang, X.; Li, M.; Xie, J.; Chen, D.; Geng, X.; Sun, S.; Liu, B.; Wang, M. *Beta* band modulation by dopamine D2 receptors in the primary motor cortex and pedunculopontine nucleus in a rat model of Parkinson's disease. *Brain Res. Bull.* **2022**, *181*, 121–128. [CrossRef] [PubMed]

33. Brittain, J.S.; Brown, P. Oscillations and the basal ganglia: Motor control and beyond. *Neuroimage* **2014**, *85 Pt 2*, 637–647. [CrossRef] [PubMed]

34. Singh, A. Oscillatory activity in the cortico-basal ganglia-thalamic neural circuits in Parkinson's disease. *Eur. J. Neurosci.* **2018**, *48*, 2869–2878. [CrossRef] [PubMed]

35. Sharott, A.; Magill, P.J.; Harnack, D.; Kupsch, A.; Meissner, W.; Brown, P. Dopamine depletion increases the power and coherence of *beta*-oscillations in the cerebral cortex and subthalamic nucleus of the awake rat. *Eur. J. Neurosci.* **2005**, *21*, 1413–1422. [CrossRef] [PubMed]

36. Prosperetti, C.; Di Giovanni, G.; Stefani, A.; Möller, J.C.; Galati, S. Acute nigro-striatal blockade alters cortico-striatal encoding: An in vivo electrophysiological study. *Exp. Neurol.* **2013**, *247*, 730–736. [CrossRef]

37. McDowell, K.A.; Shin, D.; Roos, K.P.; Chesselet, M.F. Sleep dysfunction and EEG alterations in mice overexpressing *alpha*-synuclein. *J. Park. Dis.* **2014**, *4*, 531–539. [CrossRef]

38. Caviness, J.N.; Lue, L.F.; Hentz, J.G.; Schmitz, C.T.; Adler, C.H.; Shill, H.A.; Sabbagh, M.N.; Beach, T.G.; Walker, D.G. Cortical phosphorylated α-Synuclein levels correlate with brain wave spectra in Parkinson's disease. *Mov. Disord.* **2016**, *31*, 1012–1019. [CrossRef]

39. Ninkina, N.; Papachroni, K.; Robertson, D.C.; Schmidt, O.; Delaney, L.; O'Neill, F.; Court, F.; Rosenthal, A.; Fleetwood-Walker, S.M.; Davies, A.M.; et al. Neurons expressing the highest levels of γ-synuclein are unaffected by targeted inactivation of the gene. *Mol. Cell. Biol.* **2003**, *23*, 8233–8245. [CrossRef]

40. Ninkina, N.; Connor-Robson, N.; Ustyugov, A.A.; Tarasova, T.V.; Shelkovnikova, T.A.; Buchman, V.L. A novel resource for studying function and dysfunction of α-synuclein: Mouse lines for modulation of endogenous Snca gene expression. *Sci. Rep.* **2015**, *5*, 16615. [CrossRef]

41. Franklin, K.B.J.; Paxinos, G. *The Mouse Brain in Stereotaxic Coordinates*, 3rd ed.; Academic Press: New York, NY, USA, 2007.

42. Al-Wandi, A.; Ninkina, N.; Millership, S.; Williamson, S.J.; Jones, P.A.; Buchman, V.L. Absence of *alpha*-synuclein affects dopamine metabolism and synaptic markers in the striatum of aging mice. *Neurobiol. Aging* **2010**, *31*, 796–804. [CrossRef] [PubMed]

43. Goloborshcheva, V.V.; Chaprov, K.D.; Teterina, E.V.; Ovchinnikov, R.; Buchman, V.L. Reduced complement of dopaminergic neurons in the substantia nigra pars compacta of mice with a constitutive "low footprint" genetic knockout of *alpha*-synuclein. *Mol. Brain* **2020**, *13*, 75. [CrossRef] [PubMed]

44. Hökfelt, T.; Martensson, R.; Björklund, A.; Kheinau, S.; Goldstein, M. *Handbook of Chemical Neuroanatomy*; Björklund, A., Hökfelt, T., Eds.; Elsevier Science B.V: Amsterdam, The Netherlands, 1984; Volume 2, pp. 277–379.

45. Gal'chenko, A.A.; Vorobyov, V.V. Analysis of electroencephalograms using a modified amplitude-interval algorithm. *Neurosci. Behav. Physiol.* **1999**, *29*, 157–160. [CrossRef] [PubMed]

46. Eusebio, A.; Brown, P. Synchronisation in the beta frequency-band—the bad boy of parkinsonism or an innocent bystander? *Exp. Neurol.* **2009**, *217*, 1–3. [CrossRef] [PubMed]

47. Timmermann, L.; Fink, G.R. Pathological network activity in Parkinson's disease: From neural activity and connectivity to causality? *Brain* **2011**, *134*, 332–334. [CrossRef] [PubMed]

48. Haumesser, J.K.; Beck, M.H.; Pellegrini, F.; Kühn, J.; Neumann, W.J.; Altschüler, J.; Harnack, D.; Kupsch, A.; Nikulin, V.V.; Kühn, A.A.; et al. Subthalamic beta oscillations correlate with dopaminergic degeneration in experimental parkinsonism. *Exp. Neurol.* **2021**, *335*, 113513. [CrossRef]

49. Doyle, L.M.; Kuhn, A.A.; Hariz, M.; Kupsch, A.; Schneider, G.H.; Brown, P. Levodopa-induced modulation of subthalamic beta oscillations during self-paced movements in patients with Parkinson's disease. *Eur. J. Neurosci.* **2005**, *21*, 1403–1412. [CrossRef]

50. Kühn, A.A.; Kempf, F.; Brucke, C.; Gaynor Doyle, L.; Martinez-Torres, I.; Pogosyan, A.; Trottenberg, T.; Kupsch, A.; Schneider, G.-H.; Hariz, M.I.; et al. High-frequency stimulation of the subthalamic nucleus suppresses oscillatory beta activity in patients with Parkinson's disease in parallel with improvement in motor performance. *J. Neurosci.* **2008**, *28*, 6165–6173.

51. Eusebio, A.; Chen, C.C.; Lu, C.S.; Lee, S.T.; Tsai, C.H.; Limousin, P.; Hariz, M.; Brown, P. Effects of low-frequency stimulation of the subthalamic nucleus on movement in Parkinson's disease. *Exp. Neurol.* **2008**, *209*, 125–130. [CrossRef]

*Article*

# Sex-Related Differences in Voluntary Alcohol Intake and mRNA Coding for Synucleins in the Brain of Adult Rats Prenatally Exposed to Alcohol

Viktor S. Kokhan [1,*], Kirill Chaprov [2,3], Natalia N. Ninkina [2,3], Petr K. Anokhin [1], Ekaterina P. Pakhlova [4], Natalia Y. Sarycheva [4] and Inna Y. Shamakina [1]

1   V.P. Serbsky Federal Medical Research Centre for Psychiatry and Narcology, 119034 Moscow, Russia
2   Institute of Physiologically Active Compounds RAS, 142432 Chernogolovka, Russia
3   Belgorod State National Research University, 308015 Belgorod, Russia
4   Lomonosov Moscow State University, 119991 Moscow, Russia
*   Correspondence: viktor_kohan@hotmail.com; Tel.: +7-925-462-99-48

**Abstract:** Maternal alcohol consumption is one of the strong predictive factors of alcohol use and consequent abuse; however, investigations of sex differences in response to prenatal alcohol exposure (PAE) are limited. Here we compared the effects of PAE throughout gestation on alcohol preference, state anxiety and mRNA expression of presynaptic proteins α-, β- and γ-synucleins in the brain of adult (PND60) male and female Wistar rats. Total RNA was isolated from the hippocampus, midbrain and hypothalamus and mRNA levels were assessed with quantitative RT-PCR. Compared with naïve males, naïve female rats consumed more alcohol in "free choice" paradigm (10% ethanol vs. water). At the same time, PAE produced significant increase in alcohol consumption and preference in males but not in females compared to male and female naïve groups, correspondingly. We found significantly lower α-synuclein mRNA levels in the hippocampus and midbrain of females compared to males and significant decrease in α-synuclein mRNA in these brain areas in PAE males, but not in females compared to the same sex controls. These findings indicate that the impact of PAE on transcriptional regulation of synucleins may be sex-dependent, and in males' disruption in α-synuclein mRNA expression may contribute to increased vulnerability to alcohol-associated behavior.

**Keywords:** prenatal alcohol exposure; synucleins; alcohol consumption; free-choice; hippocampus; midbrain; mRNA expression; α-synuclein; rats

**Citation:** Kokhan, V.S.; Chaprov, K.; Ninkina, N.N.; Anokhin, P.K.; Pakhlova, E.P.; Sarycheva, N.Y.; Shamakina, I.Y. Sex-Related Differences in Voluntary Alcohol Intake and mRNA Coding for Synucleins in the Brain of Adult Rats Prenatally Exposed to Alcohol. *Biomedicines* **2022**, *10*, 2163. https://doi.org/10.3390/biomedicines 10092163

Academic Editor: Laura Orio

Received: 2 July 2022
Accepted: 29 August 2022
Published: 2 September 2022

**Publisher's Note:** MDPI stays neutral with regard to jurisdictional claims in published maps and institutional affiliations.



## 1. Introduction

Alcohol use disorder (AUD) is a severe and etiologically complex disease associated with high morbidity and mortality rates. Although the development of alcohol dependence has a strong genetic component [1], there are a wide variety of environmental factors that impact the vulnerability to AUD through epigenetic modifications [2].

Maternal alcohol drinking leading to prenatal alcohol exposure (PAE) is one of the strong predictive factors of AUD [3]. Neurobehavioral disorder associated with prenatal alcohol exposure can occur in children following even low to moderate levels of maternal ethanol consumption during pregnancy and touch on the basic components of cognition—learning, memory, executive functions, affective state anxiety and depression, and addictive behavior later in life [4,5]. To date the epigenetic mechanisms underlying high risk of drug abuse which arise from maternal consumption of alcohol are unclear. It is well known that alterations in presynaptic regulation of neurotransmission are among key mechanisms involved in the long-lasting ethanol effects, excessive drinking and ethanol abuse [6]. Central component of the presynaptic neurotransmitter release machinery is soluble NSF attachment protein receptor complex, which in turn is under regulatory control of synucleins—small presynaptic proteins that are expressed from three genes (α-, β- and γ-synuclein) and may modify neurotransmitter release [7]. α-Synuclein has been previously studied as a potential player in alcohol addiction due to its role in dopamine (DA) neurotransmission [8]. Genome-wide association studies have found α-synuclein to be one of the candidate gene for alcoholism, and single-nucleotide polymorphisms in intron 4 of the α-synuclein gene (*Snca*) have been linked to alcohol craving [9]. It has been found that variability in *Snca* which affects its expression may contribute to the high risk of alcohol abuse [10]. It was shown that elevated α-synuclein mRNA levels are associated with craving in patients with alcoholism [11,12]; transgenic mice expressing human mutant A30P α-synuclein were characterized by higher motivation for ethanol and attenuated context- and cue-induced reinstatement of alcohol-seeking behavior [13]. At the same time, alcohol-dependent subjects had higher frequencies of the shortest 267 bp allele of the α-synuclein-repeat 1 marker, associated with decreased expression of α-synuclein in the autopsy samples of human prefrontal cortex [10]. These and other data have inspired the α-synuclein deficit hypothesis: low basal levels of α-synuclein in some brain regions may predispose to enhanced DA activity in response to alcohol and as a result alcohol cravings and excessive alcohol consumption, leading in turn to increases in α-synuclein [9]. This hypothesis is in accordance with the fact, that α-synuclein expression is downregulated in the frontal cortex and caudate–putamen of inbred alcohol preferring rats prior to ethanol exposure [14].

β-Synuclein—a pre-synaptic protein that co-localizes with α-synuclein, can act as α-synuclein inhibitor [15] and suggested to have neuroprotective properties [15]. Mammalian γ-synuclein is mainly expressed in the peripheral nervous system (primary sensory, sympathetic and motor neurons), but also detected in the brain [16]. The normal cellular function of γ-synuclein still remains unknown. Interestingly, γ-synuclein knockout mice were hypoactive in a novel environment [17], whereas α- and γ-synuclein double-null knockout mice were hyperactive due to the hyperdopaminergic phenotype detected by two-fold increase in the extracellular concentration of dopamine after electrical stimulation of striatum [18].

Considering that dopaminergic imbalance in the brain can be one of the determinants of high alcohol motivation [19,20], and synucleins may be critically involved in regulation of dopaminergic neurons [8] here we studied alterations in synucleins mRNA level in specific brain areas of adult male and female rats affected by prenatal alcohol exposure. It was of particular importance to focus on the sex-dependent alterations in gene expression which might provide more insight into the problem of sex differences in response to alcohol intake both in humans [21,22] and animal models of alcoholism [23,24].

## 2. Materials and Methods

### 2.1. Animals

Adult (postnatal day 60; PND60) male and female Wistar rats were supplied by Stolbovaya Animal Farm (Moscow region, Russia) and group-housed 5 per a cage in a 12/12-h light/dark cycle, 19–22 °C and 55% humidity with free access to water and standard lab chow. Following a one-week acclimation period in the animal facility two female rats were matched with one male for 72 h, thus the number of rats in one cage was 3 (2♀+ 1♂). In the current experiment conception occurred 1–2 days following male-female pairing. After confirmation of pregnancy by the presence of a vaginal plug male partners were removed and females were singly housed until giving birth. In this case, 20 female rats were distributed into two groups equally (10 animals/group): ethanol-exposed (received 10% (*v/v*) ethanol as the only fluid, E2-birth, *n* = 10) and intact (received water as the only fluid all the time, *n* = 10). A gestational period "E2-birth" is analogous to the first two trimesters of human gestation [25]. The human third trimester equivalent occurs in rodents following birth (PND 1–10) [26,27], thus, in accordance with the goal of the study (prenatal alcohol exposure study) during this period, the offspring/dams did not receive ethanol. Offspring were weaned at PND21 and housed 5 rats of the same sex in one cage. At postnatal day 60 (PND 60) male and female offspring were divided into three groups and used for: (1) test for anxiety ("light-dark box"), (2) test for alcohol preference, (3) analysis of mRNA expression. For each test we used equal number of rats per sex per litter from both PAE and control groups. Each experimental group consisted of no more than 3 littermates. A scheme of the experimental design is shown in Figure 1.

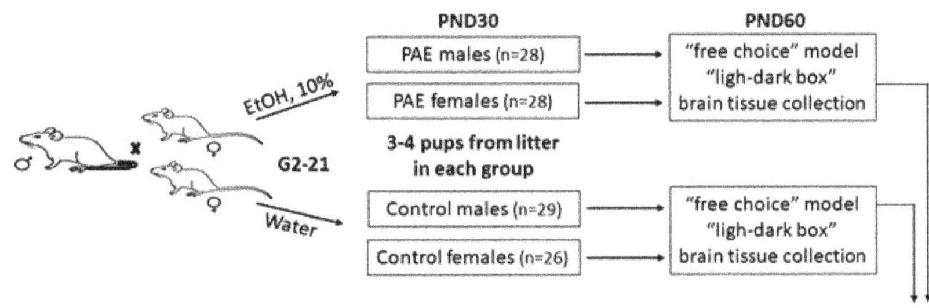


**Figure 1.** Scheme of the experimental design. EtOH—ethanol. PAE—prenatal alcohol exposure. PND—postnatal day.

### 2.2. Two-Bottle "10% Alcohol vs. Water" Choice Drinking Paradigm (Voluntary Alcohol Consumption)

At PND 60 rats were housed individually in standard cages ($43.5 \times 28 \times 16$ cm) and presented with 2 bottles containing diluted ethanol (10% $v/v$) and the other containing water, providing a 24-h continuous free-choice alcohol access. Alcohol and water intake were carefully measured by weighing the bottles every 24 h at 8 a.m. (0.1 g accuracy). The calculated parameters were daily intake levels of water, 10% Alcohol and total fluid. Alcohol preference was estimated as a ratio of 10% Alcohol vs. total daily fluid consumed (10% Alcohol/Alcohol + water). There were four groups: control male offspring from intact dams (C-m, $n = 11$), PAE male offspring (PA-m, $n = 10$), control female offspring from intact dams (C-f, $n = 8$), PAE female offspring (PA-f, $n = 10$).

### 2.3. Light-Dark Box

Anxiety-like behavior was measured via light-dark box test (TSE, Berlin, Germany). Each rat was initially placed in the middle compartment (length$\times$ width $\times$ height, $13 \times 21 \times 35$ cm) and monitored for 15 min in the box with free choice to move between and within brightly illuminated (left) and dark (right) compartments (both $21 \times 21 \times 35$ cm). The following parameters were detected for each of the box compartments: number of entrances (full-body transitions between chambers), total distance, and the total spent time. There were four groups: control male offspring from intact dams (C-m, $n = 9$), PAE male offspring (PA-m, $n = 9$), control female offspring from intact dams (C-f, $n = 9$), PAE female offspring (PA-f, $n = 9$).

### 2.4. Tissue Collection

The animals from control male offspring from intact dams (C-m, $n = 9$), PAE male offspring (PA-m, $n = 9$), control female offspring from intact dams (C-f, $n = 9$) and PAE female offspring (PA-f, $n = 9$) were euthanized by decapitation and the following brain morphological structures were isolated on ice: hippocampus (HPC), hypothalamus (HPY) and midbrain (MID). The structures were immediately frozen in liquid nitrogen.

### 2.5. RNA Extraction, cDNA Synthesis and Quantitative RT-PCR

Total RNA was extracted using Extract RNA (Evrogen, Moscow, Russia) according to the standard phenol-chloroform protocol. DNAse (Thermo Scientific, Mundelein, IL, USA) was used for RNA purification and 1 µg of total RNA was used for cDNA synthesis. Reverse transcription was performed using the MMLV-RT kit (Evrogen, Moscow, Russia) with random hexamer primers. Gene expression levels were analysed via real-time polymerase chain reaction (qRT-PCR) using qPCRmix-HS SYBR (Evrogen). The reaction was carried out in a BioRad CFX96 thermal cycler (Bio-Rad Laboratories Inc., California, USA). Relative gene expression values were obtained by using $2^{-\Delta\Delta CT}$ method with β-actin as a reference gene. Primer sequences are shown in Table 1.

**Table 1.** Oligonucleotide sequences for primers used in the qRT-PCR.

| Gene | Primers | |
| --- | --- | --- |
| | Forward | Reverse |
| β-actin | 5′-cactgccg-catcctcttcct-3′ | 5′-aaccgctcatt-gccgatagtg-3′ |
| *Snca* (α-synuclein) | 5′-tgtcaagaaggaccagatg-3′ | 5′-caggctcatagtcttggtag-3′ |
| *Sncb* (β-synuclein) | 5′-agttccccacagacctgaag-3′ | 5′-ttacgcctctggctcgtattc-3′ |
| *Sncg* (γ-synuclein) | 5′-aaacatcgtggtcaccacc-3′ | 5′-tctagtctcctccactcttg-3′ |

*2.6. Data Analysis*

The data were represented as the mean $\pm$ standard deviation (SD) and analyzed with the Statistica 12 software (StatSoft Inc., Tulsa, OK, USA). The datasets were tested for normality with the Shapiro-Wilk test (W-test) and the parametric analysis was applied if $p > 0.05$. The data obtained in the "alcohol intake" test were analyzed with repeated measures ANOVA. For the other data sets the two-way ANOVA was applied. The post-hoc Duncan's multiple range test was used when appropriate.

**3. Results**

*3.1. Observation of Pregnant Dam and Offspring*

The mean maternal ethanol consumption was recorded daily throughout pregnancy and was $2.6 \pm 0.72$ g/kg per 24 h. This moderate ethanol exposure did not affect maternal weight gain and pregnancy outcomes—litter size, number of pups born, postnatal mortality or offspring birth weight compared to control group (data not shown).

*3.2. Voluntary Alcohol Consumption*

Alcohol drinking groups had access to 10% ethanol and water in two-bottle-choice paradigm for 7 days.

Statistically significant impact of PAE ($F_{1,35} = 4.5$, $p = 0.04$), sex ($F_{1,35} = 13.7$, $p = 0.007$), and interaction of alcohol consumption dynamics and sex ($F_{6,210} = 3.8$, $p = 0.001$) on alcohol consumption were found (Figure 2a). Estimation of daily ethanol intake indicated that C-f group of rats demonstrated higher alcohol consumption at Days 1, 2, 4, 5, and 6 (264%, $p = 0.004$; 209%, $p = 0.009$; 200%, $p = 0.005$; 314%, $p = 0.004$; 152%, $p = 0.03$, respectively) compared to C-m rats. The differences between the PA-f and PA-m groups were not found. At the same time, there was apparent difference in ethanol consumption between PAE and control (naïve) males, but not between PAE and naïve females. PA-m rats demonstrated significantly higher alcohol consumption at Day 3 and Day 7 of the experiment (180%, $p = 0.04$ and 135%, $p = 0.009$, respectively) compared to C-m rats. Among all experimental groups only PA-m rats showed a 101% ($p = 0.0005$) increase in alcohol consumption from Day 1 to Day 7 of the experiment.

Effects of PAE ($F_{1,35} = 8$, $p = 0.008$), sex ($F_{1,35} = 14.5$, $p = 0.006$), and interaction of alcohol preference dynamics and sex ($F_{6,210} = 5.7$, $p = 0.00002$), alcohol consumption dynamics, PAE and sex ($F_{6,210} = 2.2$, $p = 0.04$) on alcohol preference were found (Figure 2b). Estimation of daily drinking activity indicated that female rats in both PAE (only at the beginning of testing) and C groups prefer ethanol to water to a greater extent than males. C-f group of rats demonstrated higher alcohol preference at Days 1, 2, 4, 5, and 6 (249%, $p = 0.007$; 215%, $p = 0.008$; 196%, $p = 0.008$; 331%, $p = 0.003$; 146%, $p = 0.046$, respectively) compared to C-m rats. The differences between the PA-f and PA-m groups were somewhat less pronounced: PA-f group of rats demonstrated higher alcohol preference at Day 1—on 94% ($p = 0.03$), but lower at Day 7—(41%, $p = 0.025$) compared to PA-m rats. At the same time, there was apparent difference in the percentage preference for ethanol between PAE and naïve males, but not between PAE and naïve females. PA-m rats demonstrated significantly higher alcohol preference at Day 3 and Day 7 of the experiment (200%, $p = 0.02$ and 178%, $p = 0.0005$, respectively) compared to C-m rats. Along with the detected increase in alcohol consumption, PA-m rats showed a 133% ($p = 3 \times 10^{-6}$) increase in alcohol preference from Day 1 to Day 7 of the experiment, whereas for other groups, no statistically significant change in the dynamics of alcohol preference was found.

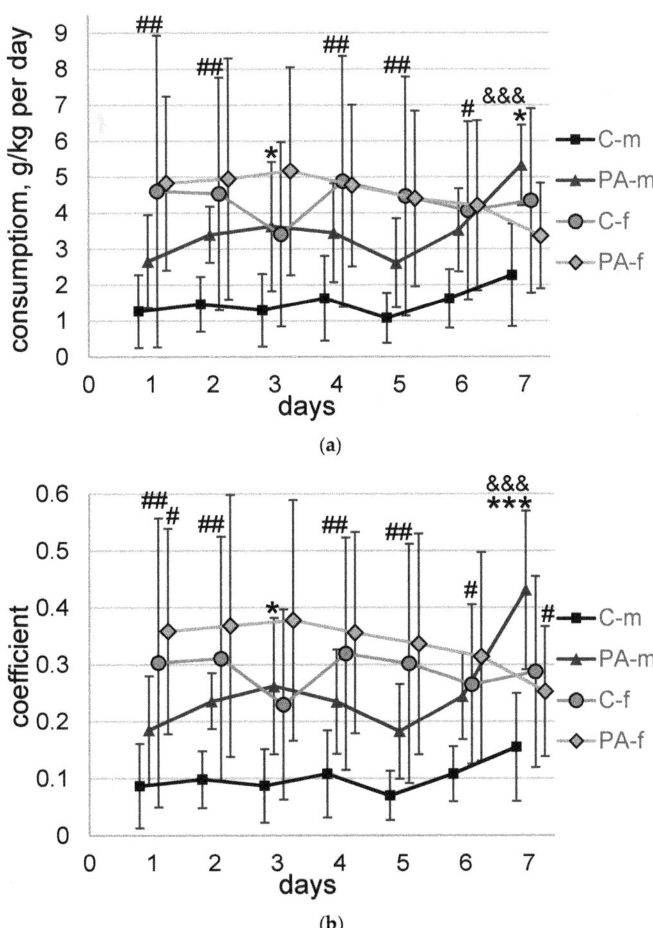

**Figure 2.** Two-bottle-choice paradigm. Data present as mean ± SD. (**a**)—alcohol consumption, g/kg of absolute ethanol per day; (**b**)—alcohol preference, coefficient: the ratio of alcohol consumed volume to total volume of alcohol and water consumed. C-m—male naïve rats, $n = 11$; PA-m—male PAE rats, $n = 10$; C-f—female naïve rats, $n = 8$; PA-f—female PAE rats, $n = 10$. Rats were provided with 24 h/day access to two-bottle choice drinking (one bottle contained 10% alcohol, the other contained water) for 7 days. Asterisk (*) indicates a significant difference between naïve and PAE rats of the same sex: *—$p < 0.05$, ***—$p < 0.001$; post hoc Duncan's test. Hash (#) indicates a significant difference between groups of the same name of different sex: #—$p < 0.05$; ##—$p < 0.01$; post hoc Duncan's test. Ampersand (&) indicates a significant difference between Day 1 and Day 7 inside a group: &&&—$p < 0.001$; post hoc Duncan's test.

### 3.3. The "Light-Dark Box" Test

The following statistically significant differences were found in the "light-dark box" test: number of entries into the dark compartment ($F_{1,32} = 6.1$, $p = 0.02$; PAE and sex factor interaction), distance in the dark compartment ($F_{1,32} = 4.2$, $p = 0.049$; PAE and sex factor interaction), number of entries into the middle compartment ($F_{1,32} = 8.2$, $p = 0.007$; PAE and sex factor interaction). An effect of PAE ($F_{1,32} = 5.8$, $p = 0.02$) and PAE and sex factor interaction ($F_{1,32} = 6.6$, $p = 0.015$) was found, when total distance in all compartments was analyzed.

There were no significant differences between PA-m and C-m rats. At the same time, PA-f rats characterized by lower number of entries (66.5%, $p = 0.03$, Figure 3a) and distance (67%, $p = 0.01$, Figure 3b) in the dark compartment, number of entries into the middle compartment (70%, $p = 0.015$, Figure 3c) and total distance in all compartments (70%, $p = 0.002$, Figure 3d) compared to C-f rats. Meanwhile, C-f rats demonstrated higher number of entries into the middle compartment (143%, $p = 0.01$) and total distance in all compartments (126%, $p = 0.03$) compared to C-m rats.

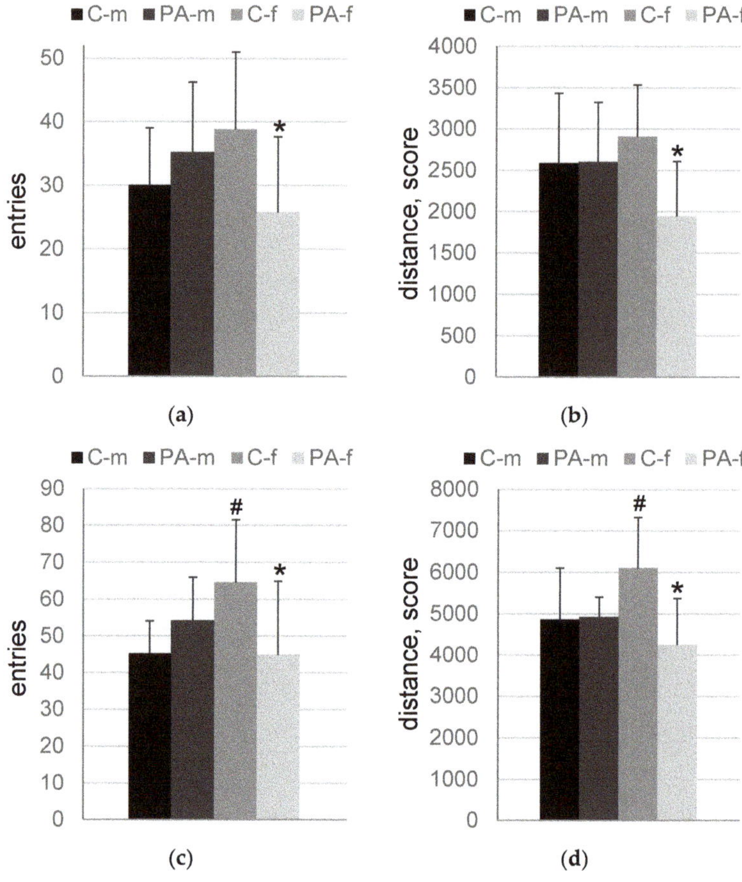


**Figure 3.** Light-dark box. Data present as mean + SD. (**a**)—entries into the dark compartment; (**b**)—distance in the dark compartment, scores; (**c**)—entries into the middle compartment; (**d**)—total distance travelled in all compartments, scores. C-m—male naïve rats, $n = 9$; PA-m—male PAE rats, $n = 9$; C-f—female naïve rats, $n = 9$; PA-f—female PAE rats, $n = 9$. Asterisk (*) indicates a significant difference between naïve and PAE rats of the same sex: *—$p < 0.05$; post hoc Duncan's test. Hash (#) indicates a significant difference between groups of the same name of different sex: #—$p < 0.05$; post hoc Duncan's test.

### 3.4. Synucleins mRNA Expression

In HPC effect of PAE ($F_{1,32} = 7.7$; $p = 0.009$) and sex ($F_{1,32} = 135$; $p = 5.29 \times 10^{-13}$) factors on $\alpha$-synuclein mRNA level was found. The expression level of *Snca* was 25.7% lower ($p = 0.0046$) in PAE-m mice compared to C-m mice. At the same time, both groups C-f and PAE-f have, respectively, 78% ($p = 0.00006$) and 60% ($p = 0.00006$) lower expression level of *Snca* compared to C-m and PAE-m mice groups, respectively. When mRNA level of $\beta$-synuclein was analyzed, only sex effect was found ($F_{1,32} = 89$; $p = 9.4 \times 10^{-11}$). Both C-f and PAE-f groups of mice had, respectively, 69% ($p = 0.00006$) and 62% ($p = 0.00006$) lower $\beta$-synuclein mRNA level compared to corresponding groups of male mice: C-m and PAE-m, respectively. The analysis of *Sncg* expression level revealed significant effects: PAE ($F_{1,32} = 21$; $p = 0.00006$), sex ($F_{1,32} = 45$; $p = 1.4 \times 10^{-7}$) and PAE-sex interaction ($F_{1,32} = 13$; $p = 0.0009$). The expression level of *Sncg* was 132% higher ($p = 0.0001$) in PAE-m mice compared to C-m mice. At the same time, both C-f and PAE-f groups of mice had, respectively, 49% ($p = 0.046$) and 71% ($p = 0.00006$) lower mRNA expression level of $\beta$-synuclein compared to corresponding groups of male mice: C-m and PA-m, respectively (Figure 4a).

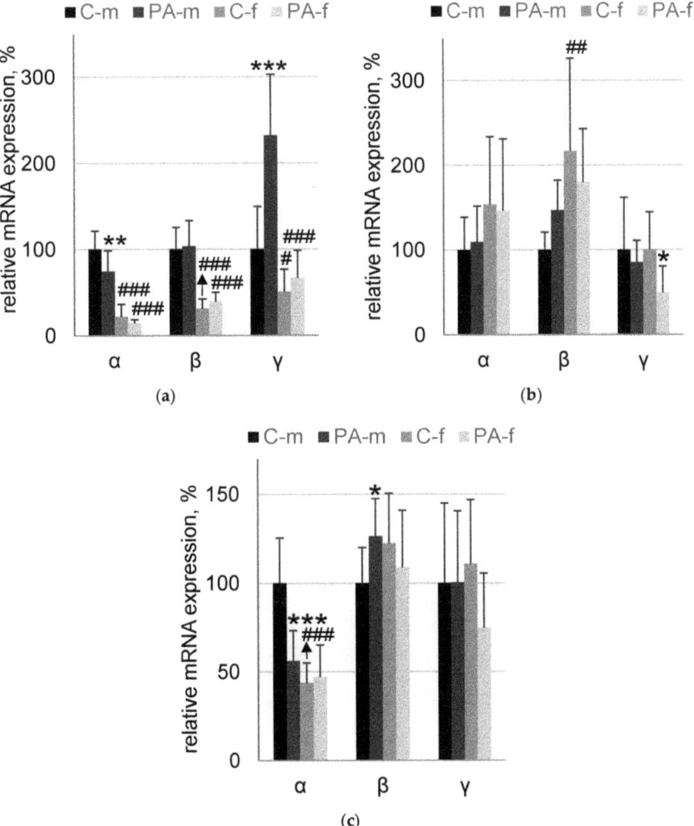


**Figure 4.** The mRNA expression levels of synucleins. (**a**)—the hippocampus; (**b**)—the hypothalamus; (**c**)—the midbrain. The corresponding members of the synuclein family are indicated by Latin letters: $\alpha$, $\beta$ and $\gamma$. C-m—male naïve rats, $n = 9$; PA-m—male PAE rats, $n = 9$; C-f—female naïve rats, $n = 9$; PA-f—female PAE rats, $n = 9$. Asterisk (*) indicates a significant difference between naïve and PAE rats of the same sex: *—$p < 0.05$; **—$p < 0.01$; ***—$p < 0.001$; a post hoc Duncan's test. Hash (#) indicates a significant difference between groups of the same name of different sex: #—$p < 0.05$; ##—$p < 0.01$; ###—$p < 0.001$; post hoc Duncan's test.

In HYP effect of sex ($F_{1,32}$ = 11.5; $p$ = 0.002) on β-synuclein mRNA expression level was found. The expression level of *Sncb* was 116% higher ($p$ = 0.001) in C-f mice compared to C-m group of mice. The analysis of *Sncg* expression level revealed significant effect of only PAE ($F_{1,32}$ = 5.4; $p$ = 0.03). γ-synuclein mRNA level was 51% ($p$ = 0.02) lower in PAE-f group compared to C-f group of mice (Figure 4b).

In MID effect of PAE ($F_{1,32}$ = 10.6; $p$ = 0.003), sex ($F_{1,32}$ = 27; $p$ = 0.00001) and their interaction ($F_{1,32}$ = 14.6; $p$ = 0.0006) on α-synuclein mRNA content was found. The expression level of *Snca* was 44% lower ($p$ = 0.0002) in PA-m mice compared to C-m mice. At the same time, the expression level of Snca was 56% lower ($p$ = 0.00006) in C-f mice compared to C-m group of mice. For β-synuclein mRNA statistically significant interaction of PAE and sex effects was found ($F_{1,32}$ = 5.3; $p$ = 0.028). A tendency to increase in β-synuclein mRNA level (26%, $p$ = 0.055, Duncan post-hoc test; $p$ = 0.038, Fisher LSD post-hoc test) in PAE-m group compared to C-m group of mice was detected (Figure 4c).

## 4. Discussion

In humans, ethanol exposure during pregnancy is one of predictive factors of future ethanol use in the offspring [28]. Rodents exposed to alcohol prenatally are also known to exhibit high ethanol preference as well as neurodevelopmental, physiological, and behavioral deficits reviewed in [25,29]. We showed that prenatal ethanol exposure increased ethanol consumption and ethanol preference in the male but not in female offspring compared to controls. Moreover, PAE males demonstrated progressive escalation of ethanol consumption and preference of male offspring. Previous studies have shown that prenatal ethanol exposure may influence the acceptance of ethanol's taste. The authors suggested that prenatal ethanol increases the risk to be engaged in alcohol abuse later in life because of increased preference for ethanol's smell and taste [30]. No effect of sex of the pups on sensory responsiveness was revealed, which allowed to pool the data for both sexes [31]. Surprisingly, we found striking differences in voluntary alcohol consumption between the sexes as a response to prenatal alcohol exposure. Importantly, females of control group drank more alcohol than males. These data are in line with previously reported results of tests for alcohol intake in rats using different experimental paradigms [32]. Earlier studies suggested that sex differences in ethanol intake in rats may be due to the sex differences in brain dopamine systems believed to mediate ethanol's reinforcing properties [33].

Here we asked whether fetal ethanol exposure can alter mRNA expression of presynaptic proteins—α-, β- and γ-synuclein known to be involved in regulation of dopamine release particularly in the mesolimbic system. We focused on three brain areas critically involved in mechanisms of alcohol-associated long-term effects—HPC, HYP and MID. Unexpectedly, PAE males had lower level of *Snca* expression in HPC and MID compared to control males, whereas no effect of PAE was found for female offspring. Interestingly, both control and PAE (to a much lesser extent) females demonstrated significantly higher levels of alcohol preference and lower levels of *Snca* expression in HIP and MID compared to males. Sex-associated difference in *Snca* expression in the brain may be explained by the effect of sex hormones including estrogen [34]. However, effect of PAE can be attributed to epigenetic alterations such as *Snca* intron 1 methylation [35]. Indeed, the inhibition of methylation in this region of *Snca* resulted in increased mRNA expression, while hypermethylation had the opposite effect [35]. One of the mechanisms involved in the regulation of *Snca* expression is dependent on the activity of DNA (cytosine-5)-methyltransferase 1 (DNMT1) [36]. This assumption is consistent with the previous results, which showed DNMT1 mRNA level is increased in the mesolimbic areas of adult PAE male rats [37]. Thus, our current data testifies that α-synuclein may be among the targets of prenatal alcohol and suggests that sex-specific long-term changes in *Snca* expression may impact the manifestation of neurobehavior disorders caused by PAE. We hypothesized several possible mechanisms behind this, which we will consider below.

It is known that α-synuclein knock-out mice have hyperdopaminergic phenotype, and knock-out of both α- and γ-synuclein genes potentiates this effect [18]. It has been estimated that the level of knockdown of endogenous α-synuclein correlates with the amount of nigral neuron loss [38]. On the contrary, decrease in α-synuclein mRNA using siRNA treatment did not cause any dramatic changes in survival of dopamine neurons or monoamine metabolism in primates [39]. These conflicting results probably reflect different levels of reduction in endogenous α-synuclein. Notably, transgenic rats overexpressing the full-length human *Snca* locus are also characterized by age-dependent degeneration of the dopaminergic system, which is preceded by the hyperdophaminergic phenotype at young age [40]. Despite early ideas about the exclusive role of the hypodopaminergic status, it is currently believed that both hypo- and hyperdopaminergia are states of vulnerability to relapse and increase context- and cue-induced alcohol-seeking behavior [41–43]. At the same time, literature data indicate that alcohol-dependent subjects are characterized by bidirectional changes in the content of α-synuclein [9,10]. Thus, a decrease in *Snca* expression and the predicted subsequent imbalance of the dopaminergic system may lead to a reduction in reward threshold and an increase in both alcohol preference at first presentation and the risk of relapse in the future.

A change in *Snca* expression was detected in HPC—one of the most-explored brain areas involved in complex processes such as learning and memory and emotional behavior [44]. Effects of alcohol on HPC are dependent on the developmental stage, producing pronounced alterations during gestation. It is well established that children with FASD show impaired cognition and intellectual abilities, deficient self-regulation, adaptive skills and low ability to apply the learned rules and information in the new context [45]. Hippocampal defects produced by maternal alcohol consumption during pregnancy had been well described in rodents [46,47] and can be due to a decrease in the number of neurons, an altered dendritic structure, and/or a reduced number of synapses [48,49]. In mice, an acute ethanol exposure during synaptogenesis produces apoptotic neurodegeneration in specific areas including HPC, cortex and striatum [50]. According to the findings from the animal studies α-synuclein plays an important role in the early development of synapses [51] and might act as a modulator of the size of the presynaptic vesicular pool in HPC [52]. Importantly, α-synuclein is highly expressed in the excitatory synapses marked by vesicular glutamate transporter-1 [53,54] and believed to be involved in mobilization of glutamate from the reserve pool [55]. Since glutamate plays a principle role in alcohol seeking, alcohol addiction and relapse [56–58] and chronic alcohol treatment promotes abnormal synaptic transmission that may lead to hippocampal cell death resulted from glutamate excitotoxicity after ethanol withdrawal [59]. Taking into account that α-synuclein knockout mice have cognitive impairments [60] we can suggest that the identified effect of downregulation of α-synuclein mRNA in HPC, perhaps, may be partly responsible for cognitive deficit which is typical for PAE offspring [61,62].

MID became another brain structure in which a change in the expression of *Snca* was detected. According to our previous data downregulation of α-synuclein mRNA expression was detected in MID of adult male rats with high initial ethanol preference [63]. It is known, a loss of α-synuclein due to the low mRNA expression or pathological aggregation may increase DA synthesis [64]. As mentioned above, this shift of the DA homeostasis to the hyperdopaminergic status may be partly responsible for hyperactivity reported in animal models of PAE [65].

We did not find any differences between PAE and control males in unconditioned anxiety-like behavior in the light-dark box. These results do not contradict the literature data, showing no significant differences in anxiety between PAE and control rats, but a pronounced increase in anxiety level in PAE males subjected to postnatal chronic mild stress [66]. At the same time, we observed decreased anxiety level in PAE females compared to control females. It should be noted that we observed this effect of PAE using the "light-dark box" test and further studies are needed to confirm the effects of PAE using a wider range of behavioral tests related to anxiety including open field and elevated plus maze. There is not much data on synucleins and anxiety behavior. It had been found that expression of α-synuclein is increased in the hippocampus HPC of rats with high levels of innate anxiety [67], however α-synuclein knock-out mice had similar to naïve mice anxiety level [68]. It looks likely that changes in α-synuclein expression are not critically involved in modulation of anxiety-related behavior. On the contrary, γ-synuclein knock-out mice characterized by decreased state anxiety and have enhancing exploratory behavior and cognitive abilities in several behavior paradigms [69], and γ-synuclein mRNA level was decreased in the HYP of PAE female rats. Thus, γ-synuclein may be responsible for alterations of emotional status of PAE female rats.

Compare to α-synuclein β- and γ-synucleins are largely understudied in the field of neuronal basis of alcohol and drug addiction. The majority of studies used overexpression or knock-out of these genes and meet difficulties in interpreting the data. It was found, that β-synuclein is upregulated in MID of mice lacking alpha-synuclein, γ-synuclein, or both α- and γ-synuclein, γ-synuclein is upregulated in mice lacking both α- and β-synuclein [70]. Even though deficiency of α-synuclein cannot be completely compensated by other members of the family [71], transcriptional activation β-synuclein in MID and γ-synuclein in HPC can be suggested as neuroadaptive mechanism. However, the opposite effect is also possible. It has been shown that increased expression of β- and γ-synuclein reduces synaptic vesicle binding of α-synuclein [72]. Relying on data that loss of α-synuclein could contribute to synaptic dysfunction in the aging brain [73], changes in the expression of β- and γ-synuclein in male PAE offspring can be interpreted as a synergistic effect leading to more pronounced synaptic dysfunction. At the same time, overexpression of γ-synuclein decreases DA neurotransmission in the nigrostriatal and mesocortical pathways, and increased γ-synuclein levels in the midbrain DA system correlated with an impaired cognitive function [74], which are typical for PAE offspring.

## 5. Conclusions

Our data show sex-specific changes in alcohol preference and mRNA expression of synucleins in rats prenatally exposed to alcohol. These facts point to the idea that alcohol consumption during gestation could alter the mechanisms of synaptic vesicle trafficking and neurotransmission in the selected brain areas, and these alterations could manifest during adulthood. Finding out how PAE changes the gene-environment interactions and determines behavioral profiles of adults including postnatal vulnerability for developing patterns of alcohol use and abuse appears to be a major goal of animal models in future research.

**Author Contributions:** Conceptualization, V.S.K. and I.Y.S.; methodology, P.K.A.; validation, N.Y.S., V.S.K. and N.N.N.; formal analysis, V.S.K.; investigation, E.P.P. and V.S.K.; resources, N.Y.S.; data curation, V.S.K.; writing—original draft preparation, I.Y.S.; writing—review and editing, N.N.N. and V.S.K.; visualization, K.C.; supervision, I.Y.S.; funding acquisition, K.C. All authors have read and agreed to the published version of the manuscript.

**Funding:** This research was funded by the grant agreement No. 075-15-2020-795, contract No. 13.1902.21.0027 of 29.09.2020 unique project ID: RF-190220X0027.

**Institutional Review Board Statement:** The study was conducted in accordance with the ethical principles stated in the European directive (86/609/EC). The research materials were reviewed and approved by the Ethics Committee of the V.P. Serbsky National Medical Research Center on Psychiatry and Addiction, Russian Ministry of Health, protocol No. 2, 10.02.2022.

**Informed Consent Statement:** Not applicable.

**Data Availability Statement:** The data presented in this study are openly available in Mendeley Data, link Kokhan, Viktor (2022), "Sex-related differences: synuclein and prenatal alcohol exposure", Mendeley Data, V1, doi: 10.17632/kmn44h85jg.1.

**Conflicts of Interest:** The authors declare no conflict of interest.

# References


1. Gupta, I.; Dandavate, R.; Gupta, P.; Agrawal, V.; Kapoor, M. Recent advances in genetic studies of alcohol use disorders. *Curr. Genet. Med. Rep.* **2020**, *8*, 27–34. [CrossRef] [PubMed]

2. Longley, M.J.; Lee, J.; Jung, J.; Lohoff, F.W. Epigenetics of alcohol use disorder-A review of recent advances in DNA methylation profiling. *Addict. Biol.* **2021**, *26*, e13006. [CrossRef] [PubMed]

3. Gaztanaga, M.; Angulo-Alcalde, A.; Chotro, M.G. Prenatal Alcohol Exposure as a Case of Involuntary Early Onset of Alcohol Use: Consequences and Proposed Mechanisms From Animal Studies. *Front. Behav. Neurosci.* **2020**, *14*, 26. [CrossRef] [PubMed]

4. Jacobson, J.L.; Akkaya-Hocagil, T.; Ryan, L.M.; Dodge, N.C.; Richardson, G.A.; Olson, H.C.; Coles, C.D.; Day, N.L.; Cook, R.J.; Jacobson, S.W. Effects of prenatal alcohol exposure on cognitive and behavioral development: Findings from a hierarchical meta-analysis of data from six prospective longitudinal U.S. cohorts. *Alcohol. Clin. Exp. Res.* **2021**, *45*, 2040–2058. [CrossRef]

5. Charness, M.E. Fetal Alcohol Spectrum Disorders: Awareness to Insight in Just 50 Years. *Alcohol Res.* **2022**, *42*, 05. [CrossRef]

6. Das, J. SNARE Complex-Associated Proteins and Alcohol. *Alcohol. Clin. Exp. Res.* **2020**, *44*, 7–18. [CrossRef]

7. Nemani, V.M.; Lu, W.; Berge, V.; Nakamura, K.; Onoa, B.; Lee, M.K.; Chaudhry, F.A.; Nicoll, R.A.; Edwards, R.H. Increased expression of alpha-synuclein reduces neurotransmitter release by inhibiting synaptic vesicle reclustering after endocytosis. *Neuron* **2010**, *65*, 66–79. [CrossRef]

8. Cahill, C.M.; Aleyadeh, R.; Gao, J.; Wang, C.; Rogers, J.T. Alpha-Synuclein in Alcohol Use Disorder, Connections with Parkinson's Disease and Potential Therapeutic Role of 5' Untranslated Region-Directed Small Molecules. *Biomolecules* **2020**, *10*, 1465. [CrossRef]

9. Levey, D.F.; Le-Niculescu, H.; Frank, J.; Ayalew, M.; Jain, N.; Kirlin, B.; Learman, R.; Winiger, E.; Rodd, Z.; Shekhar, A.; et al. Genetic risk prediction and neurobiological understanding of alcoholism. *Transl. Psychiatry* **2014**, *4*, e391. [CrossRef]

10. Janeczek, P.; MacKay, R.K.; Lea, R.A.; Dodd, P.R.; Lewohl, J.M. Reduced expression of alpha-synuclein in alcoholic brain: Influence of SNCA-Rep1 genotype. *Addict. Biol.* **2014**, *19*, 509–515. [CrossRef]

11. Bonsch, D.; Reulbach, U.; Bayerlein, K.; Hillemacher, T.; Kornhuber, J.; Bleich, S. Elevated alpha synuclein mRNA levels are associated with craving in patients with alcoholism. *Biol. Psychiatry* **2004**, *56*, 984–986. [CrossRef] [PubMed]

12. Foroud, T.; Wetherill, L.F.; Liang, T.; Dick, D.M.; Hesselbrock, V.; Kramer, J.; Nurnberger, J.; Schuckit, M.; Carr, L.; Porjesz, B.; et al. Association of alcohol craving with alpha-synuclein (SNCA). *Alcohol. Clin. Exp. Res.* **2007**, *31*, 537–545. [CrossRef]

13. Rotermund, C.; Reolon, G.K.; Leixner, S.; Boden, C.; Bilbao, A.; Kahle, P.J. Enhanced motivation to alcohol in transgenic mice expressing human alpha-synuclein. *J. Neurochem.* **2017**, *143*, 294–305. [CrossRef] [PubMed]

14. Liang, T.; Kimpel, M.W.; McClintick, J.N.; Skillman, A.R.; McCall, K.; Edenberg, H.J.; Carr, L.G. Candidate genes for alcohol preference identified by expression profiling in alcohol-preferring and -nonpreferring reciprocal congenic rats. *Genome Biol.* **2010**, *11*, R11. [CrossRef] [PubMed]

15. Janowska, M.K.; Wu, K.P.; Baum, J. Unveiling transient protein-protein interactions that modulate inhibition of alpha-synuclein aggregation by beta-synuclein, a pre-synaptic protein that co-localizes with alpha-synuclein. *Sci. Rep.* **2015**, *5*, 15164. [CrossRef]

16. Buchman, V.L.; Hunter, H.J.; Pinon, L.G.; Thompson, J.; Privalova, E.M.; Ninkina, N.N.; Davies, A.M. Persyn, a member of the synuclein family, has a distinct pattern of expression in the developing nervous system. *J. Neurosci. Off. J. Soc. Neurosci.* **1998**, *18*, 9335–9341. [CrossRef]

17. Kokhan, V.S.; Kokhan, T.Y.G.; Samsonova, A.N.; Fisenko, V.P.; Ustyugov, A.A.; Aliev, G. The Dopaminergic Dysfunction and Altered Working Memory Performance of Aging Mice Lacking Gamma-synuclein Gene. *CNS Neurol. Disord. Drug Targets* **2018**, *17*, 604–607. [CrossRef]

18. Senior, S.L.; Ninkina, N.; Deacon, R.; Bannerman, D.; Buchman, V.L.; Cragg, S.J.; Wade-Martins, R. Increased striatal dopamine release and hyperdopaminergic-like behaviour in mice lacking both alpha-synuclein and gamma-synuclein. *Eur. J. Neurosci.* **2008**, *27*, 947–957. [CrossRef]

19. Wise, R.A.; Jordan, C.J. Dopamine, behavior, and addiction. *J. Biomed. Sci.* **2021**, *28*, 83. [CrossRef]

20. Nutt, D.; Hayes, A.; Fonville, L.; Zafar, R.; Palmer, E.O.C.; Paterson, L.; Lingford-Hughes, A. Alcohol and the Brain. *Nutrients* **2021**, *13*, 3938. [CrossRef]

21. White, A.M. Gender Differences in the Epidemiology of Alcohol Use and Related Harms in the United States. *Alcohol Res.* **2020**, *40*, 01. [CrossRef] [PubMed]

22. Rossetti, M.G.; Patalay, P.; Mackey, S.; Allen, N.B.; Batalla, A.; Bellani, M.; Chye, Y.; Cousijn, J.; Goudriaan, A.E.; Hester, R.; et al. Gender-related neuroanatomical differences in alcohol dependence: Findings from the ENIGMA Addiction Working Group. *Neuroimage Clin.* **2021**, *30*, 102636. [CrossRef] [PubMed]

23. Flores-Bonilla, A.; De Oliveira, B.; Silva-Gotay, A.; Lucier, K.W.; Richardson, H.N. Shortening time for access to alcohol drives up front-loading behavior, bringing consumption in male rats to the level of females. *Biol. Sex Differ.* **2021**, *12*, 51. [CrossRef]

24. Datta, U.; Schoenrock, S.E.; Bubier, J.A.; Bogue, M.A.; Jentsch, J.D.; Logan, R.W.; Tarantino, L.M.; Chesler, E.J. Prospects for finding the mechanisms of sex differences in addiction with human and model organism genetic analysis. *Genes Brain Behav.* **2020**, *19*, e12645. [CrossRef]

25. Patten, A.R.; Fontaine, C.J.; Christie, B.R. A comparison of the different animal models of fetal alcohol spectrum disorders and their use in studying complex behaviors. *Front. Pediatr.* **2014**, *2*, 93. [CrossRef]

26. West, J.R. Fetal alcohol-induced brain damage and the problem of determining temporal vulnerability: A review. *Alcohol Drug Res.* **1987**, *7*, 423–441.

27. Quinn, R. Comparing rat's to human's age: How old is my rat in people years? *Nutrition* **2005**, *21*, 775–777. [CrossRef]

28. Baer, J.S.; Sampson, P.D.; Barr, H.M.; Connor, P.D.; Streissguth, A.P. A 21-year longitudinal analysis of the effects of prenatal alcohol exposure on young adult drinking. *Arch. Gen. Psychiatry* **2003**, *60*, 377–385. [CrossRef] [PubMed]

29. Almeida, L.; Andreu-Fernandez, V.; Navarro-Tapia, E.; Aras-Lopez, R.; Serra-Delgado, M.; Martinez, L.; Garcia-Algar, O.; Gomez-Roig, M.D. Murine Models for the Study of Fetal Alcohol Spectrum Disorders: An Overview. *Front. Pediatr.* **2020**, *8*, 359. [CrossRef]

30. Youngentob, S.L.; Glendinning, J.I. Fetal ethanol exposure increases ethanol intake by making it smell and taste better. *Proc. Natl. Acad. Sci. USA* **2009**, *106*, 5359–5364. [CrossRef]

31. Glendinning, J.I.; Simons, Y.M.; Youngentob, L.; Youngentob, S.L. Fetal ethanol exposure attenuates aversive oral effects of TrpV1, but not TrpA1 agonists in rats. *Exp. Biol. Med.* **2012**, *237*, 236–240. [CrossRef] [PubMed]

32. Li, J.; Chen, P.; Han, X.; Zuo, W.; Mei, Q.; Bian, E.Y.; Umeugo, J.; Ye, J. Differences between male and female rats in alcohol drinking, negative affects and neuronal activity after acute and prolonged abstinence. *Int. J. Physiol. Pathophysiol. Pharmacol.* **2019**, *11*, 163–176.

33. Blanchard, B.A.; Glick, S.D. Sex differences in mesolimbic dopamine responses to ethanol and relationship to ethanol intake in rats. *Recent Dev. Alcohol.* **1995**, *12*, 231–241. [CrossRef]

34. Spence, J.P.; Reiter, J.L.; Qiu, B.; Gu, H.; Garcia, D.K.; Zhang, L.; Graves, T.; Williams, K.E.; Bice, P.J.; Zou, Y.; et al. Estrogen-Dependent Upregulation of Adcyap1r1 Expression in Nucleus Accumbens Is Associated With Genetic Predisposition of Sex-Specific QTL for Alcohol Consumption on Rat Chromosome 4. *Front. Genet.* **2018**, *9*, 513. [CrossRef] [PubMed]

35. Pavlou, M.A.S.; Pinho, R.; Paiva, I.; Outeiro, T.F. The yin and yang of alpha-synuclein-associated epigenetics in Parkinson's disease. *Brain A J. Neurol.* **2017**, *140*, 878–886. [CrossRef]

36. Desplats, P.; Spencer, B.; Coffee, E.; Patel, P.; Michael, S.; Patrick, C.; Adame, A.; Rockenstein, E.; Masliah, E. Alpha-synuclein sequesters Dnmt1 from the nucleus: A novel mechanism for epigenetic alterations in Lewy body diseases. *J. Biol. Chem.* **2011**, *286*, 9031–9037. [CrossRef] [PubMed]

37. Razumkina, E.; Anokhin, P.; Sarycheva, N.; Shamakina, I. Prenatal alcohol exposure increases DNA-methyltransferases 1 and 3a its mRNA levels in the rat mesolimbic brain areas. *Eur. Neuropsychopharmacol.* **2019**, *29*, S312–S313. [CrossRef]

38. Gorbatyuk, O.S.; Li, S.; Nash, K.; Gorbatyuk, M.; Lewin, A.S.; Sullivan, L.F.; Mandel, R.J.; Chen, W.; Meyers, C.; Manfredsson, F.P.; et al. In vivo RNAi-mediated alpha-synuclein silencing induces nigrostriatal degeneration. *Mol. Ther. J. Am. Soc. Gene Ther.* **2010**, *18*, 1450–1457. [CrossRef]

39. McCormack, A.L.; Mak, S.K.; Henderson, J.M.; Bumcrot, D.; Farrer, M.J.; Di Monte, D.A. Alpha-synuclein suppression by targeted small interfering RNA in the primate substantia nigra. *PLoS ONE* **2010**, *5*, e12122. [CrossRef]

40. Polissidis, A.; Koronaiou, M.; Kollia, V.; Koronaiou, E.; Nakos-Bimpos, M.; Bogiongko, M.; Vrettou, S.; Karali, K.; Casadei, N.; Riess, O.; et al. Psychosis-Like Behavior and Hyperdopaminergic Dysregulation in Human alpha-Synuclein BAC Transgenic Rats. *Mov. Disord.* **2021**, *36*, 716–728. [CrossRef]

41. Hirth, N.; Meinhardt, M.W.; Noori, H.R.; Salgado, H.; Torres-Ramirez, O.; Uhrig, S.; Broccoli, L.; Vengeliene, V.; Rossmanith, M.; Perreau-Lenz, S.; et al. Convergent evidence from alcohol-dependent humans and rats for a hyperdopaminergic state in protracted abstinence. *Proc. Natl. Acad. Sci. USA* **2016**, *113*, 3024–3029. [CrossRef] [PubMed]

42. Hansson, A.C.; Grunder, G.; Hirth, N.; Noori, H.R.; Spanagel, R.; Sommer, W.H. Dopamine and opioid systems adaptation in alcoholism revisited: Convergent evidence from positron emission tomography and postmortem studies. *Neurosci. Biobehav. Rev.* **2019**, *106*, 141–164. [CrossRef] [PubMed]

43. Hauser, S.R.; Mulholland, P.J.; Truitt, W.A.; Waeiss, R.A.; Engleman, E.A.; Bell, R.L.; Rodd, Z.A. Adolescent Intermittent Ethanol (AIE) Enhances the Dopaminergic Response to Ethanol within the Mesolimbic Pathway during Adulthood: Alterations in Cholinergic/Dopaminergic Genes Expression in the Nucleus Accumbens Shell. *Int. J. Mol. Sci.* **2021**, *22*, 11733. [CrossRef]

44. Santangelo, V.; Cavallina, C.; Colucci, P.; Santori, A.; Macri, S.; McGaugh, J.L.; Campolongo, P. Enhanced brain activity associated with memory access in highly superior autobiographical memory. *Proc. Natl. Acad. Sci. USA* **2018**, *115*, 7795–7800. [CrossRef] [PubMed]

45. Wilhoit, L.F.; Scott, D.A.; Simecka, B.A. Fetal Alcohol Spectrum Disorders: Characteristics, Complications, and Treatment. *Community Ment Health J.* **2017**, *53*, 711–718. [CrossRef] [PubMed]

46. An, L.; Zhang, T. Spatial cognition and sexually dimorphic synaptic plasticity balance impairment in rats with chronic prenatal ethanol exposure. *Behav. Brain Res.* **2013**, *256*, 564–574. [CrossRef]

47. An, L.; Zhang, T. Prenatal ethanol exposure impairs spatial cognition and synaptic plasticity in female rats. *Alcohol* **2015**, *49*, 581–588. [CrossRef]

48. Guerri, C.; Bazinet, A.; Riley, E.P. Foetal Alcohol Spectrum Disorders and alterations in brain and behaviour. *Alcohol Alcohol.* **2009**, *44*, 108–114. [CrossRef]

49. Bon, L.I.; Zimatkin, S.M. Disruption of synaptogenesis in the rats brain cortex after antenatal alcoholisation. *J. Grodno State Med. Univ.* **2017**, *15*, 538–543. [CrossRef]

50. Sadrian, B.; Lopez-Guzman, M.; Wilson, D.A.; Saito, M. Distinct neurobehavioral dysfunction based on the timing of developmental binge-like alcohol exposure. *Neuroscience* **2014**, *280*, 204–219. [CrossRef]

51. Hsu, L.J.; Mallory, M.; Xia, Y.; Veinbergs, I.; Hashimoto, M.; Yoshimoto, M.; Thal, L.J.; Saitoh, T.; Masliah, E. Expression pattern of synucleins (non-Abeta component of Alzheimer's disease amyloid precursor protein/alpha-synuclein) during murine brain development. *J. Neurochem.* **1998**, *71*, 338–344. [CrossRef] [PubMed]

52. Murphy, D.D.; Rueter, S.M.; Trojanowski, J.Q.; Lee, V.M. Synucleins are developmentally expressed, and alpha-synuclein regulates the size of the presynaptic vesicular pool in primary hippocampal neurons. *J. Neurosci.* **2000**, *20*, 3214–3220. [CrossRef] [PubMed]

53. Taguchi, K.; Watanabe, Y.; Tsujimura, A.; Tatebe, H.; Miyata, S.; Tokuda, T.; Mizuno, T.; Tanaka, M. Differential expression of alpha-synuclein in hippocampal neurons. *PLoS ONE* **2014**, *9*, e89327. [CrossRef] [PubMed]

54. Taguchi, K.; Watanabe, Y.; Tsujimura, A.; Tanaka, M. Expression of alpha-synuclein is regulated in a neuronal cell type-dependent manner. *Anat. Sci. Int.* **2019**, *94*, 11–22. [CrossRef] [PubMed]

55. Gureviciene, I.; Gurevicius, K.; Tanila, H. Role of alpha-synuclein in synaptic glutamate release. *Neurobiol. Dis.* **2007**, *28*, 83–89. [CrossRef]

56. Cheng, H.; Kellar, D.; Lake, A.; Finn, P.; Rebec, G.V.; Dharmadhikari, S.; Dydak, U.; Newman, S. Effects of Alcohol Cues on MRS Glutamate Levels in the Anterior Cingulate. *Alcohol Alcohol.* **2018**, *53*, 209–215. [CrossRef]

57. Ceccarini, J.; Leurquin-Sterk, G.; Crunelle, C.L.; de Laat, B.; Bormans, G.; Peuskens, H.; Van Laere, K. Recovery of Decreased Metabotropic Glutamate Receptor 5 Availability in Abstinent Alcohol-Dependent Patients. *J. Nucl. Med.* **2020**, *61*, 256–262. [CrossRef]

58. Burnette, E.M.; Nieto, S.J.; Grodin, E.N.; Meredith, L.R.; Hurley, B.; Miotto, K.; Gillis, A.J.; Ray, L.A. Novel Agents for the Pharmacological Treatment of Alcohol Use Disorder. *Drugs* **2022**, *82*, 251–274. [CrossRef]

59. Gerace, E.; Landucci, E.; Bani, D.; Moroni, F.; Mannaioni, G.; Pellegrini-Giampietro, D.E. Glutamate Receptor-Mediated Neurotoxicity in a Model of Ethanol Dependence and Withdrawal in Rat Organotypic Hippocampal Slice Cultures. *Front. Neurosci.* **2018**, *12*, 1053. [CrossRef]

60. Kokhan, V.S.; Afanasyeva, M.A.; Van'kin, G.I. alpha-Synuclein knockout mice have cognitive impairments. *Behav. Brain Res.* **2012**, *231*, 226–230. [CrossRef]

61. Mattson, S.N.; Crocker, N.; Nguyen, T.T. Fetal alcohol spectrum disorders: Neuropsychological and behavioral features. *Neuropsychol. Rev.* **2011**, *21*, 81–101. [CrossRef] [PubMed]

62. Olguin, S.L.; Thompson, S.M.; Young, J.W.; Brigman, J.L. Moderate prenatal alcohol exposure impairs cognitive control, but not attention, on a rodent touchscreen continuous performance task. *Genes Brain Behav.* **2021**, *20*, e12652. [CrossRef] [PubMed]

63. Anokhin, P.K.; Shamakina, I.Y.; Ustyugov, A.A.; Bachurin, S.O.; Proskuryakova, T.V. A comparison of the expression of α-synuclein mRNA in the brain of rats with different levels of alcohol consumption. *Neurochem. J.* **2016**, *10*, 294–299. [CrossRef]

64. Tehranian, R.; Montoya, S.E.; Van Laar, A.D.; Hastings, T.G.; Perez, R.G. Alpha-synuclein inhibits aromatic amino acid decarboxylase activity in dopaminergic cells. *J. Neurochem.* **2006**, *99*, 1188–1196. [CrossRef]

65. Hausknecht, K.A.; Acheson, A.; Farrar, A.M.; Kieres, A.K.; Shen, R.Y.; Richards, J.B.; Sabol, K.E. Prenatal alcohol exposure causes attention deficits in male rats. *Behav. Neurosci.* **2005**, *119*, 302–310. [CrossRef] [PubMed]

66. Hellemans, K.G.; Verma, P.; Yoon, E.; Yu, W.; Weinberg, J. Prenatal alcohol exposure increases vulnerability to stress and anxiety-like disorders in adulthood. *Ann. N. Y. Acad. Sci.* **2008**, *1144*, 154–175. [CrossRef]

67. Chiavegatto, S.; Izidio, G.S.; Mendes-Lana, A.; Aneas, I.; Freitas, T.A.; Torrao, A.S.; Conceicao, I.M.; Britto, L.R.; Ramos, A. Expression of alpha-synuclein is increased in the hippocampus of rats with high levels of innate anxiety. *Mol. Psychiatry* **2009**, *14*, 894–905. [CrossRef]

68. Pena-Oliver, Y.; Buchman, V.L.; Stephens, D.N. Lack of involvement of alpha-synuclein in unconditioned anxiety in mice. *Behav. Brain Res.* **2010**, *209*, 234–240. [CrossRef]

69. Kokhan, V.S.; Van'kin, G.I.; Bachurin, S.O.; Shamakina, I.Y. Differential involvement of the gamma-synuclein in cognitive abilities on the model of knockout mice. *BMC Neurosci.* **2013**, *14*, 53. [CrossRef]

70. Chandra, S.; Fornai, F.; Kwon, H.B.; Yazdani, U.; Atasoy, D.; Liu, X.; Hammer, R.E.; Battaglia, G.; German, D.C.; Castillo, P.E.; et al. Double-knockout mice for alpha- and beta-synucleins: Effect on synaptic functions. *Proc. Natl. Acad. Sci. USA* **2004**, *101*, 14966–14971. [CrossRef]

71. Connor-Robson, N.; Peters, O.M.; Millership, S.; Ninkina, N.; Buchman, V.L. Combinational losses of synucleins reveal their differential requirements for compensating age-dependent alterations in motor behavior and dopamine metabolism. *Neurobiol. Aging* **2016**, *46*, 107–112. [CrossRef]

72. Carnazza, K.E.; Komer, L.E.; Xie, Y.X.; Pineda, A.; Briano, J.A.; Gao, V.; Na, Y.; Ramlall, T.; Buchman, V.L.; Eliezer, D.; et al. Synaptic vesicle binding of alpha-synuclein is modulated by beta- and gamma-synucleins. *Cell Rep.* **2022**, *39*, 110675. [CrossRef] [PubMed]

73. Mak, S.K.; McCormack, A.L.; Langston, J.W.; Kordower, J.H.; Di Monte, D.A. Decreased alpha-synuclein expression in the aging mouse substantia nigra. *Exp. Neurol.* **2009**, *220*, 359–365. [CrossRef] [PubMed]

74. Pavia-Collado, R.; Rodriguez-Aller, R.; Alarcon-Aris, D.; Miquel-Rio, L.; Ruiz-Bronchal, E.; Paz, V.; Campa, L.; Galofre, M.; Sgambato, V.; Bortolozzi, A. Up and Down gamma-Synuclein Transcription in Dopamine Neurons Translates into Changes in Dopamine Neurotransmission and Behavioral Performance in Mice. *Int. J. Mol. Sci.* **2022**, *23*, 1807. [CrossRef] [PubMed]

 *biomedicines*

*Review*

# The Role of α-Synuclein in the Regulation of Serotonin System: Physiological and Pathological Features

Lluis Miquel-Rio [1,2,3,4], Unai Sarriés-Serrano [1,2,3,5], Rubén Pavia-Collado [1,2,3,6], J Javier Meana [3,7,8] and Analia Bortolozzi [1,2,3,*]

[1] Institute of Biomedical Research of Barcelona (IIBB), Spanish National Research Council (CSIC), 08036 Barcelona, Spain
[2] Institut d'Investigacions Biomèdiques August Pi i Sunyer (IDIBAPS), 08036 Barcelona, Spain
[3] Biomedical Research Networking Center for Mental Health (CIBERSAM), Institute of Health Carlos III (ISCIII), 28029 Madrid, Spain
[4] Faculty of Medicine and Health Sciences, University of Barcelona (UB), 08036 Barcelona, Spain
[5] Department of Pharmacology, University of the Basque Country UPV/EHU, 48940 Leioa, Spain
[6] MiCure Therapeutics Ltd., Tel Aviv 6423902, Israel
[7] Department of Pharmacology, CIBERSAM, Biocruces Bizkaia Health Research Institute, University of the Basque Country UPV/EHU, 48940 Leioa, Spain
[8] Biocruces Bizkaia Health Research Institute, 48940 Leioa, Spain
* Correspondence: analia.bortolozzi@iibb.csic.es; Tel.: +34-93-363-8313

**Abstract:** In patients affected by Parkinson's disease (PD), up to 50% of them experience cognitive changes, and psychiatric disturbances, such as anxiety and depression, often precede the onset of motor symptoms and have a negative impact on their quality of life. Pathologically, PD is characterized by the loss of dopamine (DA) neurons in the substantia nigra pars compacta (SNc) and the presence of intracellular inclusions, called Lewy bodies and Lewy neurites, composed mostly of α-synuclein (α-Syn). Much of PD research has focused on the role of α-Syn aggregates in the degeneration of SNc DA neurons due to the impact of striatal DA deficits on classical motor phenotypes. However, abundant Lewy pathology is also found in other brain regions including the midbrain raphe nuclei, which may contribute to non-motor symptoms. Indeed, dysfunction of the serotonergic (5-HT) system, which regulates mood and emotional pathways, occurs during the premotor phase of PD. However, little is known about the functional consequences of α-Syn inclusions in this neuronal population other than DA neurons. Here, we provide an overview of the current knowledge of α-Syn and its role in regulating the 5-HT function in health and disease. Understanding the relative contributions to α-Syn-linked alterations in the 5-HT system may provide a basis for identifying PD patients at risk for developing depression and could lead to a more targeted therapeutic approach.

**Keywords:** depression; Parkinson's disease; α-synuclein; serotonin; raphe nuclei

**Citation:** Miquel-Rio, L.; Sarriés-Serrano, U.; Pavia-Collado, R.; Meana, J.J.; Bortolozzi, A. The Role of α-Synuclein in the Regulation of Serotonin System: Physiological and Pathological Features. *Biomedicines* **2023**, *11*, 541. https://doi.org/10.3390/biomedicines11020541

Academic Editor: Natalia Ninkina

Received: 20 December 2022
Revised: 30 January 2023
Accepted: 9 February 2023
Published: 13 February 2023



## 1. Introduction

Parkinson's disease (PD) is clinically characterized based on classic motor features including the presence of hypokinesia, rigidity, resting tremor, and impaired postural control [1–3]. A wide variety of incapacitating non-motor symptoms are also present over the course of the illness. These non-motor signs include autonomic and neuropsychiatric features such as fatigue, apathy, anxiety and depression, as well as cognitive deficits. Neuropsychiatric symptoms are inherent to the disease and are neither a result nor a side effect of long-term dopaminergic treatment [4,5]. These comorbidities are frequent and can be found in all stages of PD, from the premotor and the early untreated phases of the disease to the advanced stages of PD [6–13]. Among them, depression is one of the most prevalent neuropsychiatric symptoms, ranging from 35 to 50% of patients with PD [14–16].

Depressive disorder represents a huge burden on the quality of life in many PD patients, but is frequently undiagnosed and left untreated [17–20]. Therefore, understanding the neurobiology of depression in PD is critical to achieving the optimal care needed by patients with PD.

While the etiology of PD still remains unclear, one major neuropathological hallmark of PD is the degeneration and subsequent loss of DA neurons in the substantia nigra pars compacta (SNc) leading to prototypic motor deficits [21–23]. The SNc involves a neuronal population projecting to the caudate and putamen and is critical for the regulation of basal ganglia circuitry [24,25]. Lewy pathology (LP), which can also be observed across the central, peripheral, and enteric nervous systems (CNS, PNS, and ENS), is another major pathological finding present in about 70% of "clinically typical PD cases" [26,27]. This includes both Lewy bodies (LB) and Lewy neurites (LN), which are composed of a variety of different molecules, proteins, and organelles, including ubiquitin, tubulin, neurofilaments, lipids, and mitochondria. Among them, aggregates of $\alpha$-synuclein ($\alpha$-Syn) protein represent one of the main LP components [28–32]. To explain the widespread localization of LP and the onset of the various non-motor symptoms of PD, a critical point to consider is the dysfunction of other neuronal populations and neurotransmitter systems in regions of the CNS and PNS, other than the SNc DA neurons. Indeed, several studies reported LB-associated deficits—most likely occurring even prior to DA neurons—in cholinergic neurons in the pedunculopontine nucleus, nucleus basalis of Meynert and of the dorsal motor nucleus of the vagus, as well as in norepinephrine—NE neurons of the locus coeruleus (LC), and serotonin—5-HT (5-hydroxytryptamine) neurons of the raphe nuclei (RN) [33–35]. Furthermore, altered GABAergic and glutamatergic signaling was also reported in the amygdala and several cortical brain regions that may play important roles in the complex cognitive features of PD [34,36,37].

Lately, attention has been focused on the impaired integrity of the 5-HT system in PD, in addition to its well-known role in the pathogenesis of anxiety and depressive disorders. Notably, a growing amount of research supports a specific causal role of 5-HT system dysfunction in the progression of several PD symptoms, such as tremor and dyskinesia, but also anxiety and depression at early stages of the disease [5,38–43]. This review will discuss recent findings of the role of $\alpha$-Syn in regulating the 5-HT system. A better understanding of the relative contributions of $\alpha$-Syn-related abnormalities of the 5-HT system could lead to the identification of PD patients who are at risk of developing depression, as well as to better animal models of the disease and a more tailored therapeutic approach.

## 2. Connectivity of the Brain Serotonin System

A complete review of the brain 5-HT system is beyond the scope of the present article. The reader is referred to several reviews in the literature [44–46]. Here, we would like to highlight some features of the connectivity of the 5-HT system directly linked to its role in the neurobiology of depression.

The brain 5-HT system exerts its widespread effects from a group of relatively small brainstem nuclei known as the RN. Raphe 5-HT-producing neurons send ascending projections to the entire brain as well as descending projections to the spinal cord [47] (Figure 1A). These projections form classical synaptic connections, as well as varicosities with no associated postsynaptic structure [48,49]. Upon release, 5-HT acts primarily on G-protein coupled receptors (5-HT$_1$, 5-HT$_2$, 5-HT$_4$, 5-HT$_5$, 5-HT$_6$, 5-HT$_7$, and a single ionotropic receptor 5-HT$_3$) encoded by more than a dozen distinct genes and many more isoforms, which are differentially expressed in the brain [50,51]. Indeed, all brain regions express multiple 5-HT receptors in a receptor subtype-specific pattern [52]. In addition, individual neurons may express several 5-HT receptor subtypes. For instance, pyramidal neurons in layer V of the ventromedial prefrontal cortex (vmPFC) express 5-HT$_{1A}$ and 5-HT$_{2A}$ receptors, which exert opposite effects on neuronal firing activity [53,54]. Hence, the plethora of effects of the brain's 5-HT system is partly explained by the fact that 5-HT neurons are optimally positioned to affect the activity of a wide range of brain networks.

Among the different raphe nuclei, the dorsal raphe nucleus (DR) is the largest sero-tonergic nucleus, containing approximately one third of all 5-HT neurons in the brain [47]. As such, the human DR comprises about 250,000 neurons, out of a total of $10^{11}$ neurons in the whole brain—approximately 20,000 5-HT-producing neurons in the rat—and its axons branch widely, innervating almost all brain areas. This can be illustrated in the rat cortex, where >$10^6$ serotonergic nerve endings/mm$^3$ were noted. In addition, each cortical neuron may receive around 200 varicosities [55]. Unlike cortical and subcortical glutamatergic pro-jection neurons that exhibit precise short- or long-distance connectivity with other neuronal groups [56], DR 5-HT cells send highly divergent ascending projections connecting brain ar-eas with different functions [57]. Indeed, correlations have been reported between changes in DR 5-HT neuron activity and different cognitive processes, such as working memory [58], cognitive flexibility [59], response inhibition [60], and exploration–exploitation balance [61]. Furthermore, the raphe 5-HT system is also involved in the modulation of mood, emotion, perception, stress, reward, aggression, and social interactions, among others [62–66]. It is difficult to find a human behavior that is not regulated by a 5-HT response.

Notably, deficits in the 5-HT signaling are implicated in the neuropathology of anxiety and depression. Imbalances in the production and transmission of several neurotrans-mitters, including 5-HT, are commonly observed in the CNS of patients suffering from depressive disorder [67]. In fact, a widely accepted etiological theory is the "monoamine hypothesis of depression", which postulates that depression disorder is associated with a decreased monoamine function (NE, DA, and 5-HT) in key brain areas, such as the vmPFC, hippocampus (HPC), amygdala (AMG), nucleus accumbens (NAc), ventral tegmental area (VTA), and hypothalamus [68–70]. Notably, neuroimaging studies associate vmPFC with a broad spectrum ranging from emotion to cognitive functions, and alterations in vmPFC activity have been correlated with the biology of depression as with favorable outcomes of novel antidepressant strategies [71–73]. Likewise, structural and functional neuroimaging studies show pronounced alterations in vmPFC circuits in patients with depression and PD [74–76]. The vmPFC, which is composed of 75–80% glutamatergic pyramidal projection neurons and 20–25% GABAergic local circuit interneurons, is strongly innervated by DR 5-HT neurons [53,54]. The 5-HT fibers exert an important modulatory role of excitatory and inhibitory currents in vmPFC neurons [77,78], mainly through activation of 5-HT$_{2A}$ and 5-HT$_{1A}$ receptors, respectively. In turn, the monoamine groups, including the 5-HT neurons of the DR, are innervated by descending axons from layer V pyramidal neurons in the vmPFC [79] that control the monoamine neuron activity [80,81], thus establishing a reciprocal connectivity and mutual control (Figure 1B). Although it is beyond the scope of this review, ultimately, one can be optimistic that the functional integrity of the vmPFC–raphe nuclei circuit will play an important role in the pathophysiology of depression in early PD. New approaches are advancing in many directions to identify early PD, and the 5-HT system and its connections are an important part of these recent efforts, as will be described in the following sessions. Advanced neuroimaging techniques, next generation RNA sequencing, the recent addition of the proximity ligation assay (PLA) that specifically recognizes α-Syn aggregates and new animal models, among others, will provide support for the classification of PD based on different pathological phenotypes, leading to a more appropriate therapeutic strategy.

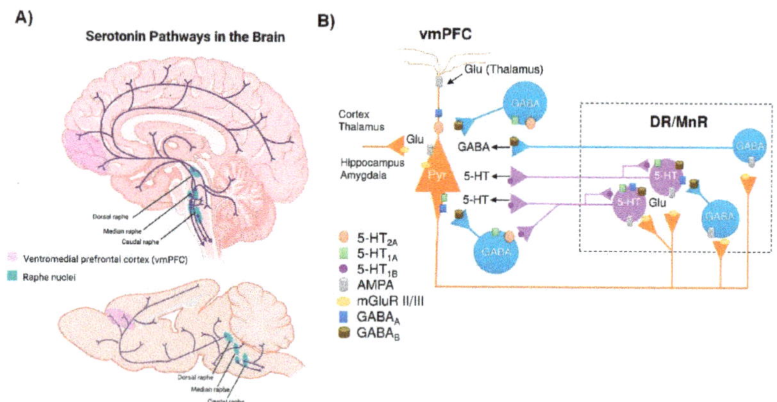

**Figure 1.** Central serotonergic pathways. (**A**) Schematic representation of the raphe nuclei in humans (**top**) and mice (**bottom**), which give rise to ascending projections to large regions of the brain, as well as descending projections predominantly innervate the cerebellum and its input structures and to the spinal cord. (**B**) Diagram showing how the ventromedial prefrontal cortex (vmPFC) and the dorsal and median raphe nuclei (DR and MrR, respectively) are anatomically and functionally connected in both directions. Pyramidal glutamatergic neurons from vmPFC send axons to raphe nuclei, where they form excitatory synapses (AMPA receptors) with 5-HT and GABAergic neurons. Stimulation of glutamatergic neurons in vmPFC primarily triggers inhibitory responses in 5-HT neurons mediated by (i) the activation of local GABAergic circuits that control the activity of 5-HT neurons in the raphe nuclei and (ii) 5-HT$_{1A}$ autoreceptor-dependent self-inhibitory responses following excitatory activation of 5-HT neurons. In addition, DR/MnR 5-HT neurons control the activity of glutamatergic neurons in the vmPFC through inhibitory 5-HT$_{1A}$ receptors and excitatory 5-HT$_{2A}$ receptors expressed in glutamatergic and GABAergic neurons. Similarly, the activity of the vmPFC-DR/MnR pathway may be affected by the activation of 5-HT$_4$ receptors on glutamatergic neurons and 5-HT$_3$ receptors on GABAergic interneurons in the outer layer of the vmPFC (not shown in the diagram). Adapted from [53,54,78].

### 3. α-Synuclein and Serotonin Neurotransmission

α-Syn is a small, natively unfolded protein belonging to the synuclein family that also encompasses β-synuclein (β-Syn) and γ-synuclein (γ-Syn). These are evolutionarily conserved proteins that have currently only been described in vertebrates, supporting the notion that they regulate some essential physiological functions [82–85]. Between them, α-Syn is the most studied protein of this family, due to its crucial role in the pathogenesis of PD and other synucleinopathies [86]. This protein is characterized by a remarkable conformational plasticity, adopting different conformations depending on the environment, i.e., neighboring proteins, lipid membranes, redox state, and local pH [87–90]. In fact, α-Syn adopts a monomeric, random coil conformation in an aqueous solution, while its interaction with lipid membranes drives the transition of the molecule part into α-helical structure. The central unstructured region of α-Syn is involved in fibril formation by converting to well-defined, β-sheet rich secondary structures. These structural and biophysical properties probably hold the key to their normal and abnormal function [91,92]. α-Syn is abundantly expressed in all neuronal types, where it localizes in presynaptic terminals [93–95] and modulates synaptic functions [96–98]. However, α-Syn is among the last presynaptic proteins to become enriched at the synapse [94] and unlike γ-Syn, it does not seem to be involved in synaptic development [99,100]. Recent studies have revealed that α-Syn is also present in different organelles, including nuclei, mitochondria, Golgi, and endoplasmic reticulum (ER) [93,101–103], although in lower concentrations than those found in synaptic locations, and its function is even less well understood [84]. This feature makes α-Syn a hub within synaptic protein interaction networks [84]. Supporting this, α-Syn was first identified at the presynaptic level as interacting with synaptic vesicle (SV)-associated proteins [93]. Indeed, it cooperates with a large number of SV surface proteins including the synapsin phosphoprotein family, complexins, and mammalian Munc 13-1, described to be affected in brain samples from PD patients and in various human α-Syn transgenic mouse lines [98,99,104–106]. Furthermore, several studies also indicated that α-Syn interacts with the SV glycoprotein 2 (SV2) family to positively modulate vesicular functions in a variety of ways, possibly by aiding in vesicular trafficking and exocytosis, as well as stabilizing stored transmitters [107,108]. In this regard, both postmortem PD brain tissue and animals overexpressing mutant α-Syn showed increases in the SV2C protein, which is abundantly expressed in the basal ganglia and selectively localizes to DA neurons [109]. Similarly, elevated levels of SV2A protein co-localizing with α-Syn were found in axonal swellings across the caudate-putamen (CPu) and cingulate cortex in a mouse model overexpressing human wild-type α-Syn in 5-HT neurons [110]. Other proteins such as Rabs, which in addition to modulating axonal traffic are also very important for the regulation of each step leading to SV release, docking and fusion at synaptic sites, interact with α-Syn [111,112]. Actually, several findings support that Rabs play a crucial role as direct mediators in the induction of synaptic alterations concerning α-Syn leading to PD pathology [112]. Taken together, the loss of α-Syn function, coupled with changes in its levels at synaptic terminals, can cause multifaceted dysregulation of many other synaptic proteins involved in neurotransmission mechanisms.

In addition to being involved in synaptic vesicular trafficking, α-Syn is also directly engaged in the regulation of monoamine (DA, NE, and 5-HT) neurotransmission homeostasis—β-Syn and γ-Syn are also involved in this regulation, although their role is less known [85,113–118]. Monoamine transporters (MAT) are transmembrane proteins solely responsible for the synaptic reuptake of DA, NE and 5-HT, and partly maintain the homeostasis of monoaminergic neurotransmission. MAT are important pharmacological targets in the therapy of various neuropsychiatric diseases, such as anxiety, depression, and suicidal behavior, among others, due to their crucial role within the brain in the replacement of monoamine neurotransmitters [119,120]. Direct interactions between α-Syn and MAT proteins have been described, indicating an important role for the synucleins in regulating MAT function, trafficking and distribution at the synapse. Even though most of the evidence is focused on DA neurotransmission and its transporter (DAT), in this review we will emphasize the role of α-Syn in the homeostasis of 5-HT neurotransmission.

Previous studies showed that the cell-surface expression and function of the 5-HT transporter (SERT) in co-transfected cells are negatively modulated by α-Syn in a non-Abeta-amyloid component (NAC) domain-dependent manner [115]. In addition, pioneering reports also showed direct interactions of α-Syn-SERT and γ-Syn-SERT proteins in cultured cells and in rat brain tissue, assessed by immunoprecipitation [115,121]. α-Syn-induced modulation of SERT trafficking is microtubule-dependent, as the microtubule-destabilizing agent nocodazole disrupts the effects of α-Syn on SERT function, reversing the inhibition of uptake in co-transfected cells [116]. More recently, in vivo studies indicated that down-regulation of α-Syn expression in raphe 5-HT neurons induced by an antisense oligonucleotide (ASO) leaves an increased synaptic 5-HT concentration, which was dependent on the reduction of SERT activity, as assessed by the selective SERT inhibitor citalopram [118]. The overexpression of α-Syn in raphe nuclei produced the opposite effects, with mice exhibiting a drop in extracellular 5-HT levels that was dependent on SERT function [110].

Moreover, α-Syn is also involved in the vesicular storage of monoamine neurotransmitters by the vesicular monoamine transporter 2 (VMAT$_2$). VMAT$_2$ mobilizes monoamines from the neuronal cytoplasm into vesicles, where they are repackaged for release at synapses [122,123]. VMAT$_2$ co-localizes with α-Syn protein in the Lewy bodies from PD brains [124], and overexpression of α-Syn negatively impairs VMAT$_2$ expression/function, leading to increased levels of cytosolic monoamine in presynaptic terminals, which in turn induce neurotoxicity [113]. These findings suggest that α-Syn may maintain high

VMAT$_2$ activity to protect monoamine neurons form cell death [125]. In support of this view, in vivo studies showed that down-regulation of $\alpha$-Syn expression in DA and 5-HT neurons increases the releasable pool of DA and 5-HT sensitive to tetrabenazine, a selective inhibitor of VMAT$_2$ [118]. Overall, the presynaptic location of $\alpha$-Syn has suggested a physiological role in neurotransmitter release and it apparently associates with the SV clustering and storage [98,99]. Furthermore, $\alpha$-Syn is abundantly expressed in DA, NA, and 5-HT neurons [96,118], defining a precise role of $\alpha$-Syn in monoamine synaptic plasticity by interacting with specific proteins that maintain monoamine homeostasis.

## 4. Dysfunction of the 5-HT System in PD Patients

The investigation of premotor pathology presents one of the most difficult problems in PD research. Although Braak and colleagues [26,126] proposed a significant premotor phase that may last as long as the symptomatic period, the identification of this phase in clinical practice is elusive. In fact, the profile of PD patients is also associated with diverse symptoms and clinical phenotypes [127]. Cumulative evidence indicates the existence of ongoing pre-SNc DA neurodegeneration during the premotor phase leading to non-motor symptoms, mainly constipation, anxiety and depression, smell loss, and rapid-eye-movement (REM) sleep behavior disorder [128]. A dysfunctional 5-HT system is generally regarded as a risk factor for depression. Consistent with this view, several reports suggest a positive correlation between decreased 5-HT neurotransmission and the severity of depression and anxiety symptoms in PD, most likely caused by pathological changes of the 5-HT neurons in the midbrain raphe nuclei [39,41,43,129,130].

By evaluating SERT availability with positron emission tomography (PET) and single photon emission computed tomography (SPECT) scans using various radioactive ligands, one can assess the integrity of the 5-HT system. The non-specific ligands [$^{123}$I]$\beta$-CIT and [$^{123}$I]FP-CIT have mostly been employed in in vivo SPECT imaging. Although these ligands have similar affinities for DAT and SERT, their thalamic and midbrain binding are considered to be SERT-specific [131]. Hence, SPECT studies using [$^{123}$I]$\beta$-CIT and [$^{123}$I]FP-CIT found decreased binding in the thalamus and midbrain of PD patients [132–136]. The PET ligands [$^{11}$C]-DASB and [$^{11}$C](+)McN5652 are highly specific for SERT. Using these ligands, several reports indicated reduced binding in different brain regions including the frontal cortex, striatum, and raphe nuclei [137–139]. Interestingly, an early study using [$^{11}$C]-DASB to map SERT changes in various PD stages based on disease duration showed reduced binding in the striatum, thalamus and anterior cingulate cortex of early-stage PD patients [140]. In the same study, decreases in SERT binding were observed in the prefrontal cortex of established PD and in the rostral and caudal raphe nuclei in advanced stages [140]. Moreover, recent SPECT and PET studies also showed 5-HT pathology in the premotor phase in mutant A53T $\alpha$-Syn gene *(SNCA)* carriers, before striatal DA loss, highlighting the early role of 5-HT pathology in the progression of PD [130].

The aforementioned studies examined PD patients without depressive symptoms. However, numerous investigations have assessed the connection between depression and SERT binding in PD. In a small cohort of depressed PD patients, early studies with [$^{11}$C]-DASB PET demonstrated that depression correlated with increased SERT binding in the dorso-lateral and prefrontal cortex [39]. Other studies also reported that the cingulate cortex and caudal raphe nuclei of depressed PD patients showed higher levels of SERT than non-depressed PD patients [41,141]. These findings are in agreement with the low levels of 5-HT and its metabolite 5-hydroxyindoleacetic acid (5-HIAA) found in the cerebrospinal fluid (CSF) of patients with PD and depression [142]. Interestingly, CSF levels of homovanillic acid (HVA), a DA metabolite, were not associated with the presence of depression in PD [142]. Likewise, in response to the functional deficit of 5-HT availability, post-synaptic 5-HT$_{1A}$ and 5-HT$_{2A}$ receptors were upregulated in cortical brain regions [143]. Furthermore, the density of 5-HT$_{2C}$ and 5-HT$_{2A}$ receptors in SN pars reticulate and striatum, respectively, appears to be increased in patients with PD [144–146]. These alterations may represent a compensatory response to a reduction of functional 5-HT levels in these nuclei.

In contrast, more recent PET studies revealed that the severe apathy in PD patients correlates with a reduction of [$^{11}$C]-DASB binding in the anterior caudate nucleus and orbitofrontal cortex, while the depression degree was exclusively related to a reduction in [$^{11}$C]-DASB binding within the bilateral subgenual anterior cingulate cortex (ACC) [5]. In another SPECT study, [$^{123}$I]FP-CIT binding was decreased in the midbrain of a cohort of PD patients with depression [147]. In light of this, it appears that, although imaging studies indicate that SERT binding in the midbrain and forebrain differs between non-depressed and depressed patients with PD, the extent to which these changes are crucial for the onset of depression is still unknown. Data also suggest that the presence of 5-HT pathology occurs at the beginning of the disease, preceding the development of the DA pathology and motor symptoms. Therefore, molecular imaging of SERT could be used to visualize the premotor pathology of PD in vivo as an adjunctive tool for screening and monitoring progression for individuals at risk of PD, thereby complementing DA imaging.

Importantly, neuropathological studies have demonstrated the presence of LBs ($\alpha$-Syn positive staining) in raphe 5-HT neurons in the early stages of the disease [148–151]. Previous studies on the propagation of $\alpha$-Syn proposed that PD begins in the medulla oblongata with LB pathology in the dorsal motor nuclei of the glossopharyngeal and vagal nerves and the adjacent intermediate reticular zone [126,152]. As PD progresses, it is proposed that the LB pathology spreads up the brainstem in an upward direction, affecting the raphe nuclei before reaching the SNc. In late stages, LBs are also found in limbic and cortical brain areas. The caudal groups of the raphe nuclei (e.g., raphe major, raphe obscure, and raphe pallidus) have been widely shown to contain LB-related lesions in the early stages of PD or even before the onset of motor symptoms [126,152,153]. The 5-HT neurons found in the caudal raphe nuclei play a role in a number of autonomic processes, including pain and decreased gastrointestinal motility, which are recognized non-motor symptoms in PD. In addition, the rostral raphe nuclei, containing the DR and median raphe nucleus (MR), also appear to be affected in PD. LB pathologies have been found in both the DR and MR of post-mortem PD brains and appear to be localized in 5-HT-containing neurons [148,149,154]. Some early studies found a significant neuronal loss within the DR from postmortem brain samples of PD patients, and this was even more pronounced in depressed PD patients [155]. However, other studies did not observe neuronal loss in DR, but did in MR [148,149,156]. Surprisingly, we found that only one of these studies used an unbiased design-based stereology method for counting cells [156]. In addition, some studies reported the absence of significant cell loss in the DR of post-mortem PD brains, but found evidence of the dysfunction of DR neurons based on reduced nucleolar volume and loss of cytoplasmic RNA [157]. Recently, it was also reported that long-range 5-HT projections from raphe are vulnerable in PD in response to hydrogen peroxide-induced cellular stress [158]. Taken together, the above findings point to neuropathological alterations in the 5-HT system in PD, comprising of the presence of LB pathology accompanied in some cases by neuronal loss in the raphe nuclei, as well as morphological changes of 5-HT fibers, which would lead to modified 5-HT neurotransmission.

## 5. Dysfunction of the 5-HT System in Animal Models with Overexpression of $\alpha$-Syn

Abundant evidence suggests that the development of PD may comprise three main phases. The onset of $\alpha$-Syn buildup in the CNS or PNS/ENS, in the absence of observable clinical symptoms, is referred to as the "preclinical PD" phase. The second phase, often known as the "pre-motor" or "prodromal," can last for more than 10 years before the disease is clinically diagnosed. It is usually accompanied by the appearance of non-motor symptoms caused in part by pre-SNc abnormalities. During this phase, PD patients may display increased anxiety as early as 16 years prior to disease diagnosis; and depression becomes significantly prevalent among PD patients in the last 3–4 years preceding diagnosis. The third phase is the "motor phase of PD", which is the one that is clinically visible and easiest to diagnose [159,160]. Understanding the pathophysiological mechanisms underlying non-motor symptoms in PD is important, but requires relevant preclinical animal models. In this sense, one of the main shortcomings of current PD-like animal models is that they focus on DA pathways, which probably do not reflect the complexity underlying the occurrence of these symptoms in patients [161,162]. In fact, there is still a paucity of studies addressing the role of brain circuits other than nigrostriatal DA systems in the early stages of the disease. In this review, we provide an overview of the current state of the field by presenting different preclinical models used in research on measures of anxiety and depressive phenotypes in rodent models of PD with emphasis on 5-HT systems.

In addition to the toxin-induced and genetic animal models of PD [162–167], in recent decades, an alternative approach to modulate the disease based on the forced expression of wild-type or mutant human $\alpha$-Syn using (1) transgenic techniques, (2) viral vector mediated transfer of $\alpha$-Syn, or (3) injection of pathogenic pre-formed $\alpha$-Syn fibrils (PFFs) has been presented. Thus, an intracellular accumulation of $\alpha$-Syn in raphe 5-HT neurons and in hippocampal 5-HT fibers, without loss of 5-HT neurons in 12-week-old transgenic mice overexpressing mutant A53T $\alpha$-Syn, was reported [168]. In parallel, mice showed a reduced 5-HT release and compromised increase in doublecortin+ neuroblasts in the dentate gyrus (DG), indicating a differential neurogenic response [168]. Another study also reported that mutant A53T $\alpha$-Syn–expressing mice (52-week-old) showed strong $\alpha$-Syn expression in the prefrontal cortex (PFC) along with reduced 5-HT innervation in layers V/VI of the PFC and enlarged axonal varicosities [169], leading to altered 5-HT signaling.

Moreover, Khol et al. [170] generated an $\alpha$-Syn transgenic rat model that displayed important features of PD such as a widespread and progressive $\alpha$-Syn aggregation pathology, DA loss and age-dependent motor decline. Notably, prior to the occurrence of the motor phenotype, rats showed profoundly impaired dendritogenesis of neuroblasts in the hippocampal DG, resulting in the severely reduced survival of adult newborn neurons. Reduced 5-HT$_{1B}$ receptor levels, lower 5-HT neurotransmitter concentration, and loss of 5-HT nerve terminals innervating the DG/CA3 subfield were indicative of decreased neurogenesis, but the number of 5-HT neurons in the raphe nuclei remained stable. The authors highlight that this transgenic rat model elicited an early anxiety-like phenotype, suggesting that $\alpha$-Syn accumulation severely impairs hippocampal neurogenesis and 5-HT neurotransmission prior to motor function [170].

Recently, the adeno-associated virus (AAV)-$\alpha$-Syn and PFFs models have been specifically adapted for study of $\alpha$-synucleinopathies using stereotaxic delivery into different brain areas, making them useful tools [171]. Therefore, a model of AAV-induced $\alpha$-synucleinopathy selectively in 5-HT neurons of rats resulted in progressive degeneration of the 5-HT axon terminals in hippocampus, without the loss of raphe 5-HT neurons [172]. Furthermore, overexpression of $\alpha$-Syn in raphe nuclei and basal forebrain cholinergic neurons of rats resulted in a more pronounced axonal pathology and significantly impaired anxiety response as assessed in the elevated plus maze [172]. Likewise, we demonstrated that AAV-induced overexpression of human $\alpha$-Syn in mouse 5-HT neurons causes a gradual accumulation and aggregation of $\alpha$-Syn in the 5-HT system. In parallel, we found alterations in axonal transport, brain-derived neurotrophic factor (BDNF) production, and 5-HT neurotransmission in 5-HT projection brain areas of PD-like mice, leading to a depressive-like phenotype (Figure 2) [110].

Other studies using rodent models of PFF also reported motor deficits and emotional and cognitive abnormalities, although the latter findings remained relatively unchanged in PFF models up to 6 months after injection [173–175]. Whether this is due to the injection site (primarily such as the striatum, SNc, or olfactory bulb) and/or the duration of the pathological spread remains to be determined. For instance, a recent study found that PFF injections in mice caused deficits in social dominance behavior and fear conditioning, two activities linked to prefrontal cortex and amygdala function, suggesting that these brain regions may be crucial in the complex emotional and cognitive traits of PD [37].

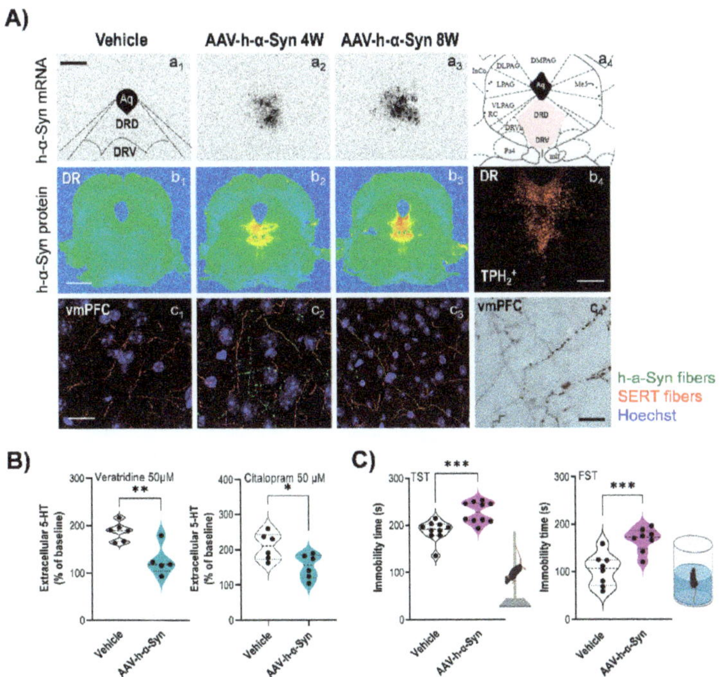



**Figure 2.** Human α-synuclein (h-α-Syn) overexpression in mouse serotonin neurons triggers a depressive-like phenotype. H-α-Syn expression was driven to raphe nuclei by a 1 µL AAV construct harboring a chicken-actin promoter (AAV-h-α-Syn) or vehicle, and mice were killed 1, 4, and 8 weeks (W) after injection. (**A**) Representative midbrain sections showing h-α-Syn mRNA levels in raphe nuclei examined by an in situ hybridization technique (**a₁–a₃**). Schematic coronal representation of mouse midbrain at −4.72 mm (AP coordinate) from bregma (**a₄**). Scale bar: 500 µm. Abbreviations: Aq (aqueduct), DRD (dorsal raphe nucleus, dorsal), and DRV (dorsal raphe nucleus, ventral). Representative coronal midbrain sections showing progressive increases of h-α-Syn protein levels in the raphe nuclei assessed by immunohistochemistry procedures (**b₁–b₃**). Signal represents the optical density (OD) of autoradiograms. Scale bar: 1 mm. Raphe serotonin (5-HT) neurons were identify using tryptophan hydroxylase (TPH₂) marker (**b₄**). Scale bar: 25 µm. Representative confocal microscopy images showing serotonin transporter (SERT) and h-α-Syn axonal co-localization in different ventromedial prefrontal cortex (vmPFC) of mice injected with AAV5 examined 4 W and 8 W later (**c₁–c₃**). The majority of the h-α-Syn-positive fibers also showed SERT staining, proving that they originate from raphe nuclei. Scale bar: 25 µm. Immunohistochemistry procedure on representative coronal brain sections reveals h-α-Syn-positive axonal swellings in the vmPFC (**c₄**). Scale bar: 25 µm. (**B**) Local infusion of veratridine (depolarizing agent, 50 µM) or citalopram (selective serotonin transporter inhibitor 50 µM) into vmPFC induced a greater effect on 5-HT release in vehicle-injected than in AAV-h-α-Syn-injected mice at 4 W post-administration. (**C**) AAV-injected mice evoked a depressive-like state in the tail suspension (TST) and forced swimming (FST) tests characterized by a longer immobility time compared to vehicle-injected mice. Values are presented as mean ± SEM. * $p < 0.05$, ** $p < 0.01$, and *** $p < 0.001$, compared to vehicle-injected mice. Adapted from [110].

In addition, some studies using cell cultures overexpressing α-Syn showed that 5-hydroxyindoleacetaldehyde (5-HIAL), a 5-HT metabolite product generated by monoamine oxidase (MAO-A), increases α-Syn oligomerization, which may explain the dysfunction of 5-HT neurons in PD [176]. Recent studies also showed the importance of maintaining the integrity of 5-HT systems, as 5-HT itself can affect the growth of amyloid-forming protein fibrils. Indeed, 5-HT or selective serotonin reuptake inhibitors (e.g., escitalopram) activate signaling that alters the processing of α-Syn fibrils as well as amyloid precursor proteins into β-amyloid (Aβ) to prevent protein aggregation by direct binding, and could be beneficial to PD and other neurodegenerative disorders [177–179].

## 6. Conclusions

The frequent occurrence of depression in PD is a prevalent and complex issue. Although often overlooked or underestimated, depression can seriously influence the course of PD and the quality of life of patients. In addition to dopaminergic depletion, several findings highlight the importance of serotonergic degeneration in PD. Thus, changes in 5-HT biochemical markers, LB pathology (α-Syn-positive staining) in raphe nuclei, and structural and functional alterations in the serotonergic system have been described, and it has been shown that these alterations in the serotonergic connectome are mainly associated with the expression of neuropsychiatric symptoms at disease onset. In support of this, the few available animal models demonstrating α-Syn-induced deficits in the serotonergic system recapitulate the mechanisms and early premotor stages of the disease. Altogether, measuring serotonergic integrity might be a useful in vivo tool to use in routines to guide the choice of the pharmacological arsenal in order to alleviate PD-related neuropsychiatric symptoms. Thus, such a measurement could serve as a sensitive marker of PD burden.

**Author Contributions:** Conceptualization, L.M.-R. and A.B.; writing—original draft preparation L.M.-R., U.S.-S., R.P.-C., J.J.M. and A.B.; writing—review and editing, A.B. All authors have read and agreed to the published version of the manuscript.

**Funding:** This research was funded by grants PID2019-105136RB-100, MCIN/AEI/10.13039/5011000 11033 and ERDF A way of making Europe, by the Euro-pean Union, and CB/07/09/0034 and CB/07/09/0008 Center for Networked Biomedical Research on Mental Health (CIBERSAM).

**Institutional Review Board Statement:** Not applicable.

**Informed Consent Statement:** Not applicable.

**Data Availability Statement:** Not applicable.

**Acknowledgments:** USS is a recipient of a fellowship from the Non-Doctor Researcher Formation Pre-doctoral Program of Basque Government, Spain. Figure 1 was created with BioRender.com.

**Conflicts of Interest:** A.B. is an inventor of the issued patents WO2011131693, WO-2014064257-A1, WO2014064258-A1 for ligand-conjugated siRNA and ASO molecules and the targeting approach for monoamine systems related to this work. The rest of the authors declare no competing interest.

## References

1. Jankovic, J. Parkinson's disease: Clinical features and diagnosis. *J. Neurol. Neurosurg. Psychiatry* **2008**, *79*, 368–376. [CrossRef]
2. Lima, M.; Martins, E.; Delattre, A.; Proenc, M.; Mori, M.; Carabelli, B. Motor and nonmotor features of Parkinson's disease: A review of clinical and experimental studies. *CNS Neurol. Disord. Drug Targets* **2012**, *11*, 439–449.
3. Postuma, R.B.; Berg, D.; Stern, M.; Poewe, W.; Olanow, C.W.; Oertel, W.; Obeso, J.; Marek, K.; Litvan, I.; Lang, A.E.; et al. MDS clinical diagnostic criteria for Parkinson's disease. *Mov. Disord.* **2015**, *30*, 1591–1601. [CrossRef]
4. Grosch, J.; Winkler, J.; Kohl, Z. Early Degeneration of both dopaminergic and serotonergic axons—A common mechanism in Parkinson's disease. *Front. Cell. Neurosci.* **2016**, *10*, 293. [CrossRef] [PubMed]
5. Maillet, A.; Krack, P.; Lhommée, E.; Météreau, E.; Klinger, H.; Favre, E.; Le Bars, D.; Schmitt, E.; Bichon, A.; Pelissier, P.; et al. The prominent role of serotonergic degeneration in apathy, anxiety and depression in de novo Parkinson's disease. *Brain* **2019**, *139*, 2486–2502. [CrossRef] [PubMed]
6. Aarsland, D.; Marsh, L.; Schrag, A. Neuropsychiatric symptoms in Parkinson's disease. *Mov. Disord.* **2009**, *24*, 2175–2186. [CrossRef]
7. Santangelo, G.; Vitale, C.; Trojano, L.; Longo, K.; Cozzolino, A.; Grossi, D.; Barone, P. Relationship between depression and cognitive dysfunctions in Parkinson's disease without dementia. *J. Neurol.* **2009**, *256*, 632–638. [CrossRef] [PubMed]
8. Chaudhuri, K.R.; Schapira, A.H. Non-motor symptoms of Parkinson's disease: Dopaminergic pathophysiology and treatment. *Lancet Neurol.* **2009**, *8*, 464–474. [CrossRef] [PubMed]
9. Thobois, S.; Ardouin, C.; Lhommée, E.; Klinger, H.; Lagrange, C.; Xie, J.; Fraix, V.; Coelho Braga, M.C.; Hassani, R.; Kistner, A.; et al. Non-motor dopamine withdrawal syndrome after surgery for Parkinson's disease: Predictors and underlying mesolimbic denervation. *Brain* **2010**, *133*, 1111–1127. [CrossRef]

10. De la Riva, P.; Smith, K.; Xie, S.X.; Weintraub, D. Course of psychiatric symptoms and global cognition in early Parkinson disease. *Neurology* **2014**, *83*, 1096–1103. [CrossRef] [PubMed]

11. Dujardin, K.; Langlois, C.; Plomhause, L.; Carette, A.S.; Delliaux, M.; Duhamel, A.; Defebvre, L. Apathy in untreated early-stage Parkinson disease: Relationship with other non-motor symptoms. *Mov. Disord* **2014**, *29*, 1796–1801. [CrossRef] [PubMed]

12. Schrag, A.; Sauerbier, A.; Chaudhuri, K.R. New clinical trials for nonmotor manifestations of Parkinson's disease. *Mov. Disord* **2015**, *30*, 1490–1504. [CrossRef] [PubMed]

13. Santos García, D.; de Deus Fonticoba, T.; Suárez Castro, E.; Borrué, C.; Mata, M.; Solano Vila, B.; Cots Foraster, A.; Álvarez Sauco, M.; Rodríguez Pérez, A.B.; Vela, L.; et al. Non-motor symptoms burden, mood, and gait problems are the most significant factors contributing to a poor quality of life in non-demented Parkinson's disease patients: Results from the COPPADIS Study Cohort. *Park. Relat. Disord.* **2019**, *66*, 151–157. [CrossRef] [PubMed]

14. Reijnders, J.S.; Ehrt, U.; Weber, W.E.; Aarsland, D.; Leentjens, F. A systematic review of prevalence studies of depression in Parkinson's disease. *Mov. Disord.* **2008**, *23*, 183–189. [CrossRef] [PubMed]

15. Aarsland, D.; Påhlhagen, S.; Ballard, C.G.; Ehrt, U.; Svenningsson, P. Depression in Parkinson disease–epidemiology, mechanisms and management. *Nat. Rev. Neurol.* **2011**, *8*, 35–47. [CrossRef]

16. Chuquilín-Arista, F.; Álvarez-Avellón, T.; Menéndez-González, M. Prevalence of Depression and Anxiety in Parkinson Disease and Impact on Quality of Life: A Community-Based Study in Spain. *J. Geriatr. Psychiatry Neurol.* **2020**, *33*, 207–213. [CrossRef]

17. Karlsen, K.H.; Tandberg, E.; Aarsland, D.; Larsen, J.P. Health related quality of life in Parkinson's disease: A prospective longitudinal study. *J. Neurol. Neurosurg. Psychiatry* **2000**, *69*, 584–589. [CrossRef]

18. Den Oudsten, B.; Van Heck, G.; De Vries, J. Quality of life and related concepts in Parkinson's disease: A systematic review. *Mov. Disord.* **2007**, *22*, 1528–1537. [CrossRef]

19. Soh, S.; Morris, M.; McGinley, J. Determinants of health-related quality of life in Parkinson's disease: A systematic review. *Park. Relat. Disord.* **2011**, *17*, 1–9. [CrossRef]

20. Hemmerle, A.; Herman, J.; Seroog, K. Stress, depression and Parkinson's disease. *Exp. Neurol.* **2012**, *233*, 79–86. [CrossRef]

21. Greffard, S.; Verny, M.; Bonnet, A.M.; Beinis, J.Y.; Gallinari, C.; Meaume, S.; Piette, F.; Hauw, J.J.; Duyckaerts, C. Motor score of the Unified Parkinson Disease Rating Scale as a good predictor of Lewy body-associated neuronal loss in the substantia nigra. *Arch. Neurol.* **2006**, *63*, 584–588. [CrossRef] [PubMed]

22. Halliday, G.; Lees, A.; Stern, M. Milestones in Parkinson's disease—Clinical and pathologic features. *Mov. Disord.* **2011**, *26*, 1015–1021. [CrossRef]

23. Hornykiewicz, O. 50 years of levodopa. *Mov. Disord.* **2015**, *30*, 1008. [CrossRef] [PubMed]

24. Foffani, G.; Obeso, J.A. A Cortical Pathogenic Theory of Parkinson's Disease. *Neuron* **2018**, *99*, 1116–1128. [CrossRef]

25. McGregor, M.M.; Nelson, A.B. Circuit Mechanisms of Parkinson's Disease. *Neuron* **2019**, *101*, 1042–1056. [CrossRef] [PubMed]

26. Braak, H.; Ghebremedhin, E.; Rub, U.; Bratzke, H.; Del Tredici, K. Stages in the development of Parkinson's disease-related pathology. *Cell Tissue Res.* **2004**, *318*, 121–134. [CrossRef]

27. Braak, H.; Del Tredici, K. Neuropathological staging of brain pathology in sporadic Parkinson's disease: Separating the wheat from the Chaff. *J. Park. Dis.* **2017**, *7*, S71–S85. [CrossRef] [PubMed]

28. Spillantini, M.G.; Schmidt, M.L.; Lee, V.M.-Y.; Trojanowski, J.Q.; Jakes, R.; Goedert, M. Alpha-Synuclein in Lewy bodies. *Nature* **1997**, *388*, 839–840. [CrossRef]

29. Dickson, D.W.; Braak, H.; Duda, J.E.; Duyckaerts, C.; Gasser, T.; Halliday, G.M.; Hardy, J.; Leverenz, J.B.; Del Tredici, K.; Wszolek, Z.K.; et al. Neuropathological assessment of Parkinson's disease: Refining the diagnostic criteria. *Lancet Neurol.* **2009**, *8*, 1150–1157. [CrossRef]

30. Spillantini, M.G.; Goedert, M. Neurodegeneration and the ordered assembly of alpha-synuclein. *Cell Tissue Res.* **2018**, *373*, 137–148. [CrossRef]

31. Shults, C.W. Lewy bodies. *Proc. Natl. Acad. Sci. USA* **2006**, *103*, 1661–1668. [CrossRef] [PubMed]

32. Shahmoradian, S.H.; Lewis, A.J.; Genoud, C.; Hench, J.; Moors, T.E.; Navarro, P.P.; Castaño-Díez, D.; Schweighauser, G.; Graff-Meyer, A.; Goldie, K.N.; et al. Lewy pathology in Parkinson's disease consists of crowded organelles and lipid membranes. *Nat. Neurosci.* **2019**, *22*, 1099–1109. [CrossRef] [PubMed]

33. Brichta, L.; Greengard, P.; Flajolet, M. Advances in the pharmacological treatment of Parkinson's disease: Targeting neurotransmitter systems. *Trends Neurosci.* **2013**, *36*, 543–554. [CrossRef] [PubMed]

34. Giguère, N.; Burke Nanni, S.; Trudeau, L.E. On Cell Loss and Selective Vulnerability of Neuronal Populations in Parkinson's Disease. *Front. Neurol.* **2018**, *9*, 455. [CrossRef]

35. Van Den Berge, N.; Ulusoy, A. Animal models of brain-first and body-first Parkinson's disease. *Neurobiol. Dis.* **2022**, *163*, 105599. [CrossRef]

36. Harding, A.J.; Stimson, E.; Henderson, J.M.; Halliday, G.M. Clinical correlates of selective pathology in the amygdala of patients with Parkinson's disease. *Brain* **2002**, *125*, 2431–2445. [CrossRef]

37. Stoyka, L.E.; Arrant, A.E.; Thrasher, D.R.; Russell, D.L.; Freire, J.; Mahoney, C.L.; Narayanan, A.; Dib, A.G.; Standaert, D.G.; Volpicelli-Daley, L.A. Behavioral defects associated with amygdala and cortical dysfunction in mice with seeded α-synuclein inclusions. *Neurobiol. Dis.* **2020**, *134*, 104708. [CrossRef]

38. Doder, M.; Rabiner, E.A.; Turjanski, N.; Lees, A.J.; Brooks, D.J. Tremor in Parkinson's disease and serotonergic dysfunction: An 11C-WAY 100635 PET study. *Neurology* **2003**, *60*, 601–605. [CrossRef]

39. Boileau, I.; Warsh, J.J.; Guttman, M.; Saint-Cyr, J.A.; McCluskey, T.; Rusjan, P.; Houle, S.; Wilson, A.A.; Meyer, J.H.; Kish, S.J. Elevated serotonin transporter binding in depressed patients with Parkinson's disease: A preliminary PET study with [11C]DASB. *Mov. Disord.* **2008**, *23*, 1776–1780. [CrossRef]

40. Pavese, N.; Metta, V.; Bose, S.K.; Chaudhuri, K.R.; Brooks, D.J. Fatigue in Parkinson's disease is linked to striatal and limbic serotonergic dysfunction. *Brain* **2010**, *133*, 3434–3443. [CrossRef]

41. Politis, M.; Wu, K.; Loane, C.; Turkheimer, F.E.; Molloy, S.; Brooks, D.J.; Piccini, P. Depressive symptoms in PD correlate with higher 5-HTT binding in raphe and limbic structures. *Neurology* **2010**, *75*, 1920–1927. [CrossRef] [PubMed]

42. Politis, M.; Wu, K.; Loane, C.; Brooks, D.J.; Kiferle, L.; Turkheimer, F.E.; Bain, P.; Molloy, S.; Piccini, P. Serotonergic mechanisms responsible for levodopa-induced dyskinesias in Parkinson's disease patients. *J. Clin. Investig.* **2014**, *124*, 1340–1349. [CrossRef] [PubMed]

43. Ballanger, B.; Klinger, H.; Eche, J.; Lerond, J.; Vallet, A.E.; Le Bars, D.; Tremblay, L.; Sgambato-Faure, V.; Broussolle, E.; Thobois, S. Role of serotonergic 1A receptor dysfunction in depression associated with Parkinson's disease. *Mov. Disord.* **2012**, *27*, 84–89. [CrossRef]

44. Adell, A.; Celada, P.; Abellán, M.T.; Artigas, F. Origin and functional role of the extracellular serotonin in the midbrain raphe nuclei. Brain research. *Brain Res. Rev.* **2002**, *39*, 154–180. [CrossRef] [PubMed]

45. Berger, M.; Gray, J.A.; Roth, B.L. The expanded biology of serotonin. *Annu. Rev. Med.* **2009**, *60*, 355–366. [CrossRef]

46. Vahid-Ansari, F.; Albert, P.R. Rewiring of the Serotonin System in Major Depression. *Front. Psychiatry* **2021**, *12*, 802581. [CrossRef]

47. Hornung, J.P. The human raphe nuclei and the serotonergic system. *J. Chem. Neuroanat.* **2003**, *26*, 331–343. [CrossRef]

48. Descarries, L.; Riad, M.; Parent, M. Ultrastructure of the Serotonin Innervation in the Mammalian Central Nervous System. In *Handbook of Behavioral Neurobiology of Serotonin*; Müller, C.P., Jacobs, B.L., Eds.; Elsevier: Amsterdam, The Netherlands, 2010.

49. Sparta, D.R.; Stuber, G.D. Cartography of serotonergic circuits. *Neuron* **2014**, *83*, 513–515. [CrossRef]

50. Bockaert, J.; Claeysen, S.A.D.; Marin, P. Classification and Signaling Characteristics of 5-HT Receptors. In *Handbook of Behavioral Neurobiology of Serotonin*; Müller, C.P., Jacobs, B.L., Eds.; Elsevier: Amsterdam, The Netherlands, 2010.

51. Weissbourd, B.; Ren, J.; DeLoach, K.E.; Guenthner, C.J.; Miyamichi, K.; Luo, L. Presynaptic partners of dorsal raphe serotonergic and GABAergic neurons. *Neuron* **2014**, *83*, 645–662. [CrossRef]

52. Mengod, G.; Palacios, J.M.; Cortés, R. Cartography of 5-HT1A and 5-HT2A Receptor Subtypes in Prefrontal Cortex and Its Projections. *ACS Chem. Neurosci.* **2015**, *6*, 1089–1098. [CrossRef]

53. Amargós-Bosch, M.; Bortolozzi, A.; Puig, M.V.; Serrats, J.; Adell, A.; Celada, P.; Toth, M.; Mengod, G.; Artigas, F. Co-expression and in vivo interaction of serotonin1A and serotonin2A receptors in pyramidal neurons of prefrontal cortex. *Cereb. Cortex* **2004**, *14*, 281–299. [CrossRef]

54. Santana, N.; Bortolozzi, A.; Serrats, J.; Mengod, G.; Artigas, F. Expression of serotonin1A and serotonin2A receptors in pyramidal and GABAergic neurons of the rat prefrontal cortex. *Cereb. Cortex* **2004**, *14*, 1100–1109. [CrossRef] [PubMed]

55. Janušonis, S. Serotonin in space: Understanding single fibers. *ACS Chem. Neurosci.* **2017**, *8*, 893–896. [CrossRef] [PubMed]

56. Gabbott, P.L.; Warner, T.A.; Jays, P.R.; Salway, P.; Busby, S.J. Prefrontal cortex in the rat: Projections to subcortical autonomic, motor, and limbic centers. *J. Comp. Neurol.* **2005**, *492*, 145–177. [CrossRef] [PubMed]

57. Muzerelle, A.; Scotto-Lomassese, S.; Bernard, J.F.; Soiza-Reilly, M.; Gaspar, P. Conditional anterograde tracing reveals distinct targeting of individual serotonin cell groups (B5-B9) to the forebrain and brainstem. *Brain Struct. Funct.* **2016**, *221*, 535–561. [CrossRef]

58. Puig, M.V.; Gulledge, A.T. Serotonin and prefrontal cortex function: Neurons, networks, and circuits. *Mol. Neurobiol.* **2011**, *44*, 449–464. [CrossRef]

59. Matias, S.; Lottem, E.; Dugué, G.P.; Mainen, Z.F. Activity patterns of serotonin neurons underlying cognitive flexibility. *ELife* **2017**, *6*, e20552. [CrossRef]

60. Pattij, T.; Schoffelmeer, A.N. Serotonin and inhibitory response control: Focusing on the role of 5-HT(1A) receptors. *Eur. J. Pharmacol.* **2015**, *753*, 140–145. [CrossRef]

61. Marquez, J.C.; Li, M.; Schaak, D.; Robson, D.N.; Li, J.M. Internal state dynamics shape brain wide activity and foraging behavior. *Nature* **2020**, *577*, 239–243. [CrossRef]

62. Maier, S.F.; Watkins, L.R. Stressor controllability and learned helplessness: The roles of the dorsal raphe nucleus, serotonin, and corticotropin-releasing factor. *Neurosci. Biobehav. Rev.* **2005**, *29*, 829–841. [CrossRef]

63. Airan, R.D.; Meltzer, L.A.; Roy, M.; Gong, Y.; Chen, H.; Deisseroth, K. High-speed imaging reveals neurophysiological links to behavior in an animal model of depression. *Science* **2007**, *317*, 819–823. [CrossRef] [PubMed]

64. Canli, T.; Lesch, K.P. Long story short: The serotonin transporter in emotion regulation and social cognition. *Nat. Neurosci.* **2007**, *10*, 1103–1109. [CrossRef] [PubMed]

65. Daut, R.A.; Fonken, L.K. Circadian regulation of depression: A role for serotonin. *Front. Neuroendocrinol.* **2019**, *54*, 100746. [CrossRef] [PubMed]

66. Hale, M.W.; Shekhar, A.; Lowry, C.A. Stress-related serotonergic systems: Implications for symptomatology of anxiety and affective disorders. *Cell. Mol. Neurobiol.* **2012**, *32*, 695–708. [CrossRef]

67. Maletic, V.; Robinson, M.; Oakes, T.; Iyengar, S.; Ball, S.G.; Russell, J. Neurobiology of depression: An integrated view of key findings. *Int. J. Clin. Pract.* **2007**, *61*, 2030–2040. [CrossRef]

68. Drevets, W.C. Neuroimaging and neuropathological studies of depression: Implications for the cognitive-emotional features of mood disorders. *Curr. Opin. Neurobiol.* **2001**, *11*, 240–249. [CrossRef]

69. Berton, O.; Nestler, E.J. New approaches to antidepressant drug discovery: Beyond monoamines. *Nat Rev. Neurosci.* **2006**, *7*, 137–151. [CrossRef]

70. Krishnan, V.; Nestler, E.J. The molecular neurobiology of depression. *Nature* **2008**, *455*, 894–902. [CrossRef]

71. Seminowicz, D.A.; Mayberg, H.S.; McIntosh, A.R.; Goldapple, K.; Kennedy, S.; Segal, Z.; Rafi-Tari, S. Limbic-frontal circuitry in major depression: A path modeling metanalysis. *NeuroImage* **2004**, *22*, 409–418. [CrossRef]

72. Mayberg, H.S.; Lozano, A.M.; Voon, V.; McNeely, H.E.; Seminowicz, D.; Hamani, C.; Schwalb, J.M.; Kennedy, S.H. Deep brain stimulation for treatment-resistant depression. *Neuron* **2005**, *45*, 651–660. [CrossRef]

73. Alexander, L.; Wood, C.M.; Roberts, A.C. The ventromedial prefrontal cortex and emotion regulation: Lost in translation? *J. Physiol.* **2022**, *601*, 37–50. [CrossRef]

74. De Schipper, L.J.; van der Grond, J.; Marinus, J.; Henselmans, J.M.L.; van Hilten, J.J. Loss of integrity and atrophy in cingulate structural covariance networks in Parkinson's disease. *Neuroimage Clin.* **2017**, *15*, 587–593. [CrossRef] [PubMed]

75. Vogt, B.A. Cingulate cortex in Parkinson's disease. *Handb. Clin. Neurol.* **2019**, *166*, 253–266.

76. Lin, H.; Cai, X.; Zhang, D.; Liu, J.; Na, P.; Li, W. Functional connectivity markers of depression in advanced Parkinson's disease. *Neuroimage Clin.* **2020**, *25*, 102130. [CrossRef]

77. Celada, P.; Puig, M.V.; Artigas, F. Serotonin modulation of cortical neurons and networks. *Front Integr. Neurosci.* **2013**, *7*, 25. [CrossRef]

78. López-Terrones, E.; Celada, P.; Riga, M.S.; Artigas, F. Preferential In Vivo Inhibitory Action of Serotonin in Rat Infralimbic versus Prelimbic Cortex: Relevance for Antidepressant Treatments. *Cereb. Cortex* **2022**, *32*, 3000–3013. [CrossRef]

79. Van Heukelum, S.; Mars, R.B.; Guthrie, M.; Buitelaar, J.K.; Beckmann, C.F.; Tiesinga, P.H.E.; Vogt, B.A.; Glennon, J.C.; Havenith, M.N. Where is Cingulate Cortex? A Cross-Species View. *Trends Neurosci.* **2020**, *43*, 285–299. [CrossRef]

80. Celada, P.; Puig, M.V.; Casanovas, J.M.; Guillazo, G.; Artigas, F. Control of dorsal raphe serotonergic neurons by the medial prefrontal cortex: Involvement of serotonin-1A, GABA(A), and glutamate receptors. *J. Neurosci.* **2001**, *21*, 9917–9929. [CrossRef] [PubMed]

81. Martín-Ruiz, R.; Puig, M.V.; Celada, P.; Shapiro, D.A.; Roth, B.L.; Mengod, G.; Artigas, F. Control of serotonergic function in medial prefrontal cortex by serotonin-2A receptors through a glutamate-dependent mechanism. *J. Neurosci.* **2001**, *21*, 9856–9866. [CrossRef]

82. George, J.M. The synucleins. *Genome Biol.* **2002**, *3*, 1–6.

83. Li, J.Y.; Henning Jensen, P.; Dahlström, A. Differential localization of α-, β-and γ-synucleins in the rat CNS. *Neuroscience* **2002**, *113*, 463–478. [CrossRef] [PubMed]

84. Longhena, F.; Faustini, G.; Spillantini, M.G.; Bellucci, A. Living in Promiscuity: The Multiple Partners of Alpha-Synuclein at the Synapse in Physiology and Pathology. *Int. J. Mol. Sci.* **2019**, *20*, 141. [CrossRef] [PubMed]

85. Pavia-Collado, R.; Rodríguez-Aller, R.; Alarcón-Arís, D.; Miquel-Rio, L.; Ruiz-Bronchal, E.; Paz, V.; Campa, L.; Galofré, M.; Sgambato, V.; Bortolozzi, A. Up and Down γ-Synuclein Transcription in Dopamine Neurons Translates into Changes in Dopamine Neurotransmission and Behavioral Performance in Mice. *Int. J. Mol. Sci.* **2022**, *23*, 1807. [CrossRef]

86. Goedert, M.; Jakes, R.; Spillantini, M.G. The Synucleinopathies: Twenty Years On. *J. Parkinsons Dis.* **2017**, *7*, S51–S69. [CrossRef] [PubMed]

87. Eliezer, D.; Kutluay, E.; Bussell, R., Jr.; Browne, G. Conformational properties of alpha-synuclein in its free and lipid-associated states. *J. Mol. Biol.* **2001**, *307*, 1061–1073. [CrossRef] [PubMed]

88. Deleersnijder, A.; Gerard, M.; Debyser, Z.; Baekelandt, V. The remarkable conformational plasticity of alpha-synuclein: Blessing or curse? *Trends Mol. Med.* **2013**, *19*, 368–377. [CrossRef] [PubMed]

89. Lv, Z.; Krasnoslobodtsev, A.V.; Zhang, Y.; Ysselstein, D.; Rochet, J.C.; Blanchard, S.C.; Lyubchenko, Y.L. Effect of acidic pH on the stability of alpha-synuclein dimers. *Biopolymers* **2016**, *105*, 715–724. [CrossRef]

90. Galvagnion, C. The Role of Lipids Interacting with alpha-Synuclein in the Pathogenesis of Parkinson's Disease. *J. Park. Dis.* **2017**, *7*, 433–450.

91. Sulzer, D.; Edwards, R.H. The physiological role of α-synuclein and its relationship to Parkinson's Disease. *J. Neurochem.* **2019**, *150*, 475–486. [CrossRef]

92. Vidović, M.; Rikalovic, M.G. Alpha-Synuclein Aggregation Pathway in Parkinson's Disease: Current Status and Novel Therapeutic Approaches. *Cells* **2022**, *11*, 1732. [CrossRef]

93. Maroteaux, L.; Campanelli, J.T.; Scheller, R.H. Synuclein: A neuron-specific protein localized to the nucleus and presynaptic nerve terminal. *J. Neurosci.* **1988**, *8*, 2804–2815. [CrossRef]

94. Withers, G.S.; George, J.M.; Banker, G.A.; Clayton, D.F. Delayed localization of synelfin (synuclein, NACP) to presynaptic terminals in cultured rat hippocampal neurons. *Brain Res. Dev. Brain Res.* **1997**, *99*, 87–94. [CrossRef]

95. Kahle, P.J.; Neumann, M.; Ozmen, L.; Muller, V.; Jacobsen, H.; Schindzielorz, A.; Okochi, M.; Leimer, U.; van Der Putten, H.; Probst, A.; et al. Subcellular localization of wild-type and Parkinson's disease-associated mutant alpha -synuclein in human and transgenic mouse brain. *J. Neurosci.* **2000**, *20*, 6365–6373. [CrossRef]

96.  Abeliovich, A.; Schmitz, Y.; Farinas, I.; Choi-Lundberg, D.; Ho, W.H.; Castillo, P.E.; Shinsky, N.; Verdugo, J.M.; Armanini, M.; Ryan, A.; et al. Mice lacking alpha-synuclein display functional deficits in the nigrostriatal dopamine system. *Neuron* **2000**, *25*, 239–252. [CrossRef]

97.  Nemani, V.M.; Lu, W.; Berge, V.; Nakamura, K.; Onoa, B.; Lee, M.K.; Chaudhry, F.A.; Nicoll, R.A.; Edwards, R.H. Increased expression of alpha-synuclein reduces neurotransmitter release by inhibiting synaptic vesicle reclustering after endocytosis. *Neuron* **2010**, *65*, 66–79. [CrossRef]

98.  Burre, J. The Synaptic Function of alpha-Synuclein. *J. Parkinsons Dis.* **2015**, *5*, 699–713. [CrossRef]

99.  Burre, J.; Sharma, M.; Sudhof, T.C. Cell Biology and Pathophysiology of alpha-Synuclein. *Cold Spring Harb. Perspect. Med.* **2018**, *8*, a024091. [CrossRef]

100.  Fernández-Nogales, M.; López-Cascales, M.T.; Murcia-Belmonte, V.; Escalante, A.; Fernández-Albert, J.; Muñoz-Viana, R.; Barco, A.; Herrera, E. Multiomic Analysis of Neurons with Divergent Projection Patterns Identifies Novel Regulators of Axon Pathfinding. *Adv Sci.* **2022**, *9*, e2200615. [CrossRef]

101.  Li, W.W.; Yang, R.; Guo, J.C.; Ren, H.M.; Zha, X.L.; Cheng, J.S.; Cai, D.F. Localization of alpha-synuclein to mitochondria within midbrain of mice. *Neuroreport* **2007**, *18*, 1543–1546. [CrossRef]

102.  Colla, E.; Jensen, P.H.; Pletnikova, O.; Troncoso, J.C.; Glabe, C.; Lee, M.K. Accumulation of toxic alpha-synuclein oligomer within endoplasmic reticulum occurs in alpha-synucleinopathy in vivo. *J. Neurosci.* **2012**, *32*, 3301–3305. [CrossRef]

103.  Pinho, R.; Paiva, I.; Jercic, K.G.; Fonseca-Ornelas, L.; Gerhardt, E.; Fahlbusch, C.; Garcia-Esparcia, P.; Kerimoglu, C.; Pavlou, M.A.; Villar-Pique, A.; et al. Nuclear localization and phosphorylation modulate pathological effects of Alpha-Synuclein. *Hum. Mol. Genet.* **2019**, *28*, 31–50. [CrossRef]

104.  Bellucci, A.; Mercuri, N.B.; Venneri, A.; Faustini, G.; Longhena, F.; Pizzi, M.; Missale, C.; Spano, P. Review: Parkinson's disease: From synaptic loss to connectome dysfunction. *Neuropathol. Appl. Neurobiol.* **2016**, *42*, 77–94. [CrossRef]

105.  Kouroupi, G.; Taoufik, E.; Vlachos, I.S.; Tsioras, K.; Antoniou, N.; Papastefanaki, F.; Chroni-Tzartou, D.; Wrasidlo, W.; Bohl, D.; Stellas, D.; et al. Defective synaptic connectivity and axonal neuropathology in a human iPSC-based model of familial Parkinson's disease. *Proc. Natl. Acad. Sci. USA* **2017**, *114*, E3679–E3688. [CrossRef]

106.  Faustini, G.; Longhena, F.; Varanita, T.; Bubacco, L.; Pizzi, M.; Missale, C.; Benfenati, F.; Bjorklund, A.; Spano, P.; Bellucci, A. Synapsin III deficiency hampers alpha-synuclein aggregation, striatal synaptic damage and nigral cell loss in an AAV-based mouse model of Parkinson's disease. *Acta Neuropathol.* **2018**, *136*, 621–639. [CrossRef]

107.  Xu, T.; Bajjalieh, S.M. SV2 modulates the size of the readily releasable pool of secretory vesicles. *Nat. Cell Biol.* **2001**, *3*, 691–698. [CrossRef]

108.  Wan, Q.F.; Zhou, Z.Y.; Thakur, P.; Vila, A.; Sherry, D.M.; Janz, R.; Heidelberger, R. SV2 acts via presynaptic calcium to regulate neurotransmitter release. *Neuron* **2010**, *66*, 884–895. [CrossRef]

109.  Dunn, A.R.; Stout, K.A.; Ozawa, M.; Lohr, K.M.; Hoffman, C.A.; Bernstein, A.I.; Li, Y.; Wang, M.; Sgobio, C.; Sastry, N.; et al. Synaptic vesicle glycoprotein 2C (SV2C) modulates dopamine release and is disrupted in Parkinson disease. *Proc. Natl. Acad. Sci. USA* **2017**, *114*, E2253–E2262. [CrossRef]

110.  Miquel-Rio, L.; Alarcón-Arís, D.; Torres-López, M.; Cóppola-Segovia, V.; Pavia-Collado, R.; Paz, V.; Ruiz-Bronchal, E.; Campa, L.; Casal, C.; Montefeltro, A.; et al. Human α-synuclein overexpression in mouse serotonin neurons triggers a depressive-like phenotype. Rescue by oligonucleotide therapy. *Transl. Psychiatry* **2022**, *12*, 79. [CrossRef]

111.  Gitler, A.D.; Bevis, B.J.; Shorter, J.; Strathearn, K.E.; Hamamichi, S.; Su, L.J.; Caldwell, K.A.; Caldwell, G.A.; Rochet, J.C.; McCaffery, J.M.; et al. The Parkinson's disease protein alpha-synuclein disrupts cellular Rab homeostasis. *Proc. Natl. Acad. Sci. USA* **2008**, *105*, 145–150. [CrossRef]

112.  Bellucci, A.; Longhena, F.; Spillantini, M.G. The Role of Rab Proteins in Parkinson's Disease Synaptopathy. *Biomedicines* **2022**, *10*, 1941. [CrossRef]

113.  Sidhu, A.; Wersinger, C.; Vernier, P. Does alpha-synuclein modulate dopaminergic synaptic content and tone at the synapse? *FASEB J.* **2004**, *18*, 637–647. [CrossRef]

114.  Wersinger, C.; Jeannotte, A.; Sidhu, A. Attenuation of the norepinephrine transporter activity and trafficking via interactions with alpha-synuclein. *Eur. J. Neurosci.* **2006**, *24*, 3141–3152. [CrossRef]

115.  Wersinger, C.; Rusnak, M.; Sidhu, A. Modulation of the trafficking of the human serotonin transporter by human alpha-synuclein. *Eur. J. Neurosci.* **2006**, *24*, 55–64. [CrossRef]

116.  Oaks, A.W.; Sidhu, A. Synuclein modulation of monoamine transporters. *FEBS Lett.* **2011**, *585*, 1001–1006. [CrossRef]

117.  Butler, B.; Saha, K.; Rana, T.; Becker, J.P.; Sambo, D.; Davari, P.; Goodwin, J.S.; Khoshbouei, H. Dopamine Transporter Activity Is Modulated by alpha-Synuclein. *J. Biol. Chem.* **2015**, *290*, 29542–29554. [CrossRef]

118.  Alarcón-Arís, D.; Recasens, A.; Galofré, M.; Carballo-Carbajal, I.; Zacchi, N.; Ruiz-Bronchal, E.; Pavia-Collado, R.; Chica, R.; Ferrés-Coy, A.; Santos, M.; et al. Selective α-Synuclein Knockdown in Monoamine Neurons by Intranasal Oligonucleotide Delivery: Potential Therapy for Parkinson's Disease. *Mol. Ther.* **2018**, *26*, 550–567. [CrossRef]

119.  Torres, G.E.; Gainetdinov, R.R.; Caron, M.G. Plasma membrane monoamine transporters: Structure, regulation and function. *Nat. Rev. Neurosci.* **2003**, *4*, 13–25. [CrossRef]

120.  Hahn, M.K.; Blakely, R.D. Monoamine transporter gene structure and polymorphisms in relation to psychiatric and other complex disorders. *Pharm. J.* **2002**, *2*, 217–235. [CrossRef]

121. Wersinger, C.; Sidhu, A. Partial regulation of serotonin transporter function by gamma-synuclein. *Neurosci. Lett.* **2009**, *453*, 157–161. [CrossRef]

122. Narboux-Nême, N.; Sagné, C.; Doly, S.; Diaz, S.L.; Martin, C.B.; Angenard, G.; Martres, M.P.; Giros, B.; Hamon, M.; Lanfumey, L.; et al. Severe serotonin depletion after conditional deletion of the vesicular monoamine transporter 2 gene in serotonin neurons: Neural and behavioral consequences. *Neuropsychopharmacology* **2011**, *36*, 2538–2550. [CrossRef]

123. Eiden, L.E.; Weihe, E. VMAT2: A dynamic regulator of brain monoaminergic neuronal function interacting with drugs of abuse. *Ann. N. Y. Acad. Sci.* **2011**, *1216*, 86–98. [CrossRef] [PubMed]

124. Yamamoto, S.; Fukae, J.; Mori, H.; Mizuno, Y.; Hattori, N. Positive immunoreactivity for vesicular monoamine transporter 2 in Lewy bodies and Lewy neurites in substantia nigra. *Neurosci. Lett.* **2006**, *396*, 187–191. [CrossRef]

125. Taylor, T.N.; Caudle, W.M.; Shepherd, K.R.; Noorian, A.; Jackson, C.R.; Iuvone, P.M.; Weinshenker, D.; Greene, J.G.; Miller, G.W. Nonmotor symptoms of Parkinson's disease revealed in an animal model with reduced monoamine storage capacity. *J. Neurosci.* **2009**, *29*, 8103–8113. [CrossRef]

126. Braak, H.; Del Tredici, K.; Rüb, U.; de Vos, R.A.; Jansen Steur, E.N.; Braak, E. Staging of brain pathology related to sporadic Parkinson's disease. *Neurobiol. Aging* **2003**, *24*, 197–211. [CrossRef]

127. De Pablo-Fernández, E.; Lees, A.J.; Holton, J.L.; Warner, T.T. Neuropathological progression of clinical Parkinson disease subtypes. *Nat. Rev. Neurol.* **2019**, *15*, 361. [CrossRef]

128. Iranzo, A.; Tolosa, E.; Gelpi, E.; Molinuevo, J.L.; Valldeoriola, F.; Serradell, M.; Sanchez-Valle, R.; Vilaseca, I.; Lomeña, F.; Vilas, D.; et al. Neurodegenerative disease status and post-mortem pathology in idiopathic rapid-eye-movement sleep behaviour disorder: An observational cohort study. *Lancet Neurol.* **2013**, *12*, 443–453. [CrossRef] [PubMed]

129. Ohno, Y.; Shimizu, S.; Tokudome, K.; Kunisawa, N.; Sasa, M. New insights into the therapeutic role of the serotonergic system in Parkinson's disease. *Prog. Neurobiol.* **2015**, *134*, 104–121. [CrossRef] [PubMed]

130. Wilson, H.; Dervenoulas, G.; Pagano, G.; Koros, C.; Yousaf, T.; Picillo, M.; Polychronis, S.; Simitsi, A.; Giordano, B.; Chappell, Z.; et al. Serotonergic pathology and disease burden in the premotor and motor phase of A53T α-synuclein parkinsonism: A cross-sectional study. *Lancet Neurol.* **2019**, *18*, 748–759. [CrossRef]

131. Tan, S.K.; Hartung, H.; Sharp, T.; Temel, Y. Serotonin-dependent depression in Parkinson's disease: A role for the subthalamic nucleus? *Neuropharmacology* **2011**, *61*, 387–399. [CrossRef]

132. Haapaniemi, T.H.; Ahonen, A.; Torniainen, P.; Sotaniemi, K.A.; Myllylä, V.V. [$^{123}$I]β-CIT SPECT demonstrates decreased brain dopamine and serotonin transporter levels in untreated parkinsonian patients. *Mov. Disord.* **2001**, *16*, 124–130. [CrossRef]

133. Kim, S.E.; Choi, J.Y.; Choe, Y.S.; Choi, Y.; Lee, W.Y. Serotonin transporters in the midbrain of Parkinson's disease patients: A study with 123I-β-CIT SPECT. *J. Nucl. Med.* **2003**, *44*, 870–876.

134. Caretti, V.; Stoffers, D.; Winogrodzka, A.; Isaias, I.U.; Costantino, G.; Pezzoli, G.; Ferrarese, C.; Antonini, A.; Wolters, E.C.; Booij, J. Loss of thalamic serotonin transporters in early drug-naïve Parkinson's disease patients is associated with tremor: An [$^{123}$I]β-CIT SPECT study. *J. Neural Transm.* **2008**, *115*, 721–729. [CrossRef]

135. Roselli, F.; Pisciotta, N.M.; Pennelli, M.; Aniello, M.S.; Gigante, A.; De Caro, M.F.; Ferrannini, E.; Tartaglione, B.; Niccoli-Asabella, A.; Defazio, G.; et al. Midbrain SERT in degenerative parkinsonisms: A 123I-FP-CIT SPECT study. *Mov. Disord.* **2010**, *25*, 1853–1859. [CrossRef] [PubMed]

136. Qamhawi, Z.; Towey, D.; Shah, B.; Pagano, G.; Seibyl, J.; Marek, K.; Borghammer, P.; Brooks, D.J.; Pavese, N. Clinical correlates of raphe serotonergic dysfunction in early Parkinson's disease. *Brain* **2015**, *138*, 2964–2973. [CrossRef]

137. Kerenyi, L.; Ricaurte, G.A.; Schretlen, D.J.; McCann, U.; Varga, J.; Mathews, W.B.; Ravert, H.T.; Dannals, R.F.; Hilton, J.; Wong, D.F.; et al. Positron emission tomography of striatal serotonin transporters in Parkinson disease. *Arch. Neurol.* **2003**, *60*, 1223–1229. [CrossRef]

138. Guttman, M.; Boileau, I.; Warsh, J.; Saint-Cyr, J.A.; Ginovart, N.; McCluskey, T.; Houle, S.; Wilson, A.; Mundo, E.; Rusjan, P.; et al. Brain serotonin transporter binding in non-depressed patients with Parkinson's disease. *Eur. J. Neurol.* **2007**, *14*, 523–528. [CrossRef]

139. Albin, R.L.; Koeppe, R.A.; Bohnen, N.I.; Wernette, K.; Kilbourn, M.A.; Frey, K.A. Spared caudal brainstem SERT binding in early Parkinson's disease. *J. Cereb. Blood Flow Metab.* **2008**, *28*, 441–444. [CrossRef] [PubMed]

140. Politis, M.; Wu, K.; Loane, C.; Kiferle, L.; Molloy, S.; Brooks, D.J.; Piccini, P. Staging of serotonergic dysfunction in Parkinson's disease: An in vivo 11C-DASB PET study. *Neurobiol. Dis.* **2010**, *40*, 216–221. [CrossRef]

141. Politis, M.; Niccolini, F. Serotonin in Parkinson's disease. *Behav. Brain Res.* **2015**, *277*, 136–145. [CrossRef]

142. Mayeux, R.; Stern, Y.; Cote, L.; Williams, J.B. Altered serotonin metabolism in depressed patients with Parkinson's disease. *Neurology* **1984**, *34*, 642–646. [CrossRef] [PubMed]

143. Chen, C.L.H.; Alder, J.; Bray, L.; Kingsbury, A.; Francis, P.; Foster, O. Post-synaptic 5-HT$_{1A}$ and 5-HT$_{2A}$ receptors are increased in Parkinson's disease neocortex. *Ann. N. Y. Acad. Sci.* **1998**, *861*, 288–289. [CrossRef] [PubMed]

144. Fox, S.H.; Brotchie, J.M. 5-HT$_{2C}$ receptor binding is increased in the substantia nigra pars reticulata in Parkinson's disease. *Mov. Disord.* **2000**, *15*, 1064–1069. [CrossRef] [PubMed]

145. Huot, P.; Johnston, T.H.; Darr, T.; Hazrati, L.N.; Visanji, N.P.; Pires, D.; Brotchie, J.M.; Fox, S.H. Increased 5-HT$_{2A}$ receptors in the temporal cortex of parkinsonian patients with visual hallucinations. *Mov. Disord.* **2010**, *25*, 1399–1408. [CrossRef]

146. Frouni, I.; Kwan, C.; Belliveau, S.; Huot, P. Cognition and serotonin in Parkinson's disease. *Prog. Brain Res.* **2022**, *269*, 373–403. [PubMed]

147. Hesse, S.; Meyer, P.M.; Strecker, K.; Barthel, H.; Wegner, F.; Oehlwein, C.; Isaias, I.U.; Schwarz, J.; Sabri, O. Monoamine transporter availability in Parkinson's disease patients with or without depression. *Eur. J. Nucl. Med. Mol. Imaging* **2009**, *36*, 428–435. [CrossRef] [PubMed]

148. Halliday, G.M.; Blumbergs, P.C.; Cotton, R.G.; Blessing, W.W.; Geffen, L.B. Loss of brainstem serotonin- and substance P-containing neurons in Parkinson's disease. *Brain Res.* **1990**, *510*, 104–107. [CrossRef]

149. Halliday, G.M.; Li, Y.W.; Blumbergs, P.C.; Joh, T.H.; Cotton, R.G.; Howe, P.R.; Blessing, W.W.; Geffen, L.B. Neuropathology of immunohistochemically identified brainstem neurons in Parkinson's disease. *Ann. Neurol.* **1990**, *27*, 373–385. [CrossRef]

150. Kish, S.J.; Tong, J.; Hornykiewicz, O.; Rajput, A.; Chang, L.J.; Guttman, M.; Furukawa, Y. Preferential loss of serotonin markers in caudate versus putamen in Parkinson's disease. *Brain* **2008**, *131*, 120–131. [CrossRef]

151. Politis, M.; Wu, K.; Loane, C.; Quinn, N.P.; Brooks, D.J.; Oertel, W.H.; Björklund, A.; Lindvall, O.; Piccini, P. Serotonin neuron loss and nonmotor symptoms continue in Parkinson's patients treated with dopamine grafts. *Sci. Transl. Med.* **2012**, *4*, 128ra41. [CrossRef]

152. Del Tredici, K.; Rüb, U.; De Vos, R.A.; Bohl, J.R.; Braak, H. Where does Parkinson's disease pathology begin in the brain? *J. Neuropathol. Exp. Neurol.* **2002**, *61*, 413–426. [CrossRef]

153. Parkkinen, L.; Pirttila, T.; Alafuzoff, I. Applicability of current staging/categorization of alpha-synuclein pathology and their clinical relevance. *Acta Neuropathol.* **2008**, *115*, 399–407. [CrossRef] [PubMed]

154. Kovacs, G.G.; Klöppel, S.; Fischer, I.; Dorner, S.; Lindeck-Pozza, E.; Birner, P.; Bötefür, I.C.; Pilz, P.; Volk, B.; Budka, H. Nucleus-specific alteration of raphe neurons in human neurodegenerative disorders. *Neuroreport* **2003**, *14*, 73–76. [CrossRef] [PubMed]

155. Paulus, W.; Jellinger, K. The neuropathologic basis of different clinical subgroups of Parkinson's disease. *J. Neuropathol. Exp. Neurol.* **1991**, *50*, 743–755. [CrossRef]

156. Cheshire, P.; Ayton, S.; Bertram, K.L.; Ling, H.; Li, A.; McLean, C.; Halliday, G.M.; O'Sullivan, S.S.; Revesz, T.; Finkelstein, D.I.; et al. Serotonergic markers in Parkinson's disease and levodopa-induced dyskinesias. *Mov. Disord.* **2015**, *30*, 796–804. [CrossRef] [PubMed]

157. Mann, D.M.; Yates, P.O. Pathological basis for neurotransmitter changes in Parkinson's disease. *Neuropathol. Appl. Neurobiol.* **1983**, *9*, 3–19. [CrossRef] [PubMed]

158. Burke, S.; Trudeau, L.E. Axonal Domain Structure as a Putative Identifier of Neuron-Specific Vulnerability to Oxidative Stress in Cultured Neurons. *eNeuro* **2022**, *9*, ENEURO.0139-22.2022. [CrossRef]

159. Akhtar, R.S.; Stern, M.B. New concepts in the early and preclinical detection of Parkinson's disease: Therapeutic implications. *Expert. Rev. Neurother.* **2012**, *12*, 1429–1438. [CrossRef]

160. Darweesh, S.K.; Verlinden, V.J.; Stricker, B.H.; Hofman, A.; Koudstaal, P.J.; Ikram, M.A. Trajectories of prediagnostic functioning in Parkinson's disease. *Brain* **2017**, *140*, 429–441. [CrossRef]

161. Faivre, F.; Joshi, A.; Bezard, E.; Barrot, M. The hidden side of Parkinson's disease: Studying pain, anxiety and depression in animal models. *Neurosci. Biobehav. Rev.* **2019**, *96*, 335–352. [CrossRef]

162. Bové, J.; Perier, C. Neurotoxin-based models of Parkinson's disease. *Neuroscience* **2012**, *211*, 51–76. [CrossRef]

163. Koprich, J.B.; Kalia, L.V.; Brotchie, J.M. Animal models of α-synucleinopathy for Parkinson disease drug development. *Nat. Rev. Neurosci.* **2017**, *18*, 515–529. [CrossRef] [PubMed]

164. Volta, M.; Melrose, H. LRRK2 mouse models: Dissecting the behavior, striatal neurochemistry and neurophysiology of PD pathogenesis. *Biochem. Soc. Trans.* **2017**, *45*, 113–122. [CrossRef] [PubMed]

165. Bastías-Candia, S.; Zolezzi, J.M.; Inestrosa, N.C. Revisiting the Paraquat-Induced Sporadic Parkinson's Disease-Like Model. *Mol. Neurobiol.* **2019**, *56*, 1044–1055. [CrossRef] [PubMed]

166. Creed, R.B.; Goldberg, M.S. New Developments in Genetic rat models of Parkinson's Disease. *Mov. Disord.* **2018**, *33*, 717–729. [CrossRef]

167. Francardo, V. Modeling Parkinson's disease and treatment complications in rodents: Potentials and pitfalls of the current options. *Behav. Brain Res.* **2018**, *352*, 142–150. [CrossRef]

168. Deusser, J.; Schmidt, S.; Ettle, B.; Plötz, S.; Huber, S.; Müller, C.P.; Masliah, E.; Winkler, J.; Kohl, Z. Serotonergic dysfunction in the A53T alpha-synuclein mouse model of Parkinson's disease. *J. Neurochem.* **2015**, *135*, 589–597. [CrossRef]

169. Wihan, J.; Grosch, J.; Kalinichenko, L.S.; Müller, C.P.; Winkler, J.; Kohl, Z. Layer-specific axonal degeneration of serotonergic fibers in the prefrontal cortex of aged A53T α-synuclein-expressing mice. *Neurobiol. Aging* **2019**, *80*, 29–37. [CrossRef]

170. Kohl, Z.; Ben Abdallah, N.; Vogelgsang, J.; Tischer, L.; Deusser, J.; Amato, D.; Anderson, S.; Müller, C.P.; Riess, O.; Masliah, E.; et al. Severely impaired hippocampal neurogenesis associates with an early serotonergic deficit in a BAC α-synuclein transgenic rat model of Parkinson's disease. *Neurobiol. Dis.* **2016**, *85*, 206–217. [CrossRef]

171. Björklund, A.; Nilsson, F.; Mattsson, B.; Hoban, D.B.; Parmar, M.A. Combined α-Synuclein/Fibril (SynFib) Model of Parkinson-Like Synucleinopathy Targeting the Nigrostriatal Dopamine System. *J. Parkinsons Dis.* **2022**, *preprint*. [CrossRef]

172. Wan, O.W.; Shin, E.; Mattsson, B.; Caudal, D.; Svenningsson, P.; Björklund, A. α-Synuclein induced toxicity in brain stem serotonin neurons mediated by an AAV vector driven by the tryptophan hydroxylase promoter. *Sci. Rep.* **2016**, *6*, 26285. [CrossRef]

173. Luk, K.C.; Kehm, V.; Carroll, J.; Zhang, B.; O'Brien, P.; Trojanowski, J.Q.; Lee, V.M. Pathological α-synuclein transmission initiates Parkinson-like neurodegeneration in nontransgenic mice. *Science* **2012**, *338*, 949–953. [CrossRef] [PubMed]

174. Rey, N.L.; Steiner, J.A.; Maroof, N.; Luk, K.C.; Madaj, Z.; Trojanowski, J.Q.; Lee, V.M.; Brundin, P. Widespread transneuronal propagation of α-synucleinopathy triggered in olfactory bulb mimics prodromal Parkinson's disease. *J. Exp. Med.* **2016**, *213*, 1759–1778. [CrossRef]

175. Chung, H.K.; Ho, H.A.; Pérez-Acuña, D.; Lee, S.J. Modeling α-Synuclein Propagation with Preformed Fibril Injections. *J. Mov. Disord.* **2019**, *12*, 139–151. [CrossRef] [PubMed]

176. Jinsmaa, Y.; Cooney, A.; Sullivan, P.; Sharabi, Y.; Goldstein, D.S. The serotonin aldehyde, 5-HIAL, oligomerizes alpha-synuclein. *Neurosci. Lett.* **2015**, *590*, 134–137. [CrossRef] [PubMed]

177. Falsone, S.F.; Leitinger, G.; Karner, A.; Kungl, A.J.; Kosol, S.; Cappai, R.; Zangger, K. The neurotransmitter serotonin interrupts α-synuclein amyloid maturation. *Biochim. Biophys. Acta* **2011**, *1814*, 553–561. [CrossRef]

178. Tin, G.; Mohamed, T.; Shakeri, A.; Pham, A.T.; Rao, P.P.N. Interactions of Selective Serotonin Reuptake Inhibitors with β-Amyloid. *ACS Chem. Neurosci.* **2019**, *10*, 226–234. [CrossRef]

179. Cirrito, J.R.; Wallace, C.E.; Yan, P.; Davis, T.A.; Gardiner, W.D.; Doherty, B.M.; King, D.; Yuede, C.M.; Lee, J.M.; Sheline, Y.I. Effect of escitalopram on Aβ levels and plaque load in an Alzheimer mouse model. *Neurology* **2020**, *95*, e2666–e2674. [CrossRef]

**MDPI**

*Review*

# Synucleins: New Data on Misfolding, Aggregation and Role in Diseases

Andrei Surguchov [1],* and Alexei Surguchev [2]

[1] Department of Neurology, Kansas University Medical Center, Kansas City, KS 66160, USA
[2] Section of Otolaryngology, Department of Surgery, Yale School of Medicine, Yale University, New Haven, CT 06520, USA
* Correspondence: asurguchov@kumc.edu

**Abstract:** The synucleins are a family of natively unfolded (or intrinsically unstructured) proteins consisting of α-, β-, and γ-synuclein involved in neurodegenerative diseases and cancer. The current number of publications on synucleins has exceeded 16.000. They remain the subject of constant interest for over 35 years. Two reasons explain this unchanging attention: synuclein's association with several severe human diseases and the lack of understanding of the functional roles under normal physiological conditions. We analyzed recent publications to look at the main trends and developments in synuclein research and discuss possible future directions. Traditional areas of peak research interest which still remain high among last year's publications are comparative studies of structural features as well as functional research on of three members of the synuclein family. Another popular research topic in the area is a mechanism of α-synuclein accumulation, aggregation, and fibrillation. Exciting fast-growing area of recent research is α-synuclein and epigenetics. We do not present here a broad and comprehensive review of all directions of studies but summarize only the most significant recent findings relevant to these topics and outline potential future directions.

**Keywords:** α-, β- and γ-synuclein; Parkinson's disease; epigenetic; protein aggregation; SNARE-complex; protein trafficking; phytochemicals; synucleinopathies; methylation; histones

**Citation:** Surguchov, A.; Surguchev, A. Synucleins: New Data on Misfolding, Aggregation and Role in Diseases. *Biomedicines* **2022**, *10*, 3241. https://doi.org/10.3390/biomedicines10123241

Academic Editor: Natalia Ninkina

Received: 15 November 2022
Accepted: 7 December 2022
Published: 13 December 2022

**Publisher's Note:** MDPI stays neutral with regard to jurisdictional claims in published maps and institutional affiliations.

## 1. Introduction

Synucleins are a family of small, prone to aggregate intrinsically disordered proteins (IDP) implicated in neurodegenerative diseases and cancer. Since synucleins are involved in severe human diseases, understandably most research is directed at unveiling their role in pathology. As a result, we know much more about their contribution to pathological mechanisms than their normal functions, which are still not fully understood.

The first synuclein was identified using antiserum against purified cholinergic synaptic vesicles as a 143 amino acid presynaptic protein in the electric organ of *Torpedo californica* [1]. Later, two additional isoforms belonging to the synuclein family were identified [2]. Three evolutionary conserved members of this family are highly expressed in the vertebrate nervous system and have been found in all vertebrates [3]. Importantly, no counterparts of synucleins were identified in invertebrates, indicating that they are vertebrate-specific proteins. Further analysis demonstrates that the number of synuclein members may differ among vertebrates. While three genes encoding α-, β-, and γ-synuclein are present in mammals and birds, a variable number of synuclein genes and corresponding proteins are expressed in fish, depending on the species. For example, four synuclein genes are identified in fugu, encoding for α, β, and two γ (g1 and g2) isoforms, but three genes are detected in zebrafish [4,5].

IDPs constitute about one-third of the human proteome [6] pointing to the importance of their structural organization. The absence of a defined structure confers members of the synuclein family conformational flexibility, allowing them to participate in dynamic and transient molecular interactions.

## 2. Three Members of The Synuclein Family

The overwhelming majority of publications on synucleins are dedicated to α-synuclein due to its involvement in Parkinson's disease (PD) and other synucleinopathies. Yet, two other proteins β- and γ-synucleins belonging to the same family are localized predominantly in neuronal terminals and implicated in the long-term modulation and preservation of nerve terminal function and dopamine homeostasis [7]. As a result, they deserve more attention to better understand the functionality of nervous system and role in carcinogenesis.

### 2.1. Common Structure of Members of the Synuclein Family

As shown in Figure 1 synucleins have a substantial degree of amino acid sequence and overall structural similarities (Figure 1).

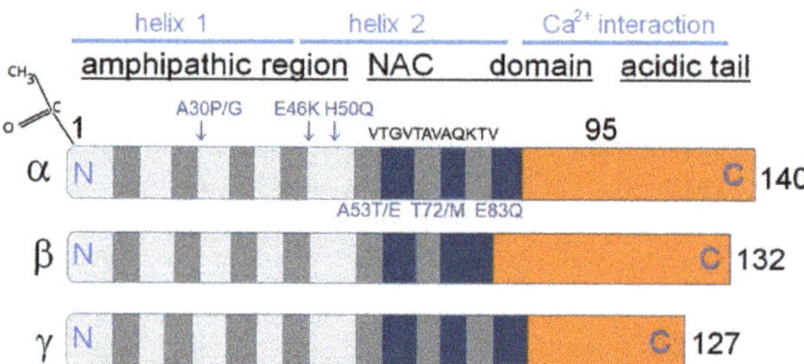


**Figure 1.** The core sections of the seven amino-terminal repeats (KTKEGV motive) are shown as dark grey bars in α and γ-synucleins. These motives are located between amino acids 7–87 of human α- and γ-synuclein. β-Synuclein has only six such repeats. Positively charged regions are shown in light gray, hydrophobic regions are blue, and negatively charged regions are light brown. In α-synuclein the non-amyloid-β component (NAC) is shown, containing amino acids 61–95, playing a key role in aggregation in vitro. The absence of this region in β-synuclein decreases its tendency to aggregate. Amino acids 71VTGVTAVAQKTV82 are the most essential for α-synuclein aggregation [8,9]; it is. This domain is partly absent in β-synuclein and to some extent conserved in γ-synucleins, which might clarify why both homologs of α-synuclein are not aggregation prone as α-synuclein. Some missense mutations predisposing to diseases ($A^{53}E$, $A^{53}T$, $A^{30}P$, $E^{46}K$, $H^{50}Q$, $T^{72}M$, and $E^{83}Q$) are displayed as blue letters above and under the α-synuclein image (Modified from [10]). At the upper left corner N-terminal acetylation of α-synuclein is shown, which slows down its aggregation and changes the morphology of the aggregates [11,12].

The members of the synuclein family mostly share a structurally similar N-terminal amino-acid sequence; α-synuclein supposedly forms a single or two distinct α-helices in this region. This transition is induced when α-synuclein interacts with the membranes via lipid bilayers. A C-terminal region contains negatively charged residues and prolines. The C-terminal domain differs in amino acid sequence among synucleins and regulates their solubility depending on its length and charge. A stretch of amino acids located between positions 61–95 in α-synuclein is called NAC (Figure 1), and a stretch of amino acids 71–82 within the α-synuclein NAC region is critical for β-sheet-rich aggregation in vitro [8,9]. This amino acid sequence is VTGVTAVAQKTV in α-synuclein, FSGAGNIAAATG in β-synuclein, and VSSVNTVATKTV in γ-synuclein. Amino acids belonging to the 71–82 stretch are necessary and sufficient for α-synuclein fibrillization [9].

Furthermore, the substitutions of amino acids (E35K + E46K + E61K = '3K') are render all synucleins considerably more toxic than their wild-type counterparts [9]. These substitutions increase α-synuclein-membrane interactions pointing to a mechanism of synuclein toxicity associated with membrane binding. Binding to lipids converts the N-terminal end to an α-helix. The N-terminal helicity negatively correlates with aggregation potential, while the C-termini differ among synucleins regulating their solubility.

Newberry et al. [13] used mutational scanning to reveal the association between its structure, activity and toxicity. They analyzed a library of protein missense variants for relative toxicity in competitive selection in yeast cells. In the headgroup region, mutations of Lys residues, which are expected to interact with acidic residues in anionic lipids, strongly decreased toxicity. Membrane binding of α-synuclein and the formation of a helix in the membrane-bound state of α-synuclein determine the level of α-synuclein toxicity. These results point to the importance of conserved lysine residues in membrane-bound α-synuclein for its toxicity. Such an approach allowed the generation of a high-resolution model for the structure and dynamics of the conformational state of α-synuclein that slows yeast growth [13].

### 2.2. Synuclein's Cellular Functions and Role in Pathology

α-Synuclein is an essential regulator of synaptic vesicle pool and trafficking, dopamine neuro-transmission, and other mechanisms involved in synaptic plasticity [14,15]. α-Synuclein performs these functions by assisting in the creation of soluble N-ethylmaleimide-sensitive-factor attachment protein receptor (SNARE)-complex [15–18]. α-Synuclein participates in chaperoning SNARE complex assembly through its interaction with vesicle-associated membrane protein 2 (VAMP2)/synaptobrevin-2 [14].

Contrary to α-synuclein, our knowledge about the role of β-synuclein and γ-synuclein in cellular functions and pathology is somewhat limited. Recent findings about β-synuclein point to its role as an important biomarker for the early stages of Alzheimer's disease (AD) [19]. The concentration of β-synuclein gradually increases in the cerebrospinal fluid beginning from the preclinical AD phase. It may be considered a promising biomarker of synaptic damage in this disease. β-Synuclein may be used as a CSF biomarker for synaptic damage in AD; its level is elevated in both dementia and pre-dementia stages of AD [19,20]. Importantly, higher CSF α-synuclein levels are reported in pre-AD subjects but not in MCI-AD and dementia AD [19], pointing to its specificity as a biomarker.

Other significant findings concerning β-synuclein and γ-synuclein have been described recently by Carnazza et al. [3]. These researchers have demonstrated that both β-synuclein and γ-synuclein have a lower affinity toward synaptic vesicles than α-synuclein. β-Synuclein and, to a lesser extent, γ-synuclein target to the synapse, despite a dramatically reduced ability to associate with membranes compared with α-synuclein. Formation of oligomers of β-synuclein or γ-synuclein with α-synuclein causes a reduction of synaptic vesicle binding of α-synuclein. These results indicate that β-synuclein and γ-synuclein are modulators of α-synuclein binding to synaptic vesicles. Therefore, they may reduce α-synuclein's physiological activity at the neuronal synapse. Another conclusion from these results is that members of the synuclein family can regulate each other's ability to bind to synaptic vesicles [3]. Inactivation of γ-synuclein affects psycho-emotional status and cognitive abilities. Moreover, in aging mice lacking γ-synuclein the dopaminergic dysfunction and altered working memory performance was demonstrated [21].

After oxidation on methionine and tyrosine located in neighboring positions, Met$^{38}$ and Tyr$^{39}$ γ-synuclein forms annular oligomers and can initiate α-synuclein aggregation [22].

Examination of synucleins in a non-human primate model of PD showed that the expression of all three members of the family was increased in the substantia nigra after treatment with prodrug to the neurotoxin-1-methyl-4-phenyl-1,2,3,6-tetrahydropyridine (MPTP). This increase correlates positively with the loss of cells and motor score [7]. The expression of the three types of synucleins was also increased in the dorsal raphe nucleus, but only α- and γ-synucleins are associated with the loss of cells and cell loss. The authors conclude that MPTP treatment mainly affects the expression level of α-synuclein and γ-synucleins and, to a lesser extent, β-synuclein levels and suggest that the three members of the synuclein family participate in the regulation of monoaminergic transmission [7].

In another study of the functional role of three members of the family, Ninkina et al. [23] found that the monoamine transporter 2–dependent dopamine uptake by synaptic vesicles of mice with β-synuclein knockout is considerably diminished. Introduction of β-synuclein recovers uptake by triple α/β/γ-synuclein–deficient striatal vesicles; however, α-synuclein or γ-synuclein do not cause such recovery. Interestingly, only β-synuclein potentiates VMAT-2–dependent uptake of dopamine and structurally similar molecules by synaptic vesicles. The authors propose that the increased presence of β-synuclein-triggered complexes at the synaptic vesicles explains the decreased sensitivity to MPTP toxicity of SNpc dopaminergic neurons in mice lacking α-synuclein and or γ-synuclein.

An association of α-synuclein and γ-synuclein with autism spectrum disorder (ASD) has recently been described. A significant reduction of plasma levels of α-synuclein and elevation of plasma γ-synuclein in children with ASD was found [24]. Plasma levels of both synucleins are significantly related to the severity of ASD and might serve as an indicator of the rigorousness of the disorder. In particular, the plasma level of γ-synuclein is lower in ASD patients than in controls, which the authors explain by disruption of the synaptic functions [24].

One of the possible mechanisms of γ-synuclein involvement in diseases may be associated with its role in the control of metabolic functions in fat cells. Rodríguez-Barrueco et al. [25] recently revealed details of the regulatory mechanism of fat mass expansion modulated by the micro-RNA [miR-424(322)/503] through γ-synuclein. This miRNA plays a crucial role in adipose tissue regulating impaired adipogenesis and fat mass via the control of γ-synuclein [25]. The authors identified and validated the direct influence of miR-424(322)/503 on the level of γ-synuclein expression and consider it a target gene in mediating the metabolic functions of adipocytes.

### 3. α-Synuclein Misfolding, Aggregation, and Fibrillation

Intracellular accumulation of prone-to-aggregate proteins associated with the formation of amyloid-like fibrils, is a common neuropathological feature of several neurodegenerative diseases. The "unfolding" of intrinsically disordered proteins is similar to the misfolding of globular proteins [26,27].

Under physiological conditions, α-synuclein in the cytoplasm exists in a dynamic equilibrium between soluble monomers and a variety of oligomers, including helically folded tetramers (Figure 2) [28–30].

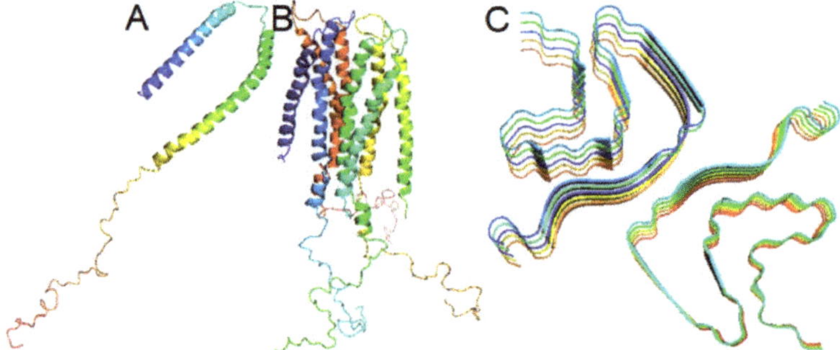


**Figure 2.** Micelle-bound, tetrameric, and fibrillar α-synuclein. (**A**)—α-Synuclein is disordered under physiological conditions in solution. Disordered form is in equilibrium with a minor α-helical tetrameric form in the cytoplasm (**B**) α-Synuclein is α-helical when bound to a cell membrane (**C**) α-Synuclein may form polymorphic amyloid fibrils with unique arrangements of cross-β-sheet motifs. From Korneev et al. [30].

Binding to lipid membranes is a key mediator of oligomerization and aggregation of α-synuclein. α-Synuclein-containing protein inclusions, aggregates, and fibrils are found in neuropathological inclusions in neurons and glia of patients with synucleinopathies [28]. Various mutations and genomic multiplications, including duplication and triplication of the α-synuclein gene cause familial PD with extramotor features [31].

It is generally supposed that α-synuclein may cause diseases via a toxic gain-of-function intrinsic for the normal protein when it surpasses a certain accumulation level. In agreement with this concept, α-synuclein knockout mice, in contrast to transgenic overexpressors, display no obvious neuropathological phenotype [14,32]. More detailed investigations reveal deficits in such null mice' dopaminergic system. The role of loss of α-synuclein functions in pathology due to the sequestration of physiological α-synuclein in deposits and inclusions cannot be completely excluded. It may damage some of the synaptic function of α-synuclein when its concentration falls below some critical level.

### 3.1. Post-Translational Modifications (PTMs) of α-Synuclein

Various PTMs regulate α-synuclein susceptibility to aggregation by changing its conformation. PTMs such as phosphorylation, ubiquitination, acetylation, nitration, SUMOylation, etc., modify α-synuclein physicochemical properties and affect the aggregation process in synucleinopathies [11,12,29,31–34]. α-Synuclein is acetylated constitutively at its N-terminal (Figure 1). As Bell et al. recently demonstrated [11,12] this PTM changes the charge and spatial structure of α-synuclein and affects its aggregation and interaction with lipid membranes. Even subtle disturbances caused by PTMs as well as mutations, can lead to remarkable alterations in the aggregation performance of this protein [11,12].

In addition to PTMs, many other factors are able to induce aggregation-prone conformations of α-synuclein such as focal changes in pH or $Ca^{2+}$ concentration. For instance, $Ca^{2+}$ binding to α-synuclein C-terminus induces N-terminal unfolding and aggregation-prone conformations [35]. At low pH, the large net negative charge at the C-terminus is reduced, diminishing charge-charge intramolecular repulsion. This change causes conversion of α-synuclein to partially unfolded conformation. Proteolytic truncation of α-synuclein C-terminus also accelerates its aggregation and increases toxicity. The C-terminus determines α-synuclein cytotoxicity and aggregation by regulating binding to membranes and chaperones [36].

### 3.2. Approaches to Reduce α-Synuclein Toxicity

One approach to neutralizing α-synuclein toxicity, which offered a lot of promise, has been targeting the pathologically aggregated form of α-synuclein. Since aggregated protein plays an essential role in synucleinopathies' pathogenesis, many attempts have been made to reduce α-synuclein accumulation and aggregation. One method has been based on the antibody specific to the aggregated form of α-synuclein [37,38]. Recently, the monoclonal antibody prasinezumab, directed at aggregated α-synuclein, has been studied as a potential treatment for PD (Pasadena trial) [39]. In a phase 2 double-blind trial, patients with earlyPD were treated by intravenous placebo or prasinezumab. The trial results were similar in the treatment groups and the placebo group indicating that prasinezumab therapy had no meaningful effect on global or imaging measures of PD progression as compared with placebo. Lang et al. [40] tested another monoclonal antibody—cinpanemab (SPARK trial) using a similar trial design with the same disappointing result indicating that both potential antibodies have no implications for current practice.

However, these negative and discouraging results should not be taken wrongly as proof that targeting α-synuclein is an unsuccessful approach. The negative result may be explained by several reasons. For example, it may be due to the failure of antibodies to enter the brain in sufficient amounts for a noticeable therapeutic effect. There are reasons to believe that antibodies that are delivered intravenously do not usually pass through intact cell membranes, including those of the blood-brain barrier [37]. Another explanation of the absence of therapeutic efficacy of these antibodies may be inability of antibodies to traverse cell membranes and penetrate intraneuronal space or cross the surface of exosomes. Thus, α-synuclein may be just inaccessible to antibodies. Potential tactics to overcome limitations of antibody therapies for synucleinopathies may be based on enhancing blood–brain barrier penetration with technologies such as magnetic resonance-guided focused ultrasound [41]. An alternative approach is developing modified antibodies that will be able to penetrate cell membranes [42]. Therefore, despite the failure of these clinical trials, there is still a chance that targeting α-synuclein will open an approach for molecular therapy for PD [37]. In other words, these results of clinical trials should not dismiss the possibility that successful results may yet be attained with the same or analogous antibody in prodromal PD or in genetic forms of this disorder. Alternative tactics to affect aggregated α-synuclein may be also beneficial [38].

### 3.3. The Quaternary Structure of α-Synuclein Fibrils Modulates α-Synuclein Pathology

Frieg et al. [43] amplified α-synuclein fibrils from brain extracts of patients with pathologically confirmed PD and multiple system atrophy, determined their 3D structures by Cryo-EM, and evaluated their potential to seed α-synuclein related pathology in oligodendrocytes. In the first step, researchers fibrilized α-synuclein by protein misfolding cyclic amplification (PMCA) (see Section 3) using brain extracts of patients. This approach showed that α-synuclein aggregated by incubation with homogenates from the brain tissues of these patients determined the fibril structures. Then, the authors analyzed fibrils by Cryo-EM and other methods and characterized their seeding potential in mouse primary oligodendroglial cultures. This examination showed distinct 3D structures and quaternary arrangement of α-synuclein fibrils.

The data demonstrate that the quaternary structure of α-synuclein fibrils is a vital factor in the seeding of pathology. Investigation of α-synuclein fibrils by Cryo-EM has shown that the dominant fibril morphology is rod polymorph [44]. Furthermore, the amplified α-synuclein fibrils differed in their inter-protofilament interface and their ability to recruit endogenous α-synuclein depending on the origin of the patient's homogenate. The fibrils were also different in their adopted helical arrangement and N-terminal region. Thus, the quaternary structure of α-synuclein fibrils modulated α-synuclein pathology inside recipient cells.

Interestingly α-synuclein monomers, oligomers and fibrils exert a differential effect on the folding and refolding of other proteins. Bagree et al. [45] examined their influence on the conformation, enzymatic activity and other properties of firefly luciferase. α-Synuclein monomers delayed the inactivation of luciferase under thermal stress conditions and enhanced the spontaneous refolding, whereas oligomers and fibrils adversely influenced luciferase activity and refolding. The oligomers suppressed spontaneous refolding, while fibrils caused a pronounced effect on the inactivation of native enzyme. Thus, various conformers of α-synuclein exert differential effect on structural modifications and misfolding of other proteins, regulating protein homeostasis.

## 4. Synuclein-Based Methods of Disease Diagnosis

Since α-synuclein misfolding and aggregation precede major clinical manifestations, detecting misfolded and aggregated α-synuclein would allow recognition of the disease at the earliest premotor phases. Further accrual and spreading of these pathological isoforms of α-synuclein usually correlate with clinical symptoms and progression of synucleinopathies. Recent findings point to a mechanism of template-mediated amplification of amyloid-like fibrils as a base of accumulation and intracellular propagation of fibril seeds by which pathological features spread along the neural circuits in the brain. Abnormal accumulation of α-synuclein [15,46], and in some cases of γ-synuclein [47,48], occurs in synucleinopathies [49]. Although relatively small α-synuclein oligomers are considered the most toxic species, it is not exactly known which α-synuclein assemblies possess the most toxic properties [26] and are the most relevant to human diseases. Accumulating data indicate that smaller-size aggregates cause a higher toxic response than filamentous aggregates (fibrils) [50]. The application of fluorophores for detecting various α-synuclein subtypes brings the hope that the initial steps of aggregation and the structure of intermediates will be better understood. A commercial aggregate-activated fluorophore Amytracker 630 (AT630) possesses photophysical properties that enable high-resolution (∼4 nm) imaging of α-synuclein and other aggregation-prone proteins. Using fluorophore AT630, the authors found that aggregates smaller than $450 \pm 60$ nm easily penetrated the plasma membrane. An important conclusion from this study is that aggregates in different synucleinopathies, i.e., PD and DLB have different potency in toxicity. Moreover, there was a linear relationship between α-synuclein aggregate size and cellular toxicity. This novel method allows for the measuring of toxic protein aggregates in biological environments quantitatively and therefore brings a better understanding of disease mechanisms under physiological conditions [50].

Sekiya et al. [51] used proximity ligation assay—a new technique to detect the distribution of α-synuclein oligomers, compare the results with immunohistochemical data and analyze the correlation between the presence of oligomers with clinical features. The results show that α-synuclein oligomers are more widespread than Lewy-related pathology (LRP, Lewy bodies, and Lewy neurites) and that α-synuclein oligomers in the hippocampus correlate with cognitive impairment. Moreover, there was discordance between the distribution of α-synuclein oligomers and LRP. In about one-half of patients, LRP were not found in the neocortex, but at least some of the α-synuclein oligomers were detected in the neocortex of all patients. More α-synuclein oligomers accumulated in the neuropil than in neurons in most brain regions. Furthermore, α-synuclein oligomers may be widespread early in the disease stage. These results may explain (at least partially) the recent failure of anti-α-synuclein therapy [52]. The unsuccessful attempts of this cure may be due to the widespread distribution of α-synuclein oligomers in earlier pathological stages of the disease. To be successful, the effective treatment of PD patients with the anti-α-synuclein approach should begin at very early PD stages or even in a preclinical phase of the disorder. Both of these stages are hard to identify based on contemporary diagnostic methods. The toxic properties of α-synuclein are intensified by cell-to-cell spread and aggregation of endogenous species in newly invaded cells [53].

Recent studies indicated an essential role of low-density lipoprotein receptor-related protein 1 (LRP1) in α-synuclein spreading between cells [54]. LPR1 is a highly expressed central nervous system protein receptor involved in intracellular signaling and endocytosis. It is located predominantly in the plasma membrane. Chen et al. [54], using LRP1 knockout (LRP1-KO) induced pluripotent stem cells (iPSCs), demonstrated that LRP1 is a vital regulator of α-synuclein neuronal uptake and mediator of its spread in the brain. LRP1 binds to the N-terminus of α-synuclein through lysine residues, being a key regulator for the endocytosis of both monomeric and oligomeric forms of α-synuclein. The new findings of LRP1's role in α-synuclein trafficking point to a potential novel target for synucleinopathies treatment.

Ozdilek and Agirbasli [55] recently demonstrated that the serum LRP1 concentration is associated with the factors influencing the prognosis of PD and disease duration and severity. Moreover, there is a correlation between LRP1 levels and abnormal aggregation of α-synuclein and other IDP between sLRP1 levels and the duration of disease [55].

Iba et al. [56] investigated the impact of aging and inflammation on the pathogenesis of synucleinopathies. They used a mouse model of DLB/PD induced by intrastriatal injection of α-synuclein preformed fibrils (PFF). Aging caused more extensive α-synuclein accumulation in the striatum and amygdala, more significant infiltration of T cells, microgliosis, and behavioral deficits. Transcriptomics analysis revealed that this network's most important upstream regulators comprised CSF2, LRG, TNFa and poly rl:rC-RNA. These results point to a key role in aging-related inflammation. More specifically, this data shows that CSF2 affects the consequences of α-synuclein spreading and suggests that targeting neuro-immune responses might be a step in developing treatments for DLB/PD. These results are also consistent with the perception that α-synuclein aggregates might lead to neurodegenerative changes by dysregulating adaptive and innate immune responses.

Various factors affect α-synuclein aggregation and fibrillation [26,27,34,44,57–59], including the inhibitory effect of β-synuclein on α-synuclein aggregation and toxicity [60,61]. In addition to this, a growing number of recent studies point to the role of defects in proteostasis, autophagy, and lysosomal function in α-synuclein intracellular accumulation and clearance. For example, impairment of lysosomal function due to mutation of PARK9 (a lysosome-related transmembrane P5 ATP transport enzyme, ATP13A2) affects the metabolism of $\alpha$-synuclein. It causes the development of PD and other synucleinopathies [62].

$\alpha$-Synuclein forms distinct spatial structures or strains at the very early stages of synucleinopathies, which are potential targets for early diagnosis and treatment. These molecular forms can be detected by various modern techniques, such as seed amplification assays (s) or protein misfolding cyclic amplification (PMCA) [63]. The combination of seed amplification assay with oligomers-specific ELISA (enzyme-linked immunosorbent assay) allows the identification of patients with PD or DLB with high sensitivity and specificity. Moreover, the methods provide vital information about the patient's clinical disease stage and the severity of the PD clinical symptoms [64].

Recent modifications of the PMCA identify the distinct $\alpha$-synuclein strains specific for different human diseases. They are successfully used for differential diagnosis of patients with PD, multiple system atrophy, DLBs and other disorders using samples of CSF, olfactory mucosa, submandibular gland biopsies, skin, and saliva even at the prodromal stages of the disease [63].

Results of a recent study by Emin et al. [65] point to the most toxic types of synuclein aggregates. The authors developed a modification of $\alpha$-synuclein aggregation reactions in the test tube with subsequent separation of aggregates by size. Wild-type $\alpha$-synuclein aggregates smaller than 200 nm in length induced inflammation and permeabilization of single-liposome membranes, whereas larger in-size aggregates were less toxic. These initial results were confirmed by characterization of soluble aggregates extracted from post-mortem brains. Aggregates from the brain had a similar size and structure as the small aggregates made in aggregation reactions in vitro. Further experiments demonstrated that the soluble aggregates from PD patients' brains were smaller and more inflammatory than the large aggregates in control brains. The authors conclude that the small non-fibrillar $\alpha$-synuclein aggregates are the critical species causing neuroinflammation and disease progression [65].

## 5. Aggregation of $\beta$-Synuclein and $\gamma$-Synuclein and Their Role in Diseases

$\alpha$-Synuclein is more aggregate-prone than $\beta$-synuclein and $\gamma$-synuclein which can be explained by differences in amino acid sequences and the intra-chain dynamics [66]. $\beta$-Synuclein is a non-amyloidogenic homolog of $\alpha$-synuclein and a natural negative regulator of $\alpha$-synuclein aggregation [60,61,67,68]. Biere et al. [69] show that while $\alpha$-synuclein forms the fibrils over time, no fibrillation could be detected for $\beta$- and $\gamma$-synuclein under the same conditions. However, under conditions that significantly speed up aggregation, $\gamma$-synuclein can form fibrils with a lag phase approximately three times slower than $\alpha$-synuclein [69]. Importantly, $\beta$-synuclein acts as a molecular chaperone elongating the lag phase of $\alpha$-synuclein aggregation [70]. $\beta$-Synuclein is capable of delaying or preventing $\alpha$-synuclein fibril formation both in vitro and in vivo.

$\beta$-synuclein may be responsible for certain forms of neurodegenerative diseases. For example, the $P^{123}H$ and $V^{70}M$ mutations in $\beta$-synuclein enhanced lysosomal pathology, play a causative role in neurodegeneration and are associated with dementia with Lewy bodies (DLB) [71,72]. In contrast to $\alpha$-synuclein, two other members of the synuclein family are not found in Lewy pathology [73].

$\beta$-Synuclein level is significantly elevated in the CSF of the pre-AD continuum since the preclinical stage. An increase of $\beta$-synuclein level in CSF occurs when t-tau and nerve fiber layer (NFL) levels are not yet significantly elevated. Therefore, it represents an emerging biomarker of synaptic damage in this disorder, which may be helpful even at the preclinical stage [19].

Due to the low level of folding, the conformation of synuclein family members is highly dynamic and they can modulate each other's aggregation propensity. Elevated temperature, low pH, and high concentrations intensify the fibrillation rate of $\alpha$- and $\gamma$-synuclein, while $\beta$-synuclein forms fibrils only at low pH. The high molar ratio of $\beta$-synuclein inhibits the fibrillation in $\alpha$- and $\gamma$-synuclein, but preformed fibrils of $\beta$- and $\gamma$-synuclein do not affect fibrillation of $\alpha$-synuclein [74].

Remarkably, increasing β-synuclein and γ-synuclein affinity to the membrane by point mutations converts these members of the synuclein family into monomers associated with the membrane, increases their cytotoxicity, and predisposes them to form round cytoplasmic inclusions [9].

The presence of γ-synuclein-positive inclusions in motor neurons of amyotrophic lateral sclerosis patients and its ability to aggregate in vitro and in vivo [47,48,75,76] suggest that a term "γ-synucleinopathy" may also be used.

Williams et al. [77] used a series of chimeras of α-synuclein and β-synuclein to probe the relative input of the N-terminal, C-terminal, and the central NAC domains to the inhibition of α-synuclein aggregation. Measurements of the rates of α-synuclein fibril formation in the presence of the chimeras indicate that the NAC is the primary determinant of self-association leading to fibril formation. However, the N- and C-terminal domains play critical roles in the fibril inhibition process. These findings provide proof that all three domains of β-synuclein together contribute to providing effective inhibition. Knowledge about such multi-site inhibitory interactions spread over the length of synuclein chains may be essential for the development of therapeutics mimicking the inhibitory effects of β-synuclein.

Interestingly, α-synuclein fibrils made in the presence of β-synuclein are less cytotoxic and possess decreased cell seeding capacity. Moreover, they are more resistant to fibril shedding compared to α-synuclein fibrils alone [23,78]. NMR studies showed that the overall structure of the core of α-synuclein when co-incubated with β-synuclein is minimally perturbed. However, the dynamics of Lys and Thr residues in the imperfect KTKEGV repeats of the α-synuclein N-terminus are increased. Thus, amyloid fibril dynamics play a crucial role in modulating synuclein toxicity and seeding.

Interestingly, β-synuclein is able to potentiate synaptic vesicle dopamine uptake and rescue dopaminergic neurons from MPTP-toxic effect in the absence of other synucleins [23]. Potentiation of dopamine uptake by β-synuclein may be explained by different protein architecture of the synaptic vesicles. These results demonstrate that of the three members of the synuclein family, only β-synuclein can potentiate VMAT-2–dependent uptake of dopamine and structurally similar molecules by synaptic vesicles. The authors hypothesize that the elevated level of β-synuclein-triggered complexes at the synaptic vesicles, and not the absence of other synucleins *per se*, causes the reduced sensitivity to MPTP toxicity in mice lacking α-synuclein and/or γ-synuclein. These results point to a critical role of β-synuclein in the resistance to MPTP toxicity by dopaminergic neurons lacking the usual balance of this protein and other members of the synuclein family [23].

## 6. Inhibitors of α-Synuclein Aggregation and Fibrillation as Potential Tools for Therapy

α-Synuclein has been identified as a key target for the development of therapeutic approaches to synucleinopathies given its central role in the pathophysiology of these diseases. Treatment strategies can be classified into several groups [79]: (1) inhibition of formation of toxic α-synuclein aggregates (aggregation inhibitors), (2) lowering α-synuclein expression (e.g., antisense therapy), (3) removal of toxic α-synuclein aggregates (immunotherapy), (4) removal of toxic forms of α-synuclein by enhancement of cellular clearance mechanisms (autophagy, lysosomal microphagy), (5) modulation of neuroinflammatory processes and (6) neuroprotective strategies. Current therapeutic approaches and elucidation of promising future treatment approaches are described in a recent article [79].

The results of recent experiments suggest that therapeutic downregulation of α-synuclein expression in PD patients may be a mostly harmless choice since it should not lead to adverse side effects on the functions of the nigrostriatal system [22].

High-throughput screening assays are efficient methods usually used to analyze a huge number of potential inhibitors of α-synuclein aggregation and to identify among them the most potent compounds. Kurnik et al. [80] used a combination of SDS-stimulated α-synuclein aggregation with Förster resonance energy transfer (FRET) to characterize the initial stages of α-synuclein aggregation. After screening 746.000 compounds, the authors identified

58 hits that markedly inhibit α-synuclein aggregation and reduce membrane permeabilization activity. The most effective aggregation inhibitors are derivatives of (4-hydroxy-naphthalen-1-yl) sulfonamide. They interact with the N-terminal part of monomeric α-synuclein with high affinity and reduce early-stage oligomer-membrane interactions. Notably, some of them also reduce α-synuclein toxicity toward neuronal cell lines.

Several approaches have been used to prevent or diminish the toxic effects of synuclein aggregation and fibrillation. One of them includes reducing α-synuclein expression, for example, by antisense oligonucleotides or nucleic acids. Other methods are based on the inhibition of toxic α-synuclein aggregates build-up using aggregation inhibitors. Some strategies are directed to dissolve and eliminate the existing intracellular or extracellular toxic α-aggregates, for instance, by immunotherapy. Various forms of α-synuclein induce an immune response including inflammation, highlighting the immune function of α-synuclein. Aggregation may render a protein "foreign" to the immune system and cause immunoresponce [81].

To reduce the toxic effect of α-synuclein, methods based on the enhancement of cellular clearance mechanisms removing toxic α-synuclein aggregates are used more often. Some of them are directed to the activation of autophagy or lysosomal microphagy.

An example of an approach based on the activation of autophagy gives hope for future therapeutic applications [82]. For instance, a small molecular inhibitor for prolyl oligopeptidase, KYP-2047, relieves α-synuclein-induced toxicity in various PD models by inducing autophagy and preventing α-synuclein aggregation, without reducing soluble α-synuclein oligomers. The authors hypothesize that KYP-2047 decreases the concentration of those toxic forms of α-synuclein that were not detected by ELISA assay specific for soluble α-synuclein. These findings indicate that when considering potential drug treatment for synucleinopathies, it is critical to characterize the α-synuclein aggregation process in detail, identifying the most toxic forms of α-synuclein. The authors also assume that previous failure to use α-synuclein aggregation inhibitors may be explained by the fact that removal of random α-synuclein aggregates forms does not provide neuroprotection [82].

Other methods, including modulation of neuroinflammatory processes and neuroprotective strategies, are currently under development. Since there is currently no method that guarantees the complete elimination of α-synuclein toxicity, new approaches and novel substances discussed below are currently under investigation.

A promising methodology for the development of new amyloid inhibitors is recently described in Murray et al. article [44]. The authors used a de novo structure-based computational design to create 35–50 residue amyloid capping inhibitors that bind to the growing ends of amyloid fibrils. The principle of the method is not restricted to α-synuclein, but can be applied to tau, beta-amyloid fibrils, and other fibril-forming proteins. Their structure was determined using Cryo-EM, ssNMR spectroscopy, and MicroED. The designed inhibitors can bind to the ends of fibrils, "capping" them and thus preventing further growth, seeding, and reducing the toxicity of such proteins both in vitro and in vivo. Since such inhibitors specifically bind to an end of an amyloid fibril, averting its elongation, other end of the fibril might be able of growing. However, analysis of this growth has shown that fibrils grow mainly unidirectionally, avoiding this issue. Importantly, inhibitors that decrease α-synuclein primary aggregation are also effective at decreasing cellular seeding.

Another promising approach to finding specific inhibitors of α-synuclein toxic species is the use of aptamers binding with high specificity to different truncated forms of α-synuclein fibrils with no cross-reactivity toward other amyloid fibrils [83]. Testing a panel of aptamers allowed to identify two of them (Apt11 and Apt15) possessing higher affinity to most C-terminally truncated forms of α-synuclein fibrils. Apt11 inhibited α-synuclein-seeded aggregation in vitro. Apt11 decreased the insoluble phosphorylated form of α-synuclein at Ser$^{129}$ (pS$^{129}$-α-syn) in a cell model. This aptamer inhibited α-synuclein aggregation assessed by RT-QuIC reactions seeded with brain homogenates isolated from PD patients. These aptamers may be used as tools for research and diagnostics or therapy.

Natural compounds also possess antiaggregational potential. Many phytochemicals inhibit α-synuclein aggregation and possess a neuroprotective effect [84–86]. For example, a polyphenolic component derived from green tea—epigallocatechin 3-gallate (EGCG) binds to α-synuclein by hydrophobic interactions and suppresses its aggregation at 100 nM in a concentration-dependent manner [85].

Curcumin is a biphenolic phytochemical compound from the root of *the Curcuma longa* efficiently inhibits α-synuclein from turning into an amyloid [87]. Curcumin prevents aggregation of both wild-type and mutant forms ($E^{46}K$ and $H^{50}Q$) of α-synuclein (Figure 1). Moreover, curcumin destabilizes preformed α-synuclein amyloid aggregates and regulates α-synuclein amyloid formation [86]. Interestingly, that curcumin induces hypermethylation of γ-synuclein gene through the upregulation of the DNA methyltransferase 3 (DNMT3) which reduces γ-synuclein expression [87]. These results show that polyphenol compounds may have therapeutic implications in PD treatment [88].

In addition to many inhibitors of synuclein aggregation, several substances enhancing α-synuclein aggregation (proaggregators) have been identified. For example, several phenyl-benzoxazol compounds, e.g., 2-arylbenzoxazole (PA86, 1) enhance the α-synuclein aggregation [80]. In another study TKD150 and TKD152 were investigated as proaggregators for α-synuclein. In comparison to a previously reported proaggregator, PA86, these two identified compounds were able to promote aggregation of α-synuclein at twice the rate [58]. These studies are important to better understand the intimate mechanism of α-synuclein aggregation.

## 7. Conclusions and Future Directions

There are several directions in synucleins studies which definitely will bring better understanding of their normal functions, association with diseases and open new strategies for the early diagnosis and timely treatment of synucleinopathies. Although the majority of investigations are still directed to α-synuclein, the functional role of two other members of the family is becoming clearer. The existing data from the literature on synucleins is descriptive, and a transition from observational to interventional study designs is currently in need. It is evident that this conversion may be slow, but hopefully it would lead to a breakthrough due to the development of multiomics and other novel research approaches.

Currently, many studies are directed to the understanding of the role of α-synuclein in the synaptic region and dopamine vesicle regulation. At the same time, its interaction with other cellular components remains poorly understood and many intriguing questions concerning synuclein functions remain unclear. For example, does α-synuclein interact with mitochondrial DNA directly and influence transcription profile? What role synucleins play in endoplasmic reticulum-Golgi traffic [89]? Recent studies investigating microtubule dynamics showed that α-synuclein facilitated the formation of short microtubules and favorably binds to 4 protofilaments. However, synuclein's role in the axonal transport and interaction with microtubules remains to be determined [90].

One of the critical directions in future work is the development of α-synuclein based biomarkers. Timely identification of PD symptoms should allow the beginning of early treatment.

### 7.1. Importance of Easily Accessible Samples for Diagnosis

Relevant biomarkers for α-synuclein pathology in PD are emerging, as well as blood-based markers of general neurodegeneration and glial activation [91,92]. Detection of misfolded α-synuclein in easily accessible fluids, e.g., blood, CSF, and skin would be a great step forward for early diagnosis of PD and other synucleinopathies [91–93]. Preliminary results indicate that skin biopsies, including nerve terminals, can be used in seeding aggregation assays to detect misfolded α-synuclein in PD, DLB, and multiple system atrophy [94]. Other potential peripheral tissues for detecting misfolded α-synuclein include the olfactory mucosa and the submandibular gland [63,94]. They can be used for both early diagnosis and also for the analysis of medication's efficiency. Biomarkers that can measure relevant treatment effects downstream of the drug target will likely be used more frequently to optimize go/no-go decisions on moving drugs forward. In the future, such biomarkers will be used in clinical routine practice to monitor the efficacy of disease-modifying therapies in individual patients.

### 7.2. Role of the Gut-Brain Axis

Recent findings point to a highly complex relationship between the gut and the brain in PD, providing the potential for the development of new biomarkers and therapeutics. The neuropathological process that leads to PD seems to start in the enteric nervous system or the olfactory bulb and spreads via rostro-cranial transmission to the substantia nigra and further into the CNS. This raises the exciting possibility that environmental substances can trigger pathogenesis [94,95]. The discovery of importance of the gut-brain axis in PD development opens a new approach for the search of therapeutic targets and biomarkers detection [95,96]. However, there are definite gaps in our knowledge and new research is needed for a better understanding of the role and interrelationships of α-synuclein with the vagus nerve, with the enteric nervous system, altered intestinal permeability, and inflammation [96,97].

Reduced activity of protein degradation systems, such as proteasomes, autophagic systems and lysosomes occur in patients with synucleinopathies. Therefore, the activation of autophagy gives hope for future therapeutic applications [36,79,82,83]. There are approaches that are not aimed at α-synuclein in a straight line but affect it indirectly. Autophagy is critically related to the anomalous accumulation of α-synuclein, and defects in lysosome-related transmembrane ATP transport enzyme ATP13A2 (also known as PARK9) impair lysosomal function and inhibit the toxicity of α-synuclein. Therefore, this enzyme may be considered a possible target in synucleinopathies [36,62]. Other autophagy activation methods also give hope for future therapeutic applications [82].

One of the promising agents causing the reduction of α-synuclein oligomer accumulation is GNE-7915, specific inhibitor of the long-term brain-penetrant leucine-rich repeat kinase 2 (LRRK2 or PARK8). This inhibitor significantly decreases oligomers in the striatum without causing adverse peripheral effects [97]. Inhibition of mutant LRRK2 hyper-kinase activity to normal physiological levels offers an encouraging and safe disease-modifying therapy to improve the course of synucleinopathies.

Small molecular inhibitors for prolyl oligopeptidase (PREP), for example, KYP-2047, relieve α-synuclein-induced toxicity, enhancing autophagy and preventing α-synuclein aggregation [83].

KYP-2047 had a significant positive impact on total oligomeric α-synuclein and particularly on proteinase K resistant α-synuclein species. There is hope that further modifications of the method will allow KYP-2047 and similar inhibitors to improve brain penetration based on the new delivery strategies. These strategies may include the fusion of effective inhibitors with a cell-penetrating peptide-based tag or conjugation to other types of delivery constructs such as brain-penetrating nanoparticles. The inhibitor designs are also genetically encodable, including viral delivery, another possible transport mechanism [82].

There is hope that scientific progress in this field will translate to the successful use of new diagnostic, prognostic, and therapeutic approaches that will ultimately improve patients' lives.

**Author Contributions:** A.S. (Andrei Surguchov) wrote the manuscript, performed the analyses and interpretation of data, gave a final approval of the version to be published, and agreed to be accountable for all aspects of the work. A.S. (Alexei Surguchev) participated in writing and editing of the manuscript and interpretation of data, made substantial contributions to the conception of the work, gave a final approval, and agreed to be accountable for all aspects of the work. All authors have read and agreed to the published version of the manuscript.

**Funding:** Some of the work by A.S. (Andrei Surguchov) was conducted at the Kansas City VA Medical Center, Kansas City, MO, United States, with support from VA Merit Review grants 1I01BX000361 and the Glaucoma Foundation grant QB42308. A.S. (Alexei Surguchev) is partially supported by YALE ENT Research grant # YD000220. The authors would like to thank Marina Nikolaev for advice and technical assistance with illustrations.

**Institutional Review Board Statement:** Not applicable.

**Informed Consent Statement:** Not applicable.

**Data Availability Statement:** Not applicable.

**Conflicts of Interest:** The authors declare no conflict of interest.

## Abbreviations

AD—Alzheimer's diseases; ATP13A2—a lysosome-related transmembrane P5 ATP transport enzyme; CryoEM—cryo-electron microscopy; CSF—cerebrospinal fluid; CSF2—colony-stimulating factor 2; DLB—dementia with Lewy bodies; DNAm—DNA methylation; EGCG—epigallocatechin 3-gallate; ELISA—enzyme-linked immunosorbent assay; IDP—intrinsically disordered proteins, iPSCs—induced pluripotent stem cells; LRP—Lewy-related pathology; LRG—lipopolysaccharide related genes; MPTP -1-methyl-4-phenyl-1,2,3,6-tetrahydropyridine; NAC—non-amyloid-β component; PD—Parkinson's disease; PFF—preformed fibrils. SNARE—soluble N-ethylmaleimide-sensitive factor attachment protein receptors; ssNMR—solid-state NMR; TNFα—Tumor Necrosis Factor-α.


## References

1. Maroteaux, L.; Campanelli, J.T.; Scheller, R.H. Synuclein: A neuron-specific protein localized to the nucleus and presynaptic nerve terminal. *J. Neurosci.* **1988**, *8*, 2804–2815. [CrossRef]
2. Jakes, R.; Spillantini, M.G.; Goedert, M. Identification of two distinct synucleins from human brain. *FEBS Lett.* **1994**, *345*, 27–32. [CrossRef]
3. Carnazza, K.E.; Komer, L.E.; Xie, Y.X.; Pineda, A.; Briano, J.A.; Gao, V.; Na, Y.; Ramlall, T.; Buchman, V.L.; Eliezer, D.; et al. Synaptic vesicle binding of α-synuclein is modulated by β- and γ-synucleins. *Cell Rep.* **2022**, *39*, 110675. [CrossRef]
4. Yoshida, H.; Craxton, M.; Jakes, R.; Zibaee, S.; Tavaré, R.; Fraser, G.; Serpell, L.C.; Davletov, B.; Crowther, R.A.; Goedert, M. Synuclein proteins of the pufferfish Fugu rubripes: Sequences and functional characterization. *Biochemistry* **2006**, *45*, 2599–2607. [CrossRef]
5. Toni, M.; Cioni, C. Fish Synucleins: An Update. *Mar. Drugs* **2015**, *13*, 6665–6686. [CrossRef] [PubMed]
6. Deiana, A.; Forcelloni, S.; Porrello, A.; Giansanti, A. Intrinsically disordered proteins and structured proteins with intrinsically disordered regions have different functional roles in the cell. *PLoS ONE* **2019**, *14*, e0217889. [CrossRef]
7. Duperrier, S.; Bortolozzi, A.; Sgambato, V. Increased Expression of Alpha-, Beta-, and Gamma-Synucleins in Brainstem Regions of a Non-Human Primate Model of Parkinson's Disease. *Int. J. Mol. Sci.* **2022**, *23*, 8586. [CrossRef]
8. Giasson, B.I.; Murray, I.V.; Trojanowski, J.Q.; Lee, V.M. A hydrophobic stretch of 12 amino acid residues in the middle of alpha-synuclein is essential for filament assembly. *J. Biol. Chem.* **2001**, *276*, 2380–2386. [CrossRef]
9. Kim, T.-E.; Newman, A.J.; Imberdis, T.; Brontesi, L.; Tripathi, A.; Ramalingam, N.; Fanning, S.; Selkoe, D.; Dettmer, U. Excess membrane binding of monomeric alpha-, beta- and gamma-synuclein is invariably associated with inclusion formation and toxicity. *Hum. Mol. Genet.* **2021**, *30*, 2332–2346. [CrossRef]
10. Goedert, M.; Spillantini, M.G. Synucleinopathies and Tauopathies. In *Basic Neurochemistry*, 8th ed.; Principles of Molecular, Cellular, and Medical Neurobiology; Scott, T.B., George, J.S., Albers, R.W., Price, D.L., Eds.; Academic Press: Cambridge, MA, USA, 2012; Chapter 47; pp. 829–843.
11. Bell, R.; Thrush, R.J.; Castellana-Cruz, M.; Oeller, M.; Staats, R.; Nene, A.; Flagmeier, P.; Xu, C.K.; Satapathy, S.; Galvagnion, C.; et al. N-Terminal Acetylation of α-Synuclein Slows down Its Aggregation Process and Alters the Morphology of the Resulting Aggregates. *Biochemistry* **2022**, *61*, 1743–1756. [CrossRef]
12. Bell, R.; Castellana-Cruz, M.; Nene, A.; Thrush, R.J.; Xu, C.K.; Kumita, J.R.; Vendruscolo, M. Effects of N-terminal acetylation on the aggregation of disease-related α-synuclein variants. *Mol. Biol.* **2022**, *10*, 167825. [CrossRef]
13. Newberry, R.W.; Leong, J.T.; Chow, E.D.; Kampmann, M.; DeGrado, W.F. Deep mutational scanning reveals the structural basis for α-synuclein activity. *Nat. Chem. Biol.* **2020**, *16*, 653–659. [CrossRef]
14. Abeliovich, A.; Schmitz, Y.; Fariñas, I.; Choi-Lundberg, D.; Ho, W.-H.; Castillo, P.; Shinsky, N.; García-Verdugo, J.M.; Armanini, M.; Ryan, A.; et al. Mice lacking alpha-synuclein display functional deficits in the nigrostriatal dopamine system. *Neuron* **2000**, *25*, 239–252. [CrossRef]
15. Burré, J.; Sharma, M.; Südhof, T.C. Cell Biology and Pathophysiology of α-Synuclein. *Cold Spring Harb. Perspect. Med.* **2018**, *8*, a024091. [CrossRef]
16. Burré, J.; Sharma, M.; Tsetsenis, T.; Buchman, V.; Etherton, M.R.; Südhof, T.C. α-Synuclein promotes SNARE-complex assembly in vivo and in vitro. *Science* **2010**, *329*, 1663–1667. [CrossRef]
17. Burré, J.; Sharma, M.; Südhof, T.C. α-Synuclein assembles into higher-order multimers upon membrane binding to promote SNARE complex formation. *Proc. Natl. Acad. Sci. USA* **2014**, *111*, E4274–E4283. [CrossRef]
18. Sulzer, D.; Edwards, R.H. The physiological role of α-synuclein and its relationship to Parkinson's disease. *J. Neurochem.* **2019**, *150*, 475–486. [CrossRef]

19. Barba, L.; Abu Rumeileh, S.; Bellomo, G.; Paoletti, F.P.; Halbgebauer, S.; Oeckl, P.; Steinacker, P.; Massa, F.; Gaetani, L.; Parnetti, L.; et al. Cerebrospinal fluid β-synuclein as a synaptic biomarker for preclinical Alzheimer's disease. *J. Neurol. Neurosurg. Psychiatry* **2022**, jnnp-2022-329124. [CrossRef]

20. Halbgebauer, S.; Oeckl, P.; Steinacker, P.; Yilmazer-Hanke, D.; Anderl-Straub, S.; von Arnim, C.; Froelich, L.; Gomes, L.A.; Hausner, L.; Huss, A.; et al. Beta-Synuclein in cerebrospinal fluid as an early diagnostic marker of Alzheimer's disease. *J. Neurol. Neurosurg. Psychiatry* **2021**, *92*, 349–356. [CrossRef]

21. Kokhan, V.; Kokhan, T.G.; Samsonova, A.N.; Fisenko, V.P.; Ustyugov, A.; Aliev, G.M. The Dopaminergic Dysfunction and Altered Working Memory Performance of Aging Mice Lacking Gamma-synuclein Gene. *CNS Neurol. Disord.-Drug Targets* **2018**, *17*, 604–607. [CrossRef]

22. Surgucheva, I.; Sharov, V.S.; Surguchov, A. γ-Synuclein: Seeding of α-synuclein aggregation and transmission between cells. Biochemistry. *Biochemistry* **2012**, *51*, 4743–4754. [CrossRef]

23. Ninkina, N.; Millership, S.J.; Peters, O.M.; Connor-Robson, N.; Chaprov, K.; Kopylov, A.T.; Montoya, A.; Kramer, H.; Withers, D.J.; Buchman, V.L. β-synuclein potentiates synaptic vesicle dopamine uptake and rescues dopaminergic neurons from MPTP-induced death in the absence of other synucleins. *J. Biol. Chem.* **2021**, *297*, 101375. [CrossRef]

24. Al-Mazidi, S.; Al-Ayadhi, L.Y. Plasma levels of alpha and gamma synucleins in autism spectrum disorder: An indicator of severity. *Med. Princ. Pract.* **2021**, *30*, 160–167. [CrossRef]

25. Rodríguez-Barrueco, R.; Latorre, J.; Devis-Jáuregui, L.; Lluch, A.; Bonifaci, N.; Llobet, F.J.; Olivan, M.; Coll-Iglesias, L.; Gassner, K.; Davis, M.L.; et al. A microRNA Cluster Controls Fat Cell Differentiation and Adipose Tissue Expansion By Regulating SNCG. *Adv. Sci.* **2022**, *9*, e2104759. [CrossRef]

26. Tofaris, G.K. Initiation and progression of α-synuclein pathology in Parkinson's disease. *Cell. Mol. Life Sci.* **2022**, *79*, 210. [CrossRef]

27. Tran, C.H.; Saha, R.; Blanco, C.; Bagchi, D.; Chen, I.A. Modulation of α-Synuclein Aggregation In Vitro by a DNA Aptamer. *Biochemistry* **2022**, *61*, 1757–1765. [CrossRef]

28. Bartels, T.; Choi, J.G.; Selkoe, D.J. Alpha-Synuclein occurs physiologically as a helically folded tetramer that resists aggregation. *Nature* **2011**, *477*, 107–110. [CrossRef]

29. Koga, S.; Sekiya, H.; Kondru, N.; Ross, O.A.; Dickson, D.W. Neuropathology and molecular diagnosis of Synucleinopathies. *Mol. Neurodegener.* **2021**, *16*, 83. [CrossRef]

30. Korneev, A.; Begun, A.; Liubimov, S.; Kachlishvili, K.; Molochkov, A.; Niemi, A.J.; Maisuradze, G.G. Exploring Structural Flexibility and Stability of α-Synuclein by the Landau-Ginzburg-Wilson Approach. *J. Phys. Chem. B* **2022**, *126*, 6878–6890. [CrossRef]

31. Chiba-Falek, O. Structural variants in SNCA gene and the implication to synucleinopathies. *Curr. Opin. Genet. Dev.* **2017**, *44*, 110–116. [CrossRef]

32. Cabin, D.E.; Shimazu, K.; Murphy, D.; Cole, N.B.; Gottschalk, W.; McIlwain, K.L.; Orrison, B.; Chen, A.; Ellis, C.E.; Paylor, R.; et al. Synaptic vesicle depletion correlates with attenuated synaptic responses to prolonged repetitive stimulation in mice lacking alpha-synuclein. *J. Neurosci.* **2002**, *22*, 8797–8807. [CrossRef]

33. Antunes, A.S.L.M. Post-translational Modifications in Parkinson's Disease. *Adv. Exp. Med. Biol.* **2022**, *1382*, 85–94. [CrossRef]

34. Yoo, H.; Lee, J.; Kim, B.; Moon, H.; Jeong, H.; Lee, K.; Song, W.J.; Hur, J.K.; Oh, J.K.H.Y. Role of post-translational modifications on the alpha-synuclein aggregation-related pathogenesis of Parkinson's disease. *BMB Rep.* **2022**, *55*, 323–335. [CrossRef]

35. Stephens, A.D.; Zacharopoulou, M.; Moons, R.; Fusco, G.; Seetaloo, N.; Chiki, A.; Woodhams, P.J.; Mela, I.; Lashuel, H.A.; Phillips, J.J.; et al. Extent of N-terminus exposure of monomeric alpha-synuclein determines its aggregation propensity. *Nat. Commun.* **2020**, *11*, 2820. [CrossRef]

36. Zhang, C.; Pei, Y.; Zhang, Z.; Xu, L.; Liu, X.; Jiang, L.; Pielak, G.J.; Zhou, X.; Liu, M.; Li, C. C-terminal truncation modulates α-Synuclein's cytotoxicity and aggregation by promoting the interactions with membrane and chaperone. *Commun. Biol.* **2022**, *5*, 798. [CrossRef]

37. Kalia, L.V. First trials test targeting of α-synuclein for Parkinson disease. *Nat. Rev. Neurol.* **2022**, *18*, 703–704. [CrossRef]

38. Whone, A. Monoclonal Antibody Therapy in Parkinson's Disease—The End? *N. Engl. J. Med.* **2022**, *387*, 466–467. [CrossRef]

39. Pagano, G.; Taylor, K.I.; Anzures-Cabrera, J.; Marchesi, M.; Simuni, T.; Marek, K.; Postuma, R.B.; Pavese, N.; Stocchi, F.; Azulay, J.-P.; et al. Trial of Prasinezumab in Early-Stage Parkinson's Disease. *N. Engl. J. Med.* **2022**, *387*, 421–432. [CrossRef]

40. Lang, A.E.; Siderowf, A.D.; Macklin, E.A.; Poewe, W.; Brooks, D.J.; Fernandez, H.H.; Rascol, O.; Giladi, N.; Stocchi, F.; Tanner, C.M.; et al. Trial of cinpanemab in early Parkinson's disease. *N. Engl. J. Med.* **2022**, *387*, 408–420. [CrossRef]

41. Meng, Y.; Pople, C.B.; Huang, Y.; Jones, R.M.; Ottoy, J.; Goubran, M.; Oliveira, L.M.; Davidson, B.; Lawrence, L.S.; Lau, A.Z.; et al. Putaminal Recombinant Glucocerebrosidase Delivery with Magnetic Resonance—Guided Focused Ultrasound in Parkinson's Disease: A Phase I Study. *Mov. Disord.* **2022**, *37*, 2134–2139. [CrossRef]

42. Shin, S.-M.; Choi, D.-K.; Jung, K.; Bae, J.; Kim, J.-S.; Park, S.-W.; Song, K.-H.; Kim, Y.-S. Antibody targeting intracellular oncogenic Ras mutants exerts anti-tumour effects after systemic administration. *Nat. Commun.* **2017**, *8*, 15090. [CrossRef]

43. Frieg, B.; Geraets, J.A.; Strohäker, T.; Dienemann, C.; Mavroeidi, P.; Jung, B.C.; Kim, W.S.; Lee, S.-J.; Xilouri, M.; Zweckstetter, M.; et al. Quaternary structure of patient-homogenate amplified α-synuclein fibrils modulates seeding of endogenous α-synuclein. *Commun. Biol.* **2022**, *5*, 1040. [CrossRef]

44. Murray, K.A.; Hu, C.J.; Griner, S.L.; Pan, H.; Bowler, J.T.; Abskharon, R.; Rosenberg, G.M.; Cheng, X.; Seidler, P.M.; Eisenberg, D.S. De novo designed protein inhibitors of amyloid aggregation and seeding. *Proc. Natl. Acad. Sci. USA* **2022**, *119*, e2206240119. [CrossRef]

45. Bagre, G.; Srivastava, T.; Mahasivam, S.; Sinha, M.; Bansal, V.; Ramanathan, R.; Priya, S.; Sharma, S.K. Differential interactions of α-synuclein conformers affect refolding and activity of proteins. *J. Biochem.* **2022**, mvac095. [CrossRef]

46. Lashuel, H.A.; Overk, C.R.; Oueslati, A.; Masliah, E. The many faces of α-synuclein: From structure and toxicity to therapeutic target. *Nat. Rev. Neurosci.* **2013**, *14*, 38–48. [CrossRef]

47. Ninkina, N.; Peters, O.; Millership, S.; Salem, H.; Van Der Putten, H.; Buchman, V.L. Gamma-synucleinopathy: Neurodegeneration associated with overexpression of the mouse protein. *Hum. Mol. Genet.* **2009**, *18*, 1779–1794. [CrossRef]

48. Surgucheva, I.; Newell, K.L.; Burns, J.; Surguchov, A. New α- and γ-synuclein immunopathological lesions in human brain. *Acta Neuropathol. Commun.* **2014**, *2*, 132. [CrossRef]

49. Goedert, M.; Jakes, R.; Spillantini, M.G. The Synucleinopathies: Twenty Years On. *J. Parkinsons Dis.* **2017**, *7*, S51–S69. [CrossRef]

50. Morten, M.J.; Sirvio, L.; Rupawala, H.; Hayes, E.M.; Franco, A.; Radulescu, C.; Ying, L.; Barnes, S.J.; Muga, A.; Yu, Y. Quantitative super-resolution imaging of pathological aggregates reveals distinct toxicity profiles in different synucleinopathies. *Proc. Natl. Acad. Sci. USA* **2022**, *119*, e2205591119. [CrossRef]

51. Sekiya, H.; Tsuji, A.; Hashimoto, Y.; Takata, M.; Koga, S.; Nishida, K.; Futamura, N.; Kawamoto, M.; Kohara, N.; Dickson, D.W.; et al. Discrepancy between distribution of alpha-synuclein oligomers and Lewy-related pathology in Parkinson's disease. *Acta Neuropathol. Commun.* **2022**, *10*, 133. [CrossRef]

52. Espay, A.J. Movement disorders research in 2021: Cracking the paradigm. *Lancet Neurol.* **2022**, *21*, 10–11. [CrossRef] [PubMed]

53. Henriques, A.; Rouvière, L.; Giorla, E.; Farrugia, C.; El Waly, B.; Poindron, P.; Callizot, N. Alpha-Synuclein: The Spark That Flames Dopaminergic Neurons, In Vitro and In Vivo Evidence. *Int. J. Mol. Sci.* **2022**, *23*, 9864. [CrossRef] [PubMed]

54. Chen, K.; Martens, Y.A.; Meneses, A.; Ryu, D.H.; Lu, W.; Raulin, A.C.; Li, F.; Zhao, J.; Chen, Y.; Jin, Y.; et al. LRP1 is a neuronal receptor for α-synuclein uptake and spread. *Mol. Neurodegener.* **2022**, *17*, 57. [CrossRef]

55. Ozdilek, B.; Agirbasli, M. Soluble LRP-1 in Parkinson's disease: Clues for paradoxical effects. *Int. J. Neurosci.* **2022**, 1–12. [CrossRef]

56. Iba, M.; McDevitt, R.A.; Kim, C.; Roy, R.; Sarantopoulou, D.; Tommer, E.; Siegars, B.; Sallin, M.; Kwon, S.; Sen, J.M.; et al. Aging exacerbates the brain inflammatory micro-environment contributing to α-synuclein pathology and functional deficits in a mouse model of DLB/PD. *Mol. Neurodegener.* **2022**, *17*, 60. [CrossRef] [PubMed]

57. Gracia, P.; Polanco, D.; Tarancón-Díez, J.; Serra, I.; Bracci, M.; Oroz, J.; Laurents, D.V.; García, I.; Cremades, N. Molecular mechanism for the synchronized electrostatic coacervation and co-aggregation of alpha-synuclein and tau. *Nat. Commun.* **2022**, *13*, 4586. [CrossRef] [PubMed]

58. Takada, F.; Kasahara, T.; Otake, K.; Maru, T.; Miwa, M.; Muto, K.; Sasaki, M.; Hirozane, Y.; Yoshikawa, M.; Yamaguchi, J. Identification of α-Synuclein Proaggregator: Rapid Synthesis and Streamlining RT-QuIC Assays in Parkinson's Disease. *ACS Med. Chem. Lett.* **2022**, *13*, 1421–1426. [CrossRef]

59. Yoo, J.M.; Lin, Y.; Heo, Y.; Lee, Y.-H. Polymorphism in alpha-synuclein oligomers and its implications in toxicity under disease conditions. *Front. Mol. Biosci.* **2022**, *9*, 959425. [CrossRef]

60. Masliah, E.; Hashimoto, M. Development of New Treatments for Parkinson's Disease in Transgenic Animal Models: A Role for β-Synuclein. *NeuroToxicology* **2002**, *23*, 461–468. [CrossRef]

61. Hashimoto, M.; Rockenstein, E.; Mante, M.; Mallory, M.; Masliah, E. beta-Synuclein inhibits alpha-synuclein aggregation: A possible role as an anti-parkinsonian factor. *Neuron* **2001**, *32*, 213–223. [CrossRef]

62. Zhang, F.; Wu, Z.; Long, F.; Tan, J.; Gong, N.; Li, X.; Lin, C. The Roles of ATP13A2 Gene Mutations Leading to Abnormal Aggregation of α-Synuclein in Parkinson's Disease. *Front. Cell. Neurosci.* **2022**, *16*, 927682. [CrossRef]

63. Bellomo, G.; De Luca, C.M.G.; Paoletti, F.P.; Gaetani, L.; Moda, F.; Parnetti, L. α-Synuclein Seed Amplification Assays for Diagnosing Synucleinopathies: The Way Forward. *Neurology* **2022**, *99*, 195–205. [CrossRef]

64. Majbour, N.; Aasly, J.; Abdi, I.; Ghanem, S.; Erskine, D.; van de Berg, W.; El-Agnaf, O. Disease-Associated α-Synuclein Aggregates as Biomarkers of Parkinson Disease Clinical Stage. *Neurology* **2022**, *99*, e2417–e2427. [CrossRef]

65. Emin, D.; Zhang, Y.P.; Lobanova, E.; Miller, A.; Li, X.; Xia, Z.; Dakin, H.; Sideris, D.I.; Lam, J.Y.L.; Ranasinghe, R.T.; et al. Small soluble α-synuclein aggregates are the toxic species in Parkinson's disease. *Nat. Commun.* **2022**, *13*, 5512. [CrossRef]

66. Ducas, V.C.; Rhoades, E. Investigation of Intramolecular Dynamics and Conformations of α-, β- and γ-Synuclein. *PLoS ONE* **2014**, *9*, e86983. [CrossRef]

67. Janowska, M.K.; Baum, J. The loss of inhibitory C-terminal conformations in disease associated P123H β-Synuclein. *Protein Sci.* **2016**, *25*, 286–294. [CrossRef]

68. Janowska, M.K.; Wu, K.-P.; Baum, J. Unveiling transient protein-protein interactions that modulate inhibition of alpha-synuclein aggregation by beta-synuclein, a pre-synaptic protein that co-localizes with alpha-synuclein. *Sci. Rep.* **2015**, *5*, 15164. [CrossRef]

69. Biere, A.L.; Wood, S.J.; Wypych, J.; Steavenson, S.; Jiang, Y.; Anafi, D.; Jacobsen, F.W.; Jarosinski, M.A.; Wu, G.-M.; Louis, J.-C.; et al. Parkinson's disease-associated alpha-synuclein is more fibrillogenic than beta- and gamma-synuclein and cannot cross-seed its homologs. *J. Biol. Chem.* **2000**, *275*, 34574–34579. [CrossRef]

70. Hayashi, J.; Carver, J.A. β-Synuclein: An Enigmatic Protein with Diverse Functionality. *Biomolecules* **2022**, *12*, 142. [CrossRef]

71. Ohtake, H.; Limprasert, P.; Fan, Y.; Onodera, O.; Kakita, A.; Takahashi, H.; Bonner, L.T.; Tsuang, D.W.; Murray, I.V.; Lee, V.M.-Y.; et al. Beta-synuclein gene alterations in dementia with Lewy bodies. *Neurology* **2004**, *63*, 805–811. [CrossRef]

72. Wei, J.; Fujita, M.; Nakai, M.; Waragai, M.; Watabe, K.; Akatsu, H.; Rockenstein, E.; Masliah, E.; Hashimoto, M. Enhanced lysosomal pathology caused by β-Synuclein mutants linked to dementia with Lewy bodies. *J. Biol. Chem.* **2007**, *282*, 28904–28914. [CrossRef] [PubMed]

73. Spillantini, M.G.; Goedert, M. Neurodegeneration and the ordered assembly of α-synuclein. *Cell Tissue Res.* **2018**, *373*, 137–148. [CrossRef] [PubMed]

74. Jain, M.K.; Singh, P.; Roy, S.; Bhat, R. Comparative Analysis of the Conformation, Aggregation, Interaction, and Fibril Morphologies of Human α-, β-, and γ-Synuclein Proteins. *Biochemistry* **2018**, *57*, 3830–3848. [CrossRef] [PubMed]

75. Peters, O.M.; Millership, S.; Shelkovnikova, T.A.; Soto, I.; Keeling, L.; Hann, A.; Marsh-Armstrong, N.; Buchman, V.L.; Ninkina, N. Selective pattern of motor system damage in gamma-synuclein transgenic mice mirrors the respective pathology in amyotrophic lateral sclerosis. *Neurobiol. Dis.* **2012**, *48*, 124–131. [CrossRef]

76. Peters, O.M.; Shelkovnikova, T.; Highley, J.R.; Cooper-Knock, J.; Hortobágyi, T.; Troakes, C.; Ninkina, N.; Buchman, V.L. Gamma-synuclein pathology in amyotrophic lateral sclerosis. *Ann. Clin. Transl. Neurol.* **2015**, *2*, 29–37. [CrossRef] [PubMed]

77. Williams, J.K.; Yang, X.; Atieh, T.B.; Olson, M.P.; Khare, S.D.; Baum, J. Multi-Pronged Interactions Underlie Inhibition of α-Synuclein Aggregation by β-Synuclein. *J. Mol. Biol.* **2018**, *430*, 2360–2371. [CrossRef]

78. Yang, X.; Williams, J.K.; Yan, R.; Mouradian, M.M.; Baum, J. Increased Dynamics of α-Synuclein Fibrils by β-Synuclein Leads to Reduced Seeding and Cytotoxicity. *Sci. Rep.* **2019**, *9*, 17579. [CrossRef] [PubMed]

79. Levin, J.; Nübling, G.; Giese, A.; Janzen, A.; Oertel, W. Neuroprotektive Therapien bei idiopathischen, genetischen und atypischen Parkinson-Syndromen mit α-Synuklein—Pathologie: Neuroprotective treatment of idiopathic, genetic and atypical Parkinson's disease with alpha-synuclein-Pathology. *Der Nervenarzt* **2021**, *92*, 1249–1259. [CrossRef]

80. Kurnik, M.; Sahin, C.; Andersen, C.B.; Lorenzen, N.; Giehm, L.; Mohammad-Beigi, H.; Jessen, C.M.; Pedersen, J.S.; Christiansen, G.; Petersen, S.V.; et al. Potent α-Synuclein Aggregation Inhibitors, Identified by High-Throughput Screening, Mainly Target the Monomeric State. *Cell Chem. Biol.* **2018**, *25*, 1389–1402.e9. [CrossRef]

81. Kline, E.M.; Houser, M.C.; Herrick, M.K.; Seibler, P.; Klein, C.; West, A.; Tansey, M.G. Genetic and Environmental Factors in Parkinson's Disease Converge on Immune Function and Inflammation. *Mov. Disord.* **2021**, *36*, 25–36. [CrossRef]

82. Eteläinen, T.S.; Kilpeläinen, T.P.; Ignatius, A.; Auno, S.; De Lorenzo, F.; Uhari-Väänänen, J.K.; Julku, U.H.; Myöhänen, T.T. Removal of proteinase K resistant αSyn species does not correlate with cell survival in a virus vector-based Parkinson's disease mouse model. *Neuropharmacology* **2022**, *218*, 109213. [CrossRef] [PubMed]

83. Hmila, I.; Sudhakaran, I.P.; Ghanem, S.S.; Vaikath, N.N.; Poggiolini, I.; Abdesselem, H.; El-Agnaf, O.M.A. Inhibition of α-Synuclein Seeding-Dependent Aggregation by ssDNA Aptamers Specific to C-Terminally Truncated α-Synuclein Fibrils. *ACS Chem. Neurosci.* **2022**, *13*, 3330–3341. [CrossRef] [PubMed]

84. Surguchov, A.; Bernal, L.; Surguchev, A.A. Phytochemicals as regulators of genes involved in synucleinopathies. *Biomolecules* **2021**, *11*, 624. [CrossRef]

85. Xu, Y.; Zhang, Y.; Quan, Z.; Wong, W.; Guo, J.; Zhang, R.; Yang, Q.; Dai, R.; McGeer, P.L.; Qing, H. Epigallocatechin Gallate (EGCG) Inhibits Alpha-Synuclein Aggregation: A Potential Agent for Parkinson's Disease. *Neurochem. Res.* **2016**, *41*, 2788–2796. [CrossRef] [PubMed]

86. Xu, B.; Chen, J.; Liu, Y. Curcumin Interacts with α-Synuclein Condensates To Inhibit Amyloid Aggregation under Phase Separation. *ACS Omega* **2022**, *7*, 30281–30290. [CrossRef]

87. Al-Yousef, N.; Shinwari, Z.; Al-Shahrani, B.; Al-Showimi, M.; Al-Moghrabi, N. Curcumin induces re expression of BRCA1 and suppression of γ synuclein by modulating DNA promoter methylation in breast cancer cell lines. *Oncol. Rep.* **2020**, *43*, 827–838. [CrossRef]

88. Wang, Y.; Wu, S.; Li, Q.; Lang, W.; Li, W.; Jiang, X.; Wan, Z.; Chen, J.; Wang, H. Epigallocatechin-3-gallate: A phytochemical as a promising drug candidate for the treatment of Parkinson's disease. *Front. Pharmacol.* **2022**, *13*, 977521. [CrossRef]

89. Gezen-Ak, D.; Yurttaş, Z.; Çamoğlu, T.; Dursun, E. Could Amyloid-β 1–42 or α-Synuclein Interact Directly with Mitochondrial DNA? A Hypothesis. ACS Chem. *ACS Chem. Neurosci.* **2022**, *13*, 2803–2812. [CrossRef]

90. Toba, S.; Jin, M.; Yamada, M.; Kumamoto, K.; Matsumoto, S.; Yasunaga, T.; Fukunaga, Y.; Miyazawa, A.; Fujita, S.; Itoh, K.; et al. Alpha-synuclein facilitates to form short unconventional microtubules that have a unique function in the axonal transport. *Sci. Rep.* **2017**, *7*, 16386. [CrossRef]

91. Hansson, O. Biomarkers for neurodegenerative diseases. *Nat. Med.* **2021**, *27*, 954–963. [CrossRef]

92. Surguchov, A. Biomarkers in Parkinson's Disease. In *Neurodegenerative Diseases Biomarkers*; Peplow, P.V., Martinez, B., Gennarelli, T.A., Eds.; Humana: New York, NY, USA, 2022; Volume 173, pp. 155–180.

93. Manne, S.; Kondru, N.; Jin, H.; Serrano, G.E.; Anantharam, V.; Kanthasamy, A.; Adler, C.H.; Beach, T.G.; Kanthasamy, A.G. Blinded RT-QuIC Analysis of α-Synuclein Biomarker in Skin Tissue From Parkinson's Disease Patients. *Mov. Disord.* **2020**, *35*, 2230–2239. [CrossRef] [PubMed]

94. Angelopoulou, E.; Paudel, Y.N.; Papageorgiou, S.G.; Piperi, C. Environmental Impact on the Epigenetic Mechanisms Underlying Parkinson's Disease Pathogenesis: A Narrative Review. *Brain Sci.* **2022**, *12*, 175. [CrossRef] [PubMed]

95. Klingelhoefer, L.; Reichmann, H. Pathogenesis of Parkinson disease—The gut-brain axis and environmental factors. *Nat. Rev. Neurol.* **2015**, *11*, 625–636. [CrossRef] [PubMed]

96. Tan, A.H.; Lim, S.Y.; Lang, A.E. The microbiome-gut-brain axis in Parkinson disease—From basic research to the clinic. *Nat. Rev. Neurol.* **2022**, *18*, 476–495. [CrossRef] [PubMed]
97. Ho, P.W.-L.; Chang, E.E.-S.; Leung, C.-T.; Liu, H.; Malki, Y.; Pang, S.Y.-Y.; Choi, Z.Y.-K.; Liang, Y.; Lai, W.S.; Ruan, Y.; et al. Long-term inhibition of mutant LRRK2 hyper-kinase activity reduced mouse brain α-synuclein oligomers without adverse effects. *NPJ Park. Dis.* **2022**, *8*, 115. [CrossRef]

*Review*

# Cellular Models of Alpha-Synuclein Aggregation: What Have We Learned and Implications for Future Study

Katrina Albert [1], Sara Kälvälä [1], Vili Hakosalo [1], Valtteri Syvänen [1], Patryk Krupa [1], Jonna Niskanen [1], Sanni Peltonen [1], Tuuli-Maria Sonninen [1] and Šárka Lehtonen [1,2,*]

[1]  A. I. Virtanen Institute for Molecular Sciences, University of Eastern Finland, 70210 Kuopio, Finland
[2]  Neuroscience Center, University of Helsinki, 00014 Helsinki, Finland
*  Correspondence: sarka.lehtonen@uef.fi

**Abstract:** Alpha-synuclein's role in diseases termed "synucleinopathies", including Parkinson's disease, has been well-documented. However, after over 25 years of research, we still do not fully understand the alpha-synuclein protein and its role in disease. In vitro cellular models are some of the most powerful tools that researchers have at their disposal to understand protein function. Advantages include good control over experimental conditions, the possibility for high throughput, and fewer ethical issues when compared to animal models or the attainment of human samples. On the flip side, their major disadvantages are their questionable relevance and lack of a "whole-brain" environment when it comes to modeling human diseases, such as is the case of neurodegenerative disorders. Although now, with the advent of pluripotent stem cells and the ability to create minibrains in a dish, this is changing. With this review, we aim to wade through the recent alpha-synuclein literature to discuss how different cell culture setups (immortalized cell lines, primary neurons, human induced pluripotent stem cells (hiPSCs), blood–brain barrier models, and brain organoids) can help us understand aggregation pathology in Parkinson's and other synucleinopathies.

**Keywords:** alpha-synuclein; SH-SY5Y; hiPSCs; organoid; aggregation; synucleinopathy; blood–brain barrier; microglia; astrocytes; overexpression; mutation; Lewy body; Parkinson's disease

**Citation:** Albert, K.; Kälvälä, S.; Hakosalo, V.; Syvänen, V.; Krupa, P.; Niskanen, J.; Peltonen, S.; Sonninen, T.-M.; Lehtonen, Š. Cellular Models of Alpha-Synuclein Aggregation: What Have We Learned and Implications for Future Study. *Biomedicines* **2022**, *10*, 2649. https://doi.org/10.3390/biomedicines10102649

Academic Editor: Natalia Ninkina

Received: 22 August 2022
Accepted: 17 October 2022
Published: 20 October 2022

**Publisher's Note:** MDPI stays neutral with regard to jurisdictional claims in published maps and institutional affiliations.

## 1. Introduction

For a scientist, entering the field of alpha-synuclein ($\alpha$-syn) research is a daunting task. Since the discoveries in 1997 that mutations in the gene encoding $\alpha$-syn (SNCA) result in an autosomal dominant form of Parkinson's disease (PD) [1] and that it is prevalent in Lewy bodies [2], aggregates linked to disease, there have been over 10,000 original articles about this enigmatic protein (Source: Scopus, Figure 1). Additionally, $\alpha$-syn also accumulates in Lewy bodies in disorders such as PD with dementia (PDD), dementia with Lewy bodies (DLB), and multiple system atrophy (MSA). These diseases are collectively termed "synucleinopathies" as they share this common pathological feature, whether the Lewy bodies are found in the cytoplasm, neurons, or glial cells. This review will focus mainly on PD, as the majority of studies attempt to model it. For a definite diagnosis of PD, Lewy bodies must be found in the substantia nigra pars compacta neurons in a post-mortem examination—although other neurons have Lewy bodies as well [3]. PDD presents similar motor symptoms to PD but with memory loss and Lewy bodies in cortical areas, and DLB is also similar to this pathologically [4]. Lastly, MSA has the unique feature that the Lewy bodies are found in the oligodendrocytes [5]. Why are these diseases, all containing neuropathological deposits of $\alpha$-syn, still somewhat distinct from one another?

$\alpha$-syn is expressed in several cell types in the central nervous system (CNS) as well as in the periphery in humans [6]. As a monomer, $\alpha$-syn forms a heterogenous population of different states in aquatic solution, which is mostly due to the fluctuating C-terminal tail and heavily dependent on the ligands it can interact with and the environment [7–10].

In terms of amino acid sequence, the first 25 amino acids are important, with the first 12 amino acids being crucial for lipid binding and docking (Figure 2). In the second α-helix are the hydrophobic areas (non-amyloid component, NAC) that are important for the oligomerization and aggregation of α-syn [11–13]. α-syn is often referred to in the literature as an "intrinsically disordered" protein, but it is rarely explained why this is important. Intrinsically disordered proteins have no unique 3D structure, which allows for greater conformational flexibility, results in a larger surface area for interacting partners, and comes with more post-translational modifications [14]. These properties are considered advantageous for the protein's function in the cell, giving it both tight control as well as the ability to quickly react to changing conditions. However, one can guess that this dynamic nature results in difficulties in wrangling intrinsically disordered proteins as drug targets. α-syn is considered mostly unstructured and belongs to a family of amyloidogenic proteins. Additionally, one more layer of complexity to the folding properties of α-syn is the recent demonstration of its association with membrane-less organelles. The association of intrinsically disordered proteins with membrane-less organelles could provide a possible explanation for the triggering of the amyloidogenesis pathway in multiple neurodegenerative diseases [15]. While in physiological conditions, the formation of α-syn-related liquid-liquid phase separation is not favored, mutations, pathological stressors such as metal ions ($Cu^{2+}$, $Fe^{3+}$), or high local concentrations of α-syn [16] may facilitate it. Additionally, this unstructured region of α-syn is converted to β-sheet-rich structures that form fibrils [17]. In short, misfolding of the soluble monomer of α-syn to temporary oligomeric structures and then into the ordered fibrillar structure is considered to be the process of aggregate formation. Further, formed aggregative structures and shapes may change based on the post-translational modifications and, importantly, depending on the experimental set-up (refer to graphical abstract).

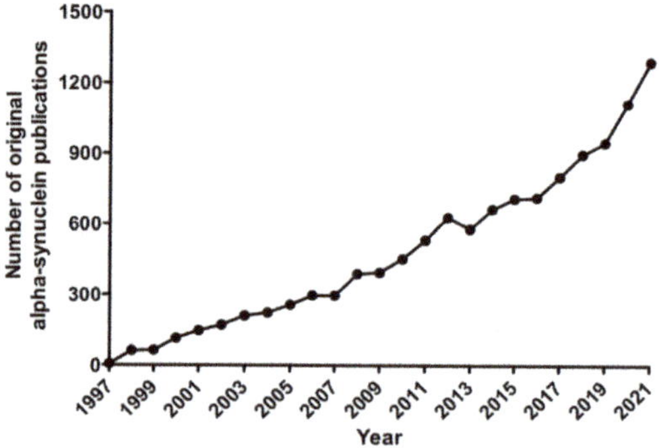



**Figure 1.** Number of original articles published per year on "alpha-synuclein" starting from 1997 when it was demonstrated that α-syn is involved in PD, up until 2021. Since 2020, over 1000 articles have been published in a year. Data was obtained from Scopus (https://www.scopus.com/, accessed on 10 January 2021). A search was made for the word "alpha-synuclein" within 'Article title, Abstract, Keywords'. This was further refined to Articles (so as to exclude Reviews and Book Chapters), then the years were narrowed from 1997 to 2021. This generated a total of 11,959 results.

It is also important to note here that for the purposes of this review when taken out of context (i.e., not associated with a particular study), we refer to monomers as the 14 kDa native protein, oligomers as larger species of α-syn that may or may not be toxic, and fibrils as the above-mentioned β-sheet structure. It is not always clear in the papers discussed below what an oligomer vs. a fibril is in some cases, particularly when taking into consideration recent research into antibodies used to label these conformations [18].

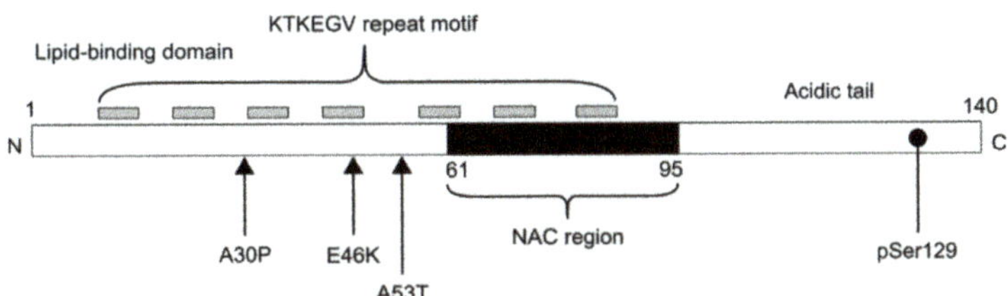



**Figure 2.** Structural diagram of alpha-synuclein protein. The N-terminal contains an amphiphilic region characterized by apolipoprotein repeat motif. The non-amyloid component (NAC) region is important in the oligomerization and aggregation of α-syn. The C-terminal domain is characterized by acidic residues Glu and Asp and assumes an inherently disordered conformation. Phosphorylation at Ser129 is a classical marker for aggregated α-syn.

## 2. Modeling Alpha-Synuclein Aggregation in Non-hiPSC Lines: What We Can Learn and Caution with Outcome Measures

Since α-syn is not naturally present in non-vertebrates, studying the propensity of this protein to aggregate has relied heavily on cellular models. As such, numerous cellular models have been generated to broaden our understanding of the role this protein has on the cellular dysfunctions observed in the disease. These models are based on the overexpression of either wild type (WT) or mutated α-syn or the addition of an exogenous protein. Multiple cell lines have been utilized to study α-syn aggregation and toxicity, ranging from differentiated immortalized cells to primary neurons of mammalian species. Aggregation can be achieved by overexpression of mutated α-syn and multiplications in the SNCA gene as well as by using exogenous α-syn preformed fibrils (PFFs) or by patient-derived aggregates. Ideal models of α-syn capture the key pathological events as seen in the clinics, such as the spread of aggregated α-syn, preferably in the form of Lewy bodies, and dysfunction of modeled cells.

There is a high amount of evidence supporting the causative role of α-syn in synucleinopathies, which has been derived from the discovery of numerous autosomal dominant SNCA mutations and multiplications [1,19–22]. Since then, these mutations have been discovered to have effects on the structure and aggregation dynamics of α-syn [23–25]. To summarize, these mutations alter the disease phenotype, severity, and age of onset, but the exact mechanism behind these aspects is still elusive. Despite this, the known mutations, such as A53T, A30P, E46K, H50Q, and G51D have been successfully used to model synucleinopathies both in vitro and in vivo.

With hundreds of studies using different cell lines, forms of α-syn, and outcome measures, it makes wading through the literature a difficult undertaking for a person new to the field. With this section of the review, we aim to summarize and highlight studies of interest in the field using immortalized and primary cells. Table 1 shows several recent studies that add to this but is not inclusive of all studies.

**Table 1.** Examples of recent studies with the use of non-iPSCs models with different $\alpha$-syn modifications. The definition of PFF and oligomeric forms of $\alpha$-syn can vary from paper to paper. PFF—preformed fibrils.

| Model | $\alpha$-syn Modification | Outcomes of the Study | References |
|---|---|---|---|
| **Cell lines** | | | |
| SH-SY5Y | Overexpression | Recipient microglia suppressed autophagy caused by enhanced expression of miR-19a-3p in exosomes, via the phosphatase and tensin homolog/AKT/mTOR signaling pathway. | [26] |
| | Overexpression | miR-335 levels are reduced in PD models and patients. Its overexpression reduced inflammation induced by LPS stimulation or LRRK2 overexpression. | [27] |
| | A53T mutation | Reduced mitochondrial oxygen flow at maximum capacity. | [28] |
| | A53T mutation | Silencing of CK2$\alpha$ results in reduced phosphorylated $\alpha$-syn at serine129 expression in cells with A53T mutation as well as functionality of dopaminergic neurons and ROS generation. | [29] |
| | PFFs | Micropinocytosis was suggested to be the main pathway of $\alpha$-syn internalization into SH-SY5Y cells and differentiated neurons. | [30] |
| | PFFs | Higher secretion levels of nanoscopic $\alpha$-syn aggregates were driven by disrupted protein homeostasis caused by PFF, but it does not lead to aggregation in the cells. | [31] |
| | Monomers, dimers, tetramers | $\alpha$-syn dimer and tetramer internalization into the cell happened primarily through endocytosis. Aggregated $\alpha$-syn from PFFs or oligomers displayed more prominent accumulation than monomers. Tetramer structures of $\alpha$-syn showed more resistance to these processes, which suggested higher infectiousness of higher oligomeric states. | [32] |
| | Monomers, polymers, $\alpha$-syn-119 | 1:4 ratio of $\alpha$-syn -119:PQQ (pyrroloquinoline quinone) resulted in neuroprotective effect, showing antioxidant effect of PQQ. PQQ can change the secondary structure of $\alpha$-syn, inhibiting oligomer formation induced by Cu(II). | [33] |
| PC12 | Overexpression | Glutamine can enhance Hsp70 expression which is able to promote degradation of $\alpha$-syn even in the presence of a proteasomal inhibitor. | [34] |
| | Overexpression | Increase in oligomerization and aggregation of $\alpha$-syn might be the result of iron accumulation in neurons. The abnormal iron levels can be caused by higher levels of $\alpha$-syn concentration in the cells. | [35] |
| | Oligomers and overexpression | Serotonin aldehyde oligomerizes $\alpha$-syn in vivo and in vitro. | [36] |
| | A53T mutation | Ubiquitin proteasome system dysfunctions caused by $\alpha$-syn in dopaminergic neurons | [37] |
| | PFFs | PC12 cell line is less resistant to $\alpha$-syn cytotoxicity than primary hippocampal neurons. | [38] |
| | PFFs | $\alpha$-syn fibril formation can be inhibited by hydroxytyrosol, and fibrils can be destabilized by hydrotyrosol. | [39] |
| | (MPP$^+$)-treated cells | Low-intensity ultrasounds stimulation results in ROS generation inhibition in MPP+ treated cells, lowering levels of $\alpha$-syn aggregation. | [40] |
| LUHMES | Overexpression | Transcriptome and proteasome analyses identified differential regulation of genes associated with PD. Vesicular transport and the lysosome were leading mechanisms. | [41] |
| | SNCA knockout | 401 genes associated to the cell cycle had reduced expression after SNCA knockout in dopaminergic neurons. | [42] |
| | Overexpression | During drug screening, PDE1A inhibition showed the most effective results against $\alpha$-syn toxicity. | [43] |
| | A30P mutation | Overexpression of WT and A30P m $\alpha$-syn had a significant effect in DNA methylation of genes related to glutamate signaling and locomotor pathways | [44] |

**Table 1.** *Cont.*

| Model | α-syn Modification | Outcomes of the Study | References |
|---|---|---|---|
| MN9D | Normal expression | α-syn accumulation was inhibited by suppression of prolonged adenosine A1 receptor activation. | [45] |
| | Normal expression | Damage induced by 6-OHDA can be reduced by icaritin (ICT). ICT increases SOD activity, TH expression, but decreased ROS production and α-syn expression. | [46] |
| BV2 | A53T mutation | Polygala saponins fractions inhibited NLRP3 inflammasome by AMPK/mTOR and PINK1/parkin pathways, which contribute to the regulation of neuroinflammation decrease and neuronal death via mitophagy | [47] |
| | α-syn and MPP+ co-treatment | α-syn and MPP+ co-treatment induced activation of NLRP3 inflammasome. | [48] |
| | Monomers, oligomers | Monomeric α-syn promotes microglial inflammatory phenotype by ERK, NF-κB, and PPARγ pathways. | [49] |
| | α-syn -enriched conditioned media | Neuroinflammation caused by impairment in microglial autophagy is disrupted by α-syn on the Tlr4-dependent p38 and Akt-mTOR pathway. | [50] |
| | PFFs | α-syn fibrils caused a strong inflammatory response. Level of fibrilization is a main trait for its intake. | [51] |
| | A53T mutation | Norepinephrine release caused dopaminergic neuron viability disruption in the noradrenergic system. | [52] |
| Primary neurons | A53T α-syn T22N Rab7A mutations | Wild type Ras-related in brain 7 (Rab7) reduced α-syn decreased α-syn toxicity, e.g., oxidative stress, mitochondrial perturbations, and DNA damage. | [53] |
| | PFFs and α-syn E35K E46K E61K mutants | Lysophosphatidylcholine acyltransferase 1 regulates α-syn pathology. Utilization of α-syn E35K E46K E61K model. | [54] |
| | LRRK2 inhibition | α-syn localization at the presynaptic terminal is connected to the kinase activity of LRRK2. | [55] |
| | PFFs | α-syn aggregation induced by PFFs is not influenced by insulin-related signaling in primary dopamine neurons. | [56] |
| | PFFs | Tannic acid showed the best results in two-step screening for α-syn aggregation inhibitors. | [57] |
| | PFFs and AS69 protein | There is no change in PFF uptake in the presence to AS69 protein, that binds to α-syn, but AS69 decreases α-syn pathology. | [58] |
| | Monomer | GLP-1R-associated neuroprotective and neurotrophic cell signaling can be activated by GLP-1 (9–36). | [59] |
| Human neural stem cell line (ReNcell) | PFF or overexpression | α-syn and its aggregate degradation can happen with miR-7 use. miR-7 can also decrease α-syn expression. | [60] |

Several cellular studies have demonstrated the capacity for overexpressed or mutated α-syn to cause neuronal dysfunction as seen in PD, as follows: disruption of the ubiquitin-dependent proteolytic system in PC12 cells (an immortalized cell line of rat pheochromocytoma often used in neuroscience research [61]) with the A53T mutation [62], disruption in calcium signaling in SH-SY5Y cells with the A53T mutation [29], impaired dopamine (DA) release in stable PC12 cells expressing non-toxic amounts of A30P mutant or WT α-syn [63] as well as vesicular DA storage impairments in a human mesencephalic cell line (MESC2.10) expressing A53T mutant α-syn [64]. In addition to these immortalized cell models, murine primary neurons have also been broadly used to study the effects of α-syn overexpression. In a rat midbrain DA primary neuron culture, A30P and A53T mutations have been shown to have a decreasing effect on the regenerative capacity of the neurons [65]. Furthermore, primary cortical neurons of mice expressing A53T mutant α-syn have displayed reduced mitochondrial mobility as well as a decrease in mitochondrial membrane potential and in respiratory capacity [66]. There are also multiple different approaches to achieving expression of mutated *SNCA* or overexpression of the WT version. A few examples include using plasmid constructs [67,68] or viral vectors [69].

While these studies and others contribute important findings to the field of PD research, particularly in terms of using mutations that confer the disease. These mutations often have a more severe phenotype than an expression of the WT, which matches the clinical presentation of, for example, the A53T mutation, which leads to an earlier onset and more severe progression of the disease [70]. However, overexpression/mutation of α-syn may not be the best way to model sporadic disease in cell monoculture. DA neurons in particular are known to be sensitive to stressors [71], and therefore an abnormal level of protein expression at an acute timepoint would likely cause dysfunction regardless of the protein being expressed. Sure enough, already more than 20 years ago, researchers showed that green fluorescent protein (GFP), commonly used as a reporter and/or control in studies modeling neurodegeneration, can be toxic to cell lines with GFP overexpression [72] and also to primary cortical neurons [73,74]. Therefore, caution needs to be taken in interpreting the results from cell studies where α-syn is overexpressed. Which control protein or mutation was used to compare the effects to? What was the level of gene expression of α-syn versus the control protein/mutation? These questions need to be answered before one can conclude that the cell death or dysfunction was the result of α-syn specifically.

In contrast to overexpression/mutation models, the following one major advance has emerged in recent years in the field of α-syn research: in 2009, Luk et al., were able to demonstrate that the addition of α-syn PFFs to SH-SY5Y culture (an immortalized cell line of human neuroblastoma, often used as a neuronal model, including as a PD model since DA markers have been observed [75]) induced recruitment and phosphorylation of endogenous α-syn into insoluble pathogenic forms, which leads to the formation of Lewy body-like intracellular inclusions. In most studies using α-syn PFFs the protocol to generate the fibrils is similar where recombinant mouse or human α-syn monomer is purified, put at 37 °C, and shaken for 7 days. This results in the fibrillar form of α-syn which is then sonicated (either using a probe tip or water bath sonicator) and applied to the culture [76].

Similar findings have also been observed in primary neuron cultures seeded with exogenous α-syn fibrils [77,78]. In addition, these studies reported synaptic dysfunction and cell death induced by the Lewy body-like inclusions. Findings from α-syn PFF cellular models have raised questions about whether the exogenous pathological forms of α-syn could induce synucleinopathies in vivo as well. Remarkably, brain injection of α-syn PFFs is sufficient to cause cell-to-cell transmission of pathological α-syn, leading to Lewy body-like inclusion formation in A53T α-syn transgenic [79] and in non-transgenic mice [80,81], and countless others). More recently, seeded α-syn PFFs induced the aggregation of endogenous α-syn while significantly increasing the secretion of nanoscopic α-syn aggregates by disrupting the protein homeostasis of SH-SY5Y cells [31] and in Cascella et al. (2021) [82], where they used rat primary hippocampal cells and SH-SY5Y cells (as well as hiPSC-derived DA neurons) to inspect the spread of differently produced α-syn fibrils/oligomers and outer lipid membrane dysfunction and intake of oligomers. Their findings show that by comparing two different α-syn fibril oligomeric structures, one with a disordered secondary structure and the other with a β-sheet rich core, the latter can cause major membrane disruption and, in their model, be toxic to the neurons, whereas the former one is biologically inert.

Although the α-syn PFF model is not perfect, as one is adding exogenous material that is produced in *E. coli* or in HEK293 cells, it has strongly demonstrated the possibility of endogenous α-syn aggregation as seen in patients. This can be used to model different synucleinopathies by adding the PFFs to different cell types and observing the effects. Of course, researchers also need to take into consideration the amount of α-syn PFFs added—adding a too-high concentration and causing unfettered cell death, but too low concentration may not result in any significant event. Additionally, there may be batch-to-batch variation between produced α-syn fibrils, as well as differences in sonication steps [83]. This can affect the length of the α-syn fibrils and consequently their seeding efficiency [84]. Therefore, a standardized protocol would be best suited so that studies between labs could be compared side by side. There are good protocols for producing α-syn fibrils [78] or using them in vivo [83,85], which could serve as starting material for 2D cultures as well, if researchers were to agree on the concentration of PFFs and the outcome measures used.

Another important factor to be taken into consideration while assessing the relevance of an α-syn PFF model is that the Lewy body pathology is sometimes achieved with experimental manipulations such as additional factors assisting the intake of fibrils [76]. This is in contrast with the majority of PD cases since they are sporadic and express normal levels of α-syn. That being said, the ease of access and biological relevance of the α-syn PFF model in relation to both sporadic and genetic PD is high since it can recapitulate the uptake and seeding of α-syn in WT cells as well as those with overexpressed or mutated α-syn. Here, in particular, the propagation of α-syn is worth mentioning, since this has long been considered a key event in sporadic PD [86,87]. Interestingly, both α-syn PFFs, as well as tau fibrils, have been shown to be taken up in a similar manner in mouse-derived multipotent neural stem cells [88] and in rat neuroblastoma cells and oligodendrocyte-like cells [89].

One option for α-syn fibrils to become internalized and at the same time propagate from cell to cell that has been shown recently is the tunneling of nanotubes between cells. Work from a mouse brain tumor, neuron-like CAD cells, and mouse primary cortical neurons to hiPSC-derived models has shown these tracks, or at least the close proximity of acceptor and donor cells, between cells to be a potential spreading mechanism for α-syn fibrils [90–92]. These papers also demonstrate the importance of lysosomal-mediated vesicle propagation of α-syn fibrils in neuronal-like CAD cells. These results are fascinating and relate strongly to intracellular lysosomal degradation pathways that have been shown to be crucial in neuronal cells.

After the initial intake of α-syn fibrils, they have been shown to be transported into the neurons themselves. Transport can happen in the retrograde (towards cell soma) or anterograde direction, but with α-syn fibrils, the retrograde direction is more prominent, which has been widely demonstrated in mouse primary cortical neurons [93–95]. This is an advantage in terms of disease modeling since for the disease progression intracellular α-syn aggregation and the formation of Lewy body inclusions in the area of the cell is a key hallmark of PD and other synucleinopathies [96]. Specifically, in mouse primary neurons using α-syn PFFs the authors of [97] were able to verify that endosomal trafficking and lysosomal intake are the main events after PFF intake. While not the first observations of such events with primary neurons the above study shows exhaustive analysis of differentially conjugated or modified α-syn PFFs to determine important parameters regarding intracellular handling and the pathological events that follow. In the future, using a battery of analysis methods that differentiate between exogenously added α-syn fibrils and endogenous aggregated α-syn can serve to help researchers understand the aggregation process.

While in this review we discuss a lot about endogenous α-syn aggregation and how it can cause the formation of Lewy bodies, we do not get too deep into the topic and debate about the nature of Lewy bodies themselves. While it is important to point out that the term "Lewy body" is sometimes used indiscriminately in the literature; therefore, it would be in the best interest to have standardized ways to detect Lewy bodies in models of synucleinopathy. However, due to their heterogeneous quality from patient to patient or even cell to cell within a single cell line, the task has been difficult. Not a single antibody alone or three together are undisputed ways to prove the presence of Lewy bodies, although they may be convincing in a certain context. Usually, microscopes are limited by three different fluorophores when using fluorescent imaging and antibodies are commonly against ubiquitination, phosphorylated α-syn at ser129, and mitochondrial membrane fragment or lipids. So far, advances in imaging technologies and the adaptions to researchers' needs have addressed this problem by both recognizing the Lewy bodies with classical antibody markers and the correlation of inclusion morphology analyzed by high magnification imaging, achieved e.g., by electron microscopy. One such approach is the correlative light and electron microscopy imaging (CLEM), demonstrated in the PD research field by [98], which reported the composition of Lewy bodies from PD patients. Later also [99] with CLEM and primary mouse hippocampal neurons or [100] with cryo-electron tomography and labeled α-syn fibrils in primary cortical neurons, were able to use these approaches while modeling the synucleinopathies with α-syn PFFs. These papers clearly demonstrate the benefit of characterization of the achieved Lewy bodies from the model used with multiple approaches, but it also puts pressure on small labs which need yet another validation assay for their α-syn-related paper to show biological relevance. Finding accessible ways to observe and categorize Lewy bodies is therefore also a priority in the synucleinopathy field.

### 3. Modeling Alpha-Synuclein Aggregation in hiPSC-Derived Neurons: Opportunities and Challenges

Human induced pluripotent stem cells (hiPSCs) provide a unique platform for cellular modeling as they can be directly isolated from patients with disease and they maintain the genetic information of the donor, including chromosome abnormalities and gene mutations. This can be utilized when studying synucleinopathies since there are multiple known pathogenic mutations, as discussed previously. The majority of hiPSC models are focused on the triplication and A53T point mutation of SNCA due to their clear phenotype, both in the cell and in the patient (Table 2). However, SNCA multiplications in patients are rare [101], so this needs to be taken into account when applying findings to other forms of PD. HiPSC-derived DA neurons with SNCA triplication display elevated α-syn mRNA levels, which leads to increased expression of the protein itself [102]. Furthermore, the lines manifesting the SNCA triplication exhibit a decreased capacity to differentiate into DA neurons while also having lower neuronal activity when compared to control lines. HiPSC-derived neurons with this mutation have also shown increased levels of phosphorylated α-syn [103] as well as an increased accumulation of α-syn into aggregates, resembling Lewy body pathology [104]. In other studies using hiPSC-derived neurons [105] showed that overexpression of WT or pathological genetic variances of α-syn (A53T, A30P) can induce mitochondrial-related dysfunction, where WT and A53T were more potent to cause a difference, and A30P was not due to poor mitochondrial lipid layer binding. Similarly, [106] showed that other pathogenic variants which can produce oligomeric α-syn (E46K and E57K) and duplication of WT SNCA also impair mitochondrial anterograde trafficking in an hiPSC-derived neuronal model. This is in line with previously reported α-syn-related mitochondrial dysfunction as reviewed in [107–109] and seems to be through MAPK signaling, which is triggered by α-syn [110].

**Table 2.** Examples of studies with the use of iPSCs models with different α-syn modifications. The definition of PFF and oligomeric forms of α-syn can vary from paper to paper. PFF—preformed fibrils.

| iPSCs Model | α-syn Modification | Outcomes of the Study | References |
|---|---|---|---|
| iPSC- derived neurons | Overexpression | α-syn binds directly to the bTubIII and it is linked to the neuritic integrity in PD. | [111] |
| | A53T mutation | Higher levels (mRNA) of α-syn, early changes in expression of genes related to metabolism, differentiation/development, ion transport, cytoskeleton, extracellular matrix organization, and synaptogenesis. | [112] |
| | A53T mutation | Abnormal accumulation of α-syn disrupts mRNA stability in PD iPSC neurons, disturbs the decapping module in PD brain. | [113] |

**Table 2.** *Cont.*

| iPSCs Model | α-syn Modification | Outcomes of the Study | References |
|---|---|---|---|
| | A53T mutation and isogenic line | Increase in SNCA/α-syn can be enhanced by recombinant pro-cathepsin D. | [114] |
| | A53T mutation | Neuron degradation, protein aggregates, increase in protein synthesis. | [28] |
| | SNCA A53T and GBA1 mutation | ER fragmentation can be caused by the α-syn accumulation in midbrain neurons. | [115] |
| | LRRK2 mutation | Carbosilane dendrimers use can counteract abnormal α-syn accumulation. | [116] |
| | A53T mutation and PFFs | Generation of reliable humanized seeding model for pharmacological research. | [117] |
| | PFFs and ribbons | Spreading of α-syn fibrils and ribbons, aggregation of endogenous α-syn. | [118] |
| | PFFs | α-syn aggregation in the form of phosphorylated α-syn. | |
| iPSC-derived astrocytes | LRRK2 mutation | LRRK2 and GBA mutations in astrocytes contribute to PD development, manifesting several disease hallmarks. | [119] |
| | ATP13A2 mutation | ATP13A2 mutation in astrocytes results in α-syn accumulation in dopaminergic neurons and ATP13A2 deficiency compromises protective astrocytes function from α-syn aggregation. | [120] |
| | LRRK2 mutation | Astrocytes play role in dopaminergic cell death in PD pathogenesis, by dysfunctions in pathway of protein degradation. | [121] |
| | PFFs | Astrocytes and microglia revealed synchronous activity in processing α-syn aggregates. | [122] |
| | PFFs | Exposure of α-syn to PFFs leads to antigen presenting phenotype in astrocytes with upregulation of major histocompatibility complex and antigen molecules, while TNF-α activates pro-inflammatory pathway. | [123] |
| | PFFs | Astrocytic α-syn uptake can be limited by binding to clusterin. A-syn clearance can be improved with lower clusterin levels. | [124] |
| iPSCs-derived endothelial cells | PFFs | Generation of a substantia nigra brain chip, reproducing α-syn pathology in vivo during PFFs exposure. | [125] |
| iPSCs-derived oligodendrocytes and midbrain spheroids | Overexpression and A53T mutation | PD and MSA can affect oligodendrocytes in early cellular pathways and alterations. Epigenetic, genetic changes, and immune reactivity in MSA can be connected to each other by immune component triggered by α-syn. | [126] |
| Organoids (iPSCs) | Overexpression | SNCA triplication in midbrain organoids revealed pathological hallmarks of synucleinopathies in glial and neuronal cells. | [127] |
| | PINK1 mutation | 2-Hydroxypropyl-β-Cyclodextrin treatment resulted in mitophagy improvement and better dopaminergic differentiation by protein levels modifications. | [128] |

**Table 2.** *Cont.*

| iPSCs Model | α-syn Modification | Outcomes of the Study | References |
|---|---|---|---|
| | PARKIN mutation | Mutation in PARKIN results in reduced IF activity. | [129] |
| | PINK1 mutation | Decreased amount of dopamine in vesicles, higher expression of α-syn. | [130] |
| | LRRK2 mutation | Midbrain organoids with LRRK2 mutation showed 3D pathological hallmarks of sporadic PD in patients. Thiol oxidoreductase functions are important in the LRRK2-associated PD development. | [131] |
| | APOE knockout | Lower levels of apoE induce aggregation of insoluble α-syn and phosphorylated α-syn, increased synapse loss, excess lipid droplet formation (hence GBA reduction and endo-lysosomal dysregulation). | [132] |
| | PFFs | Enteroendocrine cells are a key component of gut-brain hypothesis for the outcome and α-syn pathology development, and they show uptake and propagation of PFFs to neurons. | [133] |
| Organoids (ESC-derived) | GBA1 knockout and overexpression | β-sheet–rich α-syn aggregates can be the result of loss of glucocerebrosidase, linked with α-syn overexpression. | [134] |
| | DNAJC6 mutation | Loss of function in DNAJ6 gene results in α-syn aggregation caused by impairment in autophagy. | [135] |

So far there are only a few publications where synucleinopathies are modeled in hiPSC-derived neurons with exogenous α-syn fibrils alone. Gribaudo et al., (2019) [118] demonstrated that aggregation of endogenous α-syn can be achieved in a hiPSC-derived cortical neuron model by the addition of either exogenous α-syn PFFs or ribbons, made as in Bousset et al., (2013) [136]. This work showed that while the intake of exogenous α-syn fibrils happens relatively fast, even within one day, the seeding of endogenous α-syn can take more than 30 days. In their model, they used human cortical neurons matured from hiPSC-derived dorsal telencephalic neuronal progenitors. They showed that α-syn ribbons and fibrils possess different seeding properties, where ribbons are faster to cause prion-like propagation. They simultaneously showed the propagation of labeled α-syn fibrils through receptor cell axons towards acceptor cells using distal chamber microchips. An interesting result from the study is that these cortical neurons transport took α-syn fibrils at a velocity of ~2.6 μm/s intracellularly and the ribbons at a similar pace. The authors interpret the result of different strains propagating at a similar rate as an indication that the transport is mediated in the same way. Perhaps this method of measuring the propagation velocity of α-syn can aid in the estimation of the disease progression [137], although it is currently unclear what this means in terms of relevance to the patient. Lessons for future experiments can also be taken from the fact that they did not use any α-syn genetic variant cell line, but rather a moderately high α-syn concentration of 0.5 μM and were able to show mild accumulation of endogenously aggregated α-syn and Lewy body-like structures as follows: halo-like staining of phosphorylated α-syn, co-localizing with ubiquitination, p62-sequestosome1, and HSP-70. This pathological model was also shown to cause mitochondrial fragmentation and $Ca^{2+}$ oscillation frequency changes, related to neuronal dysfunction, with ribbons being more potent. Previously, similar $Ca^{2+}$ changes have been reported in hiPSC-derived neuronal models [138,139]. Recently, [140] witnessed similar higher potency with α-syn ribbons over fibrils in a screening study using hiPSC-derived DA neurons from multiple sources. In their approach, they used biologically relevant, more mature DA neurons (>45 days since differentiation), which were shown to be electrophysiologically active. Importantly, Tanudjojo et al. [140] were able to show

PFF-induced endogenous α-syn aggregation in hiPSC-derived DA neurons in healthy control lines using 1 μM concentration. The authors used α-syn fibrils generated in *E. coli* and also observed that these fibrils amplify in the presence of brain homogenates derived from both PD and MSA patients.

In the future, hiPSC-derived PD models can be used to introduce base-level questions by first conducting omic studies, which can then be answered in a more detailed study setup. Using hiPSCs to study α-syn aggregation has the advantage that, in addition to using human and, in some cases, PD patient-derived neurons, we can have a more biologically relevant model than immortalized or primary cells while keeping the tight experimental control present using cell lines.

However, despite the numerous advantages that patient-derived hiPSCs provide (pluripotency, supply of material, fewer ethical issues associated with human sample use), there are still limitations hindering their potential. One major issue is the heterogeneity of hiPSCs derived from different donors or even in clones from the same donor. It is also difficult to intervene with the variability as, in some instances, it can be either a protocol effect or a patient effect. Secondly, it is difficult to generate hiPSC-derived DA neurons with an aging phenotype since the reprogramming process will reset the cell into a more youthful state [141,142]. This is an issue when the goal is to model a disease that is heavily associated with aging.

## 4. Effects of Alpha-Synuclein on Microglia

Lately, PD research has begun to focus on how the body's immune system may contribute to disease pathogenesis [143]. This means not only looking at neurons, and in terms of CNS cells, microglia have emerged as an important study target.

The mouse microglia BV2 cell line has been extensively used for different microglial studies with α-syn. The BV2 cell line is generated by harvesting primary microglia from 1-week-old female mice (*Mus musculus*) and then transformed to a cell line by infecting the cells with retrovirus J2 carrying a v-raf/v-myc oncogene [144]. As mentioned before, α-syn can be found in monomeric, oligomeric, or fibrillar forms. The definition of oligomeric and fibrillar forms varies from paper to paper. To recapitulate an inflammatory state of microglia, lipopolysaccharide (LPS) and/or interferon-gamma (IFNγ) are used. LPS is an endotoxin that induces inflammatory activation of microglia, mouse microglia in particular, whereas IFNγ is an internal signaling protein that particularly stimulates human microglia.

The aggregation state of α-syn affects the activation of BV2 cells [51]. LPS (0.1, 1, 10 μg/mL) induces the release of cytokines TNF-α and IL-1β in a dose dependent-manner after 6 h incubation when inspected with enzyme-linked immunosorbent assay (ELISA). When BV2 cells are treated with different aggregation states and different concentrations (0, 0.1, 1, 2.5 μM) of α-syn, the fibrillic α-syn has the strongest effect on TNF-α and IL-1β excretion. TNF-α excretion hits a plateau at 1 μM dose, whereas IL-1β increases significantly only at the highest dose (2.5 μM). Monomeric α-syn induces a lower response, but interestingly, oligomeric α-syn fails to induce any activity. The aggregation state also affects α-syn internalization—monomers and oligomers are not taken into the cells as easily, whereas the fibrillar form gets phagocytosed. Depending on the concentration of α-syn fibrils, 60–80% of the cells internalized the fibrils. Moreover, the cells treated with α-syn monomers and fibrils appear to proliferate more (MTS-assay).

Further to this, [145] tested several in vitro preparations of α-syn in BV2 cells. They also found that α-syn fibrils (compared to monomers, oligomers, and ribbons) showed the most robust release of TNF-α. Interestingly, they observed that several familial mutations in the monomeric form of α-syn also induced TNF-α release compared to the WT (A53T, A53E, E46K, H50Q, G51D). Conversely, when BV2 are primed with monomeric α-syn (for 2, 6, or 12 h), the microglia take an anti-inflammatory phenotype through ERK, NF-κB, and PPARγ pathways [49]. This appears to attenuate the pro-inflammatory and neurotoxic effects caused by fibrillic α-syn.

In relation to microglial uptake of α-syn as mentioned above, using primary microglia cells isolated from an 8-month-old mouse brain, [146] demonstrated that the cells had decreased uptake of α-syn oligomers from the media of α-syn-overexpressing HEK293 cells compared to microglia from a 1 to 3-day-old mouse brain.

Only a couple of studies have been published using hiPSC-derived microglia in vitro in concert with exogenous α-syn. Similarly to BV2 cells, hiPSC-derived microglia treated with oligomeric α-syn (high molecular weight, purified from *E. coli*) had a significant amount of Il-1β secreted compared to monomeric treated microglia, as well as caspase-1 activation [147]. In the other study, the authors co-cultured microglia with astrocytes to understand if there was a synergistic effect between the two cell types on aggregates [122]. Using α-syn fibrils (along with amyloid-β fibrils) they found that both cell types take up the extracellular α-syn, but that it appears the microglia were able to degrade it more efficiently than astrocytes. When the cells were in culture together, the microglia took up more α-syn over time versus the astrocytes; however, the total accumulation of the protein was reduced compared to the monoculture setup. This relationship between microglia and astrocytes and how they process α-syn aggregates needs further study.

With more and more research pointing to microglia as a major player in neurodegenerative diseases, it is important to explore their role in synucleinopathy. Current studies show interesting results in terms of α-syn causing an inflammatory response in microglia and also how microglia may interact with it and take it up or clear it away. However, much more research is needed as there are still central questions that remain unanswered as it is still unclear whether microglia play a causative role in PD or whether they are only reacting to α-syn accumulation and neuronal loss.

## 5. Modeling Alpha-Synuclein Pathology Using Astrocytes

Astrocytes are the most abundant cells in the CNS, where they have many different functions [148,149]. They have an important role in the development of the CNS but also work as neuroprotective cells, providing both structural and metabolic support. In addition, they regulate the blood flow in microvessels and the permeability of the blood–brain barrier (BBB). They also contribute to the immune response via their cytokine and chemokine production [150]. In general, they possess other functions that are necessary for the normal function of the whole CNS.

A lot of knowledge of synucleinopathy in astrocytes has been achieved in the last two decades with animal in vivo and in vitro studies, but also in human in vitro studies. The development in hiPSC technology has provided a new source of human astrocytes for research use, which has made astrocyte studies more practicable and ethically less problematic compared to studies made with primary- or embryonic stem cell (ESC)-derived astrocytes. This development is clearly visible in studies, as most of the human astrocyte studies related to synucleinopathy have been made during the last few years and with hiPSC-derived astrocytes.

As early as 2006, the effect of α-syn on astrocytes was studied with human primary astrocytes by [151] to see how exposure to α-syn affects the astrocytes. The study showed that human α-syn fibrils (though it is not explicitly mentioned or measured in the study) purified from *E. coli* (WT, and mutated forms A30P, A53T, and E46K) induces an inflammatory response in astrocytes. A-syn induced the expression of inflammatory mediator ICAM-1 and secretion of IL-6, and it was also shown that ERK1/2, JNK, and p38 pathways that were associated with the actions of α-syn, were activated in the presence of α-syn, and their inhibition decreased the IL-6 levels. Much later in 2021, Russ et al. [123] showed with hiPSC-derived astrocytes that α-syn fibril exposure leads astrocytes towards an antigen-presenting phenotype with upregulation of major histocompatibility complex and human leukocyte antigen molecules whereas TNF-α initiates a strong pro-inflammatory cascade with activation of nuclear factor κB (NF-κB) pathway and release of pro-inflammatory cytokines.

In 2017, the transfer of α-syn between astrocytes was studied by Rostami et al. [152] using hESC-derived astrocytes. They found out that after exposure of astrocytes to oligomeric α-syn, the astrocytes failed to degrade the oligomeric α-syn during the studied period even though the monomeric form was degraded rapidly. The accumulation of α-syn leads to swelling of the endoplasmic reticulum and mitochondrial disturbances. They also observed that stressed astrocytes transferred intracellular α-syn via tunneling nanotubes to healthy astrocytes (similarly to studies in neurons discussed above) and healthy astrocytes transferred mitochondria to stressed astrocytes in the same way. However, it must be kept in mind that the transfer of α-syn happens also via other ways.

To see how astrocytes differ in PD patients and healthy individuals, the differences between healthy and PD astrocytes with the LRRK2 mutation were studied by Sonninen et al. 2020 [119]. We compared hiPSC-derived astrocytes from three control lines, one isogenic control, and four PD lines with the LRRK2 G2019S mutation. The study showed that astrocytes from PD patients expressed higher levels of α-syn and possessed a more reactive phenotype compared to healthy astrocytes. The PD astrocytes also had altered calcium signaling and metabolism.

As mentioned earlier, astrocytes have neuroprotective functions in the CNS. These functions have been studied by Domenico et al. 2019 [121] and Tsunemi et al. 2020 [120] using hiPSC-derived DA neurons and astrocytes. The studies show that healthy astrocytes are able to protect the neurons (also neurons from PD patients) from neurodegeneration. However, in astrocytes from PD patients, the neuroprotective capability is compromised and may even induce neurodegeneration. In the study by Domenico et al. 2019 [121], they co-cultured the astrocytes and neurons from controls (from healthy individuals) and PD patients (from patients with the LRRK2 G2019S mutation). They found that in healthy neurons cultured with PD astrocytes, the neurodegeneration was clearly detectable when compared to cultures with control astrocytes. Importantly, the accumulation of α-syn was increased in neurons cultured with PD astrocytes. Instead, when PD neurons were cultured with control astrocytes, the astrocytes decreased the neurodegeneration of PD neurons compared to co-cultures with PD astrocytes. The results also show that in PD astrocytes, autophagy was impaired, leading to α-syn accumulation in the astrocytes. In the study from Tsunemi et al. 2020 [120], neuronal α-syn was observed to be cleared by astrocytes. However, the ATP13A2 mutations reduced the astrocytic uptake and clearance of α-syn leading to α-syn accumulation in DA neurons. They also noticed that ATP13A2 deficiency reduced phagocytosis and the endosomal pathway in both neurons and astrocytes.

Besides co-cultures with neurons, the co-culture with microglia and its effect on α-syn pathology has been conducted. As described earlier with microglia, the study from Rostami et al. (2021) [122] showed that co-culture of astrocytes and microglia increases the degradation of α-syn.

Human astrocyte models have provided a lot of information on the effects of α-syn on astrocytes. However, the research is still limited, and more knowledge must be gained with human cells. All of the studies reviewed above have used the 2D culture of astrocytes to study the effects of α-syn. The 2D culture models are easy to use, but they are not as physiologically relevant as 3D models, in which the cells are able to form an in vivo-like morphology. In co-culture models of astrocytes and neurons, both contact [120] and non-contact [121] cultures were used. In the in vivo environment, the astrocytes and neurons are in close contact with each other, and part of the signaling pathways requires close contact between the cell types. Thus, the contact co-culture setup mimics the situation in vivo better, but it might restrict the analysis that can be made. When modeling processes such as synucleinopathy, this becomes particularly relevant as the interplay between neurons and glia is important.

## 6. Role of the Blood–Brain Barrier in Alpha-Synuclein Pathology

Endothelial cells (ECs) form the walls of all the vessels in the body [153]. However, in the CNS, the microvasculature has specific properties that are not found in the periphery. In the CNS, the microvasculature restricts the movement of molecules due to a higher number of tight junctions and transporters. This property of microvasculature in the CNS is called

BBB. In addition to the ECs, astrocytes and pericytes are needed to maintain and regulate the function of the BBB. The BBB is important for maintaining homeostasis in the CNS and protecting it from harmful substances such as inflammatory agents and toxins. However, the BBBs ability to not allow most molecules through also makes drug delivery to the CNS a difficult endeavor.

Dysfunction of the BBB is associated with different neurodegenerative diseases such as Alzheimer's disease, PD, Huntington's disease, and amyotrophic lateral sclerosis [154]. However, its role in the progression of the disease is still largely unknown. Both in vivo and in vitro models have been developed to study the role of the BBB in α-syn pathology. In this part, we go through the research on the role of the BBB in α-syn pathology. As there is so little research on the BBB and α-syn using human cells, in addition to human in vitro models, we will also review the research made with in vivo and in vitro animal models.

### 6.1. In Vivo Models

Several in vivo studies about α-syn and the BBB have been conducted to study different aspects of α-syn pathology. Some of them have concentrated on the transport of α-syn and others on the effect of α-syn on the BBB. Some of the studies have used in vitro models besides in vivo models to ensure findings and to further investigate the possible mechanisms behind the effect.

The transport of α-syn has been studied from different perspectives by Sui et al., and Matsumoto et al., using in vivo models. In 2014, Sui et al. [155] studied the transport of α-syn through the BBB in mice. They showed that α-syn can be transported into and out of the brain by the BBB, and that the transport could at least in part happen via low-density lipoprotein receptor-related protein-1 (LRP-1). Later, Matsumoto et al. (2017) [156] also studied α-syn transport to the brain, but via extracellular vesicles (EVs). The study was conducted by injecting labeled α-syn-rich EVs from humans (PD patients and controls) into the periphery of mice. The study showed that the EVs are transported to the brain, especially with pre-treatment of peripheral LPS, and that EVs from PD patients trigger a stronger inflammatory response in microglia compared to EVs from controls. These results suggest that systemic inflammation might promote the access of α-syn containing EVs into the brain by initiating an inflammatory response that might also further enhance neurodegeneration.

The effect of α-syn on the BBB in vivo has been studied with mice overexpressing human α-syn by Elabi et al., 2021 and Lan et al. 2021 [157,158]. In the study by Elabi et al., (2021), they observed that the BBB was compromised and pericytes were activated already at the early stages of the disease in mice with overexpression. Moreover, the density of microvessels was altered in mice with overexpression of α-syn. Instead, Lan et al., 2021 studied the mechanism leading to BBB disruption using an in vitro model in addition to an in vivo model. They saw a decrease in tight junction protein expression in ECs and an increase in BBB leakage caused by α-syn accumulation in and activation of astrocytes in vivo. Further in vitro studies showed that oligomeric α-syn induced BBB disruption mediated by the release of VEGFA and nitric oxide. The inhibition of the VEGFA signaling pathway protected the BBB from disruption in vivo and in vitro, so VEGFA seemed to be the primary mediator of BBB disruption.

Dohgu et al. (2019) [159] used an in vitro model using cells derived from animals to study α-syn pathology. They studied rat ECs and pericytes to elucidate how α-syn affects the BBB. They found that monomeric α-syn initiates the release of inflammatory mediators from pericytes, inducing BBB disruption.

### 6.2. Human in Vitro Models

Only two studies [125,160] of the role of BBB on α-syn pathology have been made using human cells and even in these the other also used mouse primary cells. These two studies were conducted to study the effects of α-syn on the permeability of the BBB and the functions of different cell types.

In 2016, Kuan et al., studied the effect of α-syn PFFs on immortalized EC line hCMEC/D3 in monoculture and co-culture with mouse primary astrocytes or primary human fetal cortical cells. The study showed the impairment in the expression of several tight junction-related proteins in the presence of α-syn, a result also found in PD patient brains, but still, the barrier function of the ECs was not affected except transiently in neuronal co-cultures [160].

Research from Pediaditakis et al., 2021 tested the effect of a α-syn PFFs on an organ-on-chip model of the neurovascular unit. In the research, they used hiPSC-derived brain microvascular Ecs and DA neurons, primary human astrocytes, pericytes, and microglia. The study showed that exposure of the neurovascular unit model to α-syn leads to the death of DA neurons, activation of microglia and astrocytes, as well as disruption of the BBB, mainly observed by increased permeability. However, the study did not distinguish whether the disruption of the BBB was caused directly by the exposure of ECs to α-syn, or indirectly through other cell types [125].

In BBB research, the in vivo models are still important for studying CNS functions as the cell models are still relatively simple compared to tissues and whole organisms and thus do not recapitulate the complex functions found in vivo. However, the development of in vitro models has enabled more physiologically relevant modeling that mimics the in vivo situation better than earlier models. One example of this is the organ-on-chip model that was used in the above-mentioned study from [125]. Before the development of hiPSC technology, in vitro studies were mostly made with animal cells, as they were easily available, unlike human samples, the availability of which is very limited. HiPSC technology has provided an unlimited source of human cells, which has made in vitro studies with human cells more practicable.

It is clear from these studies across different species and models that there is still a considerable amount to learn about synucleinopathy and α-syn aggregation related to the BBB. Models that use multiple cell types to understand how the BBB may be disrupted in disease are however bringing us closer to understanding this aspect of neurodegeneration.

## 7. Oligodendrocytes and Synucleinopathy

Oligodendrocytes are myelinating cells in the CNS. Defects in oligodendrocyte function can impair signal transduction due to the loss of the insulating myelin sheaths on the axons of the neurons and aggravate degeneration through the loss of neurotrophic factors. Collectively, these conditions are called demyelinating diseases, of which multiple sclerosis is the most common. The role of oligodendrocytes in synucleinopathies is not well understood, but emerging evidence has implicated their dysfunction as a possible factor in these diseases. As mentioned, MSA α-syn accumulates in oligodendrocytes and forms glial cytoplasmic inclusions (GCIs) [5], which has been recognized as a core pathological feature of the disease. Mature oligodendrocytes have not been found to express *SNCA*, leading to speculation that the GCIs are formed by exogenous α-syn being taken up by the cells. However, *SNCA* expression is transiently upregulated during oligodendrocyte maturation [161,162], leading to an alternative hypothesis that the gene may be dysregulated in MSA.

Oligodendrocyte progenitor cells (OPCs) are retained throughout adulthood. In the case of injury to the myelin sheath, OPCs can mature into myelinating oligodendrocytes and repair the damage. Importantly, it has been noted that the number of OPCs is increased in the brains of MSA patients. Despite the abundant OPC pool, there is considerable degeneration of myelin that is spatially associated with a high GCI burden [163]. Mice overexpressing human wild-type *SNCA* under the *MBP* promoter showed an increased number of OPCs in conjunction with myelin loss, mirroring the pathology found in brain tissue samples of MSA patients [164].

In vitro, overexpression of *SNCA* disrupted oligodendroglial maturation in the CG4 cell line [164], with similar findings reported in primary rat OPCs [165]. Transcriptomic profiling of OPCs derived from MSA and PD patient iPSCs has also shown impaired maturation into myelinating oligodendrocytes, with the phenotype shifting more towards an antigen-presenting cell type [126]. Contrary to previous studies using rat cells, Azevedo et al., found that *SNCA* overexpression increased the expression of maturation and myelination-associated genes in healthy iPSC-derived OPCs. However, the authors propose that α-syn expression may be necessary for early lineage commitment of precursor cells but impedes maturation at later stages [126].

While mature oligodendrocytes have not been found to uptake exogenous α-syn fibrils, OPCs and intermediate oligodendrocytes are competent. Primary rat OPCs treated with human α-syn PFFs develop intracellular inclusions containing endogenous rat α-syn, which are retained by the cell through maturation. Furthermore, α-syn PFFs induce the expression of endogenous α-syn in OPCs but not in mature cells [126]. Fibrillary α-syn perturbs normal oligodendrocyte maturation and has been shown to have cytotoxic effects if introduced at intermediate stages of development [126]. In healthy iPSC-derived OPCs, exposure to exogenous α-syn species resulted in the upregulation of genes associated with immune functions. In agreement with the findings by Kaji et al., (2020), the exposure also resulted in considerable toxicity that was more pronounced for fibrillar α-syn [126].

Whether oligodendrocyte dysfunction in MSA is the result of dysregulation of α-syn expression in maturating cells, uptake of aggregated α-syn by OPCs or a mix of both factors remains to be settled. While bona fide neuronal inclusions are rarely present in MSA, a study utilizing α-syn proximity-ligation revealed a high neuronal oligomeric α-syn load in MSA patient brain tissue [166]. The authors suggest that the oligomeric α-syn may have direct neurotoxic effects, while the uptake of neuron-derived oligomers by OPCs results in the characteristic GCI formation and demyelination. Still, further work is needed to understand the complex etiology of α-syn pathology in oligodendrocytes and how it may contribute to neurodegeneration in MSA and other synucleinopathies.

## 8. HiPSC-Derived Brain Organoids as an Emerging Modeling Platform for Synucleinopathies

Brain organoids have emerged as a promising new in vitro cell model for various neurological disorders. They are especially attractive for modeling developmental and familial diseases, as they have been shown to recapitulate human neurodevelopment and can be generated from patient-derived hiPSCs [167–169]. Directed differentiation of brain organoids enables the generation of more specific models for studying diseases affecting only certain brain regions or cell populations. In 2016, Jo et al., successfully generated midbrain-like organoids from hiPSCs, which could be used as PD models [170]. In addition to providing a human brain-like model system, brain organoids can be generated in large numbers for high-throughput experiments used in toxicity screening and drug development [171,172]. Additionally, they give the advantage of the surrounding cell types, not just DA neurons alone.

Along these lines, while brain organoids have been shown to contain a variety of ectoderm-derived neuronal and glial cell types, microglia are typically absent. During brain development in vivo, mesoderm-derived erythromyeloid progenitors migrate from the yolk sac and colonize the fetal CNS, giving rise to the adult microglia population [173–175]. Most protocols for deriving brain organoids include the use of dual-SMAD inhibitors to guide differentiation towards the neuroectoderm [176]. Although it is possible for microglia to emerge from residual mesodermal progenitors in unguided brain organoid differentiation, batch-to-batch variation in the distribution and numbers of microglia is a major limitation of this approach [177]. A possible solution to this is to derive the microglia progenitors separately and incorporate them into the organoids [178–181].

Brain organoids lack ECs and therefore do not develop vasculature. This limits the transfer of nutrients and typically results in a necrotic core, which can compromise neuronal survival and functionality. In addition, necrotic cells inadvertently activate microglia and astrocytes. Vascularizing brain organoids improves nutrient diffusion and allows for the construction of neurovascular models as well as studying BBB function. Transplanted brain organoids can be vascularized in vivo by invading vessels from the host [182], although the nonhuman origin of these vessels is a limitation. According to Pham et al., (2018), the vasculature can be humanized by incorporating human ECs into the organoids prior to transplantation [183]. Fully in vitro approaches include direct conversion to ECs [184], co-culture with human umbilical vein ECs [185], and assembloids composed of the brain and vascular organoids [186–188].

Toxin-based animal models of PD have been a standard in the field for decades. Still, they do not represent the true progression of the disease as follows: the loss of DA neurons is rapid and there is no synucleinopathy involved. It should be noted, however, that midbrain organoids can recapitulate the loss of TH+ neurons in vitro when exposed to toxic agents like MPTP or 6-OHDA [189]. While the relevance of these models to PD as seen in humans is debatable, the susceptibility of organoids to MPTP toxicity is an important finding. MPTP itself is not toxic and must be converted to MPP+ by the MAO-B enzyme, which is highly expressed in astrocytes. Recapitulation of this toxic cascade in midbrain organoids supports the notion that 3D cell models can mimic the physiological microenvironment of brain tissue.

As discussed earlier, mutations affecting the SNCA gene confer a high risk of developing synucleinopathy. Recently, Mohamed et al., (2021) generated midbrain organoids from hiPSCs carrying the SNCA triplication. They reported that these aged SNCA triplication organoids develop pS129 α-syn+ aggregates, which co-localize with both astrocytes and neurons. They also found the organoids to have decreased numbers of DA neurons [127]. Recent findings also suggest that combined SNCA overexpression and glucocerebridase deficiency can further promote the induction of synucleinopathy-like pathology in midbrain organoids [134]. While midbrain organoids are a natural choice for modeling PD, cerebral organoids could provide a platform for studying DLB and PDD, which share many of the genetic risk factors with PD. ApoE variants are known to affect Alzheimer's disease risk, with the APOE4 isoform conferring a significantly increased risk. Interestingly, APOE4 has also been associated with the risk of DLB and PDD [190]. In an article published in 2021, Zhao et al., reported increased α-syn aggregation in ApoE$^{-/-}$ cerebral organoids. Abnormal lipid metabolism and decreased glucocerebrosidase activity were also noted. Furthermore, it was shown that treatment with ApoE2 or ApoE3, but not ApoE4, could partially rescue the phenotype. Organoids generated from hiPSC lines homozygous for *APOE4* also showed increased levels of insoluble α-syn when compared to those derived from *APOE3* homozygote lines [132].

Mutations affecting the LRRK2 gene are the most common risk factor for developing familial PD. Furthermore, the presentation is often very similar to that of idiopathic PD. In 2019, Kim et al., showed midbrain organoids with inserted LRRK2-G2019S mutation to have lower gene expression of DA neuron markers TH, AADC, and DAT in comparison to the isogenic control. In addition, mutant organoids were reported to be more susceptible to MPTP toxicity. They also identified thioredoxin-interacting protein (TXNIP) as an intermediate actor in the pathology [131]. TXNIP has previously been reported to modulate α-syn accumulation via autophagy inhibition [191].

The current focus of hiPSC-derived midbrain organoid models for synucleinopathy has been rather narrow, with the effort being put into studying familial forms of the disease. With multiple genetic factors affecting disease risk, midbrain organoids carrying well-known risk variants have garnered the most interest. Indeed, there have been multiple reports of PD-like pathology arising organically in mutant organoids. In addition, some variants seem to increase the susceptibility of DA neurons to toxicity by MPTP or 6-OHDA, suggesting that these mutations can make DA neurons more vulnerable to stressors. In 2022, Rodriguez et al., showed that fibrillar α-syn is internalized by hiPSC-derived brain organoids and induces the formation of intracellular inclusions [133]. It should be emphasized, however, that no studies exposing midbrain organoids specifically to α-syn PFFs have been published yet. Furthermore, midbrain organoids intrinsically lack microglia. While there have been reports of microglia successfully incorporated into brain organoids [178,179], including a recent report on midbrain organoids [181], studies on α-syn processing and pathology in microglia-enhanced organoids have not been published.

Brain organoid technology has progressed rapidly in recent years. Improved culture systems, optimized differentiation protocols, and the development of new scaffold materials can help mitigate notable issues such as variability in represented cell populations and necrosis due to insufficient nutrient supply. More work is needed to create a truly comprehensive in vitro model of the human brain, including vasculature and microglia. There is a need for more focus on the function of microglia and how they interact with astrocytes and neurons and whether they contribute to the spreading of pathological α-syn in the brain. In addition, hiPSC-derived models of late-onset disease are inherently limited by the immature cell phenotype, which is not representative of the aged brain. This limitation was also highlighted in a recent review by Nogueira et al., (2022). They suggest that post-mortem human brain tissue samples or ex vivo organotypic culture should be used to validate the findings obtained from organoid models [192]. Nevertheless, with future advancements and standardization of methodology, brain organoids could provide a powerful platform for studying molecular and cellular disease mechanisms driving synucleinopathy in a biologically relevant, human-based system.

## 9. Conclusions and Future

Although α-syn is involved in several diseases after thousands of publications, it is still not fully known how it contributes to disease initiation or progression. While α-syn is present normally in the human brain, not everyone develops neurodegenerative disorders such as PD. Additionally, while aging is a major factor in sporadic PD, it is likely that other factors such as inflammation, the environment, and genetics contribute as well. Therefore, studying α-syn in a single cell type in a dish (whether of human origin or not) without further perturbation will not unequivocally answer these questions.

On the other hand, while studying the human brain in vivo using imaging techniques has given us useful knowledge, it is not always possible to gain detailed information on the pathophysiology of the disease. Additionally, obtaining patient post-mortem samples is not available to all labs, sample sizes are not large for rare diseases, and this only gives endpoint information. Here, cellular models have been instrumental in helping researchers understand the mechanism of α-syn aggregation and how it works in the cascade of disease initiation and progression. Immortalized cell lines and primary cells from rodents have been used robustly in the field, and with the advent of hiPSC technology, researchers hope to gain better insight into these human diseases. In particular, models that incorporate multiple cell types through co-culture, chips, or organoids are exciting for understanding synucleinopathies and other neurodegenerative diseases that can affect several cells and pathways in the CNS. With this comes some loss of experimental control, but their relevance to the human condition is higher than cell lines or animals alone.

In terms of α-syn itself, as discussed above, better models of Lewy bodies that would be observed in patients are necessary to bring the field forward. We have seen that there are several ways to model α-syn aggregation in vitro using a combination of techniques. Each of these has pros and cons, but we must always be clear and certain of what the outcome measures are and how the model is relevant to finding new therapies and understanding disease mechanisms. Perhaps currently, there is no "one size fits all" model of synucleinopathy. Furthermore, there can be a variety of types of aggregates, and whether and which aggregates are causing disease is still not fully clear. A recent study demonstrated that it is actually small soluble α-syn aggregates that cause toxicity to cells, and these resemble those found in the post-mortem patient's brain [193]. A major question is also how a single protein is causing a spectrum of diseases. α-syn is a protein with different conformations and post-translational modifications, both also depending on conditions. Lately, there have been interesting publications studying conformations of α-syn using cryo-EM, where the authors, for example, have shown that α-syn filaments from MSA patients differ from those with DLB [194]. (See [195] for a review of recent research about different α-syn strains in relation to disease). This also has the potential to bring us toward more disease-relevant models. Importantly, not every PD patient develops Lewy bodies [196], nor does amount of α-syn-rich Lewy bodies always correlate with cell loss [197]. All of this leads to the idea that personalized medicine is the future of treatment for patients with synucleinopathy. Which kinds of aggregates are present and at what level is important if we are treating patients with pathology that likely began several years before symptoms appear. Here, hiPSCs are a powerful tool—cells can be grown from the patient and drugs can be tested on how they affect aggregation and cell function, for example.

Therefore, to conclude, using in vitro cellular models, regardless of source, is still a useful and viable way to test conditions that would be more time-consuming, expensive, and labor-intensive in vivo. The types of models discussed in this review all have their ways of contributing to further understanding of α-syn aggregation as long as we also take into account their limitations.

**Author Contributions:** K.A. and Š.L. conceived of the review idea, commented on and prepared the final manuscript. K.A. wrote the introduction and conclusions, contributed to the microglia part, and added text to the entire manuscript. S.K. made the graphical abstract and wrote the oligodendrocyte and organoid parts. V.H. and V.S. wrote the non-hiPSC and hiPSC neuronal parts. P.K. prepared the tables. J.N. contributed to the microglia part. S.P. wrote the astrocytes and BBB parts. T.-M.S. commented on the astrocytes and BBB parts. All authors have read and agreed to the published version of the manuscript.

**Funding:** K.A. was funded by the Orion Foundation and the Päivikki and Sakari Sohlberg Foundation. V.H. was funded through the Finnish Parkinson Foundation. Š.L. was funded through grants from Sigrid Jusélius Foundation, Janne and Aatos Erkko Foundation and the Päivikki and Sakari Sohlberg Foundation.

**Institutional Review Board Statement:** Not applicable.

**Informed Consent Statement:** Not applicable.

**Conflicts of Interest:** The authors declare no conflict of interest.

## References

1. Polymeropoulos, M.H.; Lavedan, C.; Leroy, E.; Ide, S.E.; Dehejia, A.; Dutra, A.; Pike, B.; Root, H.; Rubenstein, J.; Boyer, R.; et al. Mutation in the α-Synuclein Gene Identified in Families with Parkinson's Disease. *Science* **1997**, *276*, 2045–2047. [CrossRef] [PubMed]

2. Spillantini, M.G.; Schmidt, M.L.; Lee, V.M.; Trojanowski, J.Q.; Jakes, R.; Goedert, M. Alpha-Synuclein in Lewy Bodies. *Nature* **1997**, *388*, 839–840. [CrossRef] [PubMed]

3. Galvin, J.E.; Lee, V.M.-Y.; Trojanowski, J.Q. Synucleinopathies: Clinical and Pathological Implications. *Arch. Neurol.* **2001**, *58*, 186–190. [CrossRef] [PubMed]

4. Lippa, C.F.; Duda, J.E.; Grossman, M.; Hurtig, H.I.; Aarsland, D.; Boeve, B.F.; Brooks, D.J.; Dickson, D.W.; Dubois, B.; Emre, M.; et al. DLB and PDD Boundary Issues: Diagnosis, Treatment, Molecular Pathology, and Biomarkers. *Neurology* **2007**, *68*, 812–819. [CrossRef] [PubMed]

5. Spillantini, M.G.; Crowther, R.A.; Jakes, R.; Cairns, N.J.; Lantos, P.L.; Goedert, M. Filamentous Alpha-Synuclein Inclusions Link Multiple System Atrophy with Parkinson's Disease and Dementia with Lewy Bodies. *Neurosci. Lett.* **1998**, *251*, 205–208. [CrossRef]

6. Kasen, A.; Houck, C.; Burmeister, A.R.; Sha, Q.; Brundin, L.; Brundin, P. Upregulation of α-Synuclein Following Immune Activation: Possible Trigger of Parkinson's Disease. *Neurobiol. Dis.* **2022**, *166*, 105654. [CrossRef] [PubMed]

7. Ferreon, A.C.M.; Gambin, Y.; Lemke, E.A.; Deniz, A.A. Interplay of α-Synuclein Binding and Conformational Switching Probed by Single-Molecule Fluorescence. *Proc. Natl. Acad. Sci. USA* **2009**, *106*, 5645–5650. [CrossRef]

8. Frimpong, A.K.; Abzalimov, R.R.; Uversky, V.N.; Kaltashov, I.A. Characterization of Intrinsically Disordered Proteins with Electrospray Ionization Mass Spectrometry: Conformational Heterogeneity of α-Synuclein. *Proteins* **2010**, *78*, 714–722. [CrossRef]

9. Theillet, F.-X.; Binolfi, A.; Bekei, B.; Martorana, A.; Rose, H.M.; Stuiver, M.; Verzini, S.; Lorenz, D.; van Rossum, M.; Goldfarb, D.; et al. Structural Disorder of Monomeric α-Synuclein Persists in Mammalian Cells. *Nature* **2016**, *530*, 45–50. [CrossRef]

10. Alam, P.; Bousset, L.; Melki, R.; Otzen, D.E. α-Synuclein Oligomers and Fibrils: A Spectrum of Species, a Spectrum of Toxicities. *J. Neurochem.* **2019**, *150*, 522–534. [CrossRef]

11. Giasson, B.I.; Murray, I.V.; Trojanowski, J.Q.; Lee, V.M. A Hydrophobic Stretch of 12 Amino Acid Residues in the Middle of Alpha-Synuclein Is Essential for Filament Assembly. *J. Biol. Chem.* **2001**, *276*, 2380–2386. [CrossRef]

12. El-Agnaf, O.M.; Jakes, R.; Curran, M.D.; Wallace, A. Effects of the Mutations Ala30 to Pro and Ala53 to Thr on the Physical and Morphological Properties of Alpha-Synuclein Protein Implicated in Parkinson's Disease. *FEBS Lett.* **1998**, *440*, 67–70. [CrossRef]

13. Lashuel, H.A.; Overk, C.R.; Oueslati, A.; Masliah, E. The Many Faces of α-Synuclein: From Structure and Toxicity to Therapeutic Target. *Nat. Rev. Neurosci.* **2013**, *14*, 38–48. [CrossRef]

14. Babu, M.M.; van der Lee, R.; de Groot, N.S.; Gsponer, J. Intrinsically Disordered Proteins: Regulation and Disease. *Curr. Opin. Struct. Biol.* **2011**, *21*, 432–440. [CrossRef] [PubMed]

15. Ryan, V.H.; Fawzi, N.L. Physiological, Pathological, and Targetable Membraneless Organelles in Neurons. *Trends Neurosci.* **2019**, *42*, 693–708. [CrossRef] [PubMed]

16. Ray, S.; Singh, N.; Kumar, R.; Patel, K.; Pandey, S.; Datta, D.; Mahato, J.; Panigrahi, R.; Navalkar, A.; Mehra, S.; et al. α-Synuclein Aggregation Nucleates through Liquid–Liquid Phase Separation. *Nat. Chem.* **2020**, *12*, 705–716. [CrossRef] [PubMed]

17. Breydo, L.; Wu, J.W.; Uversky, V.N. α-Synuclein Misfolding and Parkinson's Disease. *Biochim. Et Biophys. Acta (BBA)–Mol. Basis Dis.* **2012**, *1822*, 261–285. [CrossRef] [PubMed]

18. Kumar, S.T.; Jagannath, S.; Francois, C.; Vanderstichele, H.; Stoops, E.; Lashuel, H.A. How Specific Are the Conformation-Specific α-Synuclein Antibodies? Characterization and Validation of 16 α-Synuclein Conformation-Specific Antibodies Using Well-Characterized Preparations of α-Synuclein Monomers, Fibrils and Oligomers with Distinct Structures and Morphology. *Neurobiol. Dis.* **2020**, *146*, 105086. [CrossRef]

19. Krüger, R.; Kuhn, W.; Müller, T.; Woitalla, D.; Graeber, M.; Kösel, S.; Przuntek, H.; Epplen, J.T.; Schöls, L.; Riess, O. Ala30Pro Mutation in the Gene Encoding Alpha-Synuclein in Parkinson's Disease. *Nat. Genet.* **1998**, *18*, 106–108. [CrossRef]

20. Pasanen, P.; Myllykangas, L.; Siitonen, M.; Raunio, A.; Kaakkola, S.; Lyytinen, J.; Tienari, P.J.; Pöyhönen, M.; Paetau, A. Novel α-Synuclein Mutation A53E Associated with Atypical Multiple System Atrophy and Parkinson's Disease-Type Pathology. *Neurobiol. Aging* **2014**, *35*, 2180.e1-5. [CrossRef]

21. Chartier-Harlin, M.-C.; Kachergus, J.; Roumier, C.; Mouroux, V.; Douay, X.; Lincoln, S.; Levecque, C.; Larvor, L.; Andrieux, J.; Hulihan, M.; et al. Alpha-Synuclein Locus Duplication as a Cause of Familial Parkinson's Disease. *Lancet* **2004**, *364*, 1167–1169. [CrossRef]

22. Singleton, A.B.; Farrer, M.; Johnson, J.; Singleton, A.; Hague, S.; Kachergus, J.; Hulihan, M.; Peuralinna, T.; Dutra, A.; Nussbaum, R.; et al. α-Synuclein Locus Triplication Causes Parkinson's Disease. *Science* **2003**, *302*, 841–841. [CrossRef]

23. Sahay, S.; Ghosh, D.; Singh, P.K.; Maji, S.K. Alteration of Structure and Aggregation of α-Synuclein by Familial Parkinson's Disease Associated Mutations. *Curr. Protein Pept. Sci.* **2017**, *18*, 656–676. [CrossRef]

24. Lázaro, D.F.; Rodrigues, E.F.; Langohr, R.; Shahpasandzadeh, H.; Ribeiro, T.; Guerreiro, P.; Gerhardt, E.; Kröhnert, K.; Klucken, J.; Pereira, M.D.; et al. Systematic Comparison of the Effects of Alpha-Synuclein Mutations on Its Oligomerization and Aggregation. *PLoS Genet.* **2014**, *10*, e1004741. [CrossRef]

25. Ranjan, P.; Kumar, A. Perturbation in Long-Range Contacts Modulates the Kinetics of Amyloid Formation in α-Synuclein Familial Mutants. *ACS Chem. Neurosci.* **2017**, *8*, 2235–2246. [CrossRef]

26. Zhou, T.; Lin, D.; Chen, Y.; Peng, S.; Jing, X.; Lei, M.; Tao, E.; Liang, Y. α-Synuclein Accumulation in SH-SY5Y Cell Impairs Autophagy in Microglia by Exosomes Overloading MiR-19a-3p. *Epigenomics* **2019**, *11*, 1661–1677. [CrossRef]

27. Oliveira, S.R.; Dionísio, P.A.; Gaspar, M.M.; Correia Guedes, L.; Coelho, M.; Rosa, M.M.; Ferreira, J.J.; Amaral, J.D.; Rodrigues, C.M.P. MiR-335 Targets LRRK2 and Mitigates Inflammation in Parkinson's Disease. *Front. Cell Dev. Biol.* **2021**, *9*, 661461. [CrossRef]

28. Risiglione, P.; Cubisino, S.A.M.; Lipari, C.L.R.; De Pinto, V.; Messina, A.; Magrì, A. α-Synuclein A53T Promotes Mitochondrial Proton Gradient Dissipation and Depletion of the Organelle Respiratory Reserve in a Neuroblastoma Cell Line. *Life* **2022**, *12*, 894. [CrossRef]

29. Potdar, C.; Kaushal, A.; Raj, A.; Mallick, R.; Datta, I. Reduction of Phosphorylated α-Synuclein through Downregulation of Casein Kinase 2α Alleviates Dopaminergic-Neuronal Function. *Biochem. Biophys. Res. Commun.* **2022**, *615*, 43–48. [CrossRef]

30. Hivare, P.; Gadhavi, J.; Bhatia, D.; Gupta, S. α-Synuclein Fibrils Explore Actin-Mediated Macropinocytosis for Cellular Entry into Model Neuroblastoma Neurons. *Traffic* **2022**, *23*, 391–410. [CrossRef]

31. Sang, J.C.; Hidari, E.; Meisl, G.; Ranasinghe, R.T.; Spillantini, M.G.; Klenerman, D. Super-Resolution Imaging Reveals α-Synuclein Seeded Aggregation in SH-SY5Y Cells. *Commun. Biol.* **2021**, *4*, 613. [CrossRef]

32. Shearer, L.J.; Petersen, N.O.; Woodside, M.T. Internalization of α-Synuclein Oligomers into SH-SY5Y Cells. *Biophys. J.* **2021**, *120*, 877–885. [CrossRef]

33. Li, S.; Raja, A.; Noroozifar, M.; Kerman, K. Understanding the Inhibitory and Antioxidant Effects of Pyrroloquinoline Quinone (PQQ) on Copper(II)-Induced α-Synuclein-119 Aggregation. *ACS Chem. Neurosci.* **2022**, *13*, 1178–1186. [CrossRef]

34. Wang, H.; Tang, C.; Jiang, Z.; Zhou, X.; Chen, J.; Na, M.; Shen, H.; Lin, Z. Glutamine Promotes Hsp70 and Inhibits α-Synuclein Accumulation in Pheochromocytoma PC12 Cells. *Exp. Med.* **2017**, *14*, 1253–1259. [CrossRef]

35. Ortega, R.; Carmona, A.; Roudeau, S.; Perrin, L.; Dučić, T.; Carboni, E.; Bohic, S.; Cloetens, P.; Lingor, P. α-Synuclein Over-Expression Induces Increased Iron Accumulation and Redistribution in Iron-Exposed Neurons. *Mol. Neurobiol.* **2016**, *53*, 1925–1934. [CrossRef]

36. Jinsmaa, Y.; Cooney, A.; Sullivan, P.; Sharabi, Y.; Goldstein, D.S. The Serotonin Aldehyde, 5-HIAL, Oligomerizes Alpha-Synuclein. *Neurosci. Lett.* **2015**, *590*, 134–137. [CrossRef]

37. Zondler, L.; Kostka, M.; Garidel, P.; Heinzelmann, U.; Hengerer, B.; Mayer, B.; Weishaupt, J.H.; Gillardon, F.; Danzer, K.M. Proteasome Impairment by α-Synuclein. *PLoS ONE* **2017**, *12*, e0184040. [CrossRef] [PubMed]

38. Heravi, M.; Dargahi, L.; Parsafar, S.; Tayaranian Marvian, A.; Aliakbari, F.; Morshedi, D. The Primary Neuronal Cells Are More Resistant than PC12 Cells to α-Synuclein Toxic Aggregates. *Neurosci. Lett.* **2019**, *701*, 38–47. [CrossRef] [PubMed]
39. Hornedo-Ortega, R.; Cerezo, A.B.; Troncoso, A.M.; Garcia-Parrilla, M.C. Protective Effects of Hydroxytyrosol against α-Synuclein Toxicity on PC12 cells and Fibril Formation. *Food Chem. Toxicol.* **2018**, *120*, 41–49. [CrossRef] [PubMed]
40. Karmacharya, M.B.; Hada, B.; Park, S.R.; Choi, B.H. Low-Intensity Ultrasound Decreases α-Synuclein Aggregation via Attenuation of Mitochondrial Reactive Oxygen Species in MPP(+)-Treated PC12 Cells. *Mol. Neurobiol.* **2017**, *54*, 6235–6244. [CrossRef] [PubMed]
41. Höllerhage, M.; Stepath, M.; Kohl, M.; Pfeiffer, K.; Chua, O.W.H.; Duan, L.; Hopfner, F.; Eisenacher, M.; Marcus, K.; Höglinger, G.U. Transcriptome and Proteome Analysis in LUHMES Cells Overexpressing Alpha-Synuclein. *Front. Neurol.* **2022**, *13*, 787059. [CrossRef]
42. Prahl, J.; Pierce, S.E.; Coetzee, G.A.; Tyson, T. Alpha-Synuclein Negatively Controls Cell Proliferation in Dopaminergic Neurons. *Mol. Cell. Neurosci.* **2022**, *119*, 103702. [CrossRef]
43. Höllerhage, M.; Moebius, C.; Melms, J.; Chiu, W.-H.; Goebel, J.N.; Chakroun, T.; Koeglsperger, T.; Oertel, W.H.; Rösler, T.W.; Bickle, M.; et al. Protective Efficacy of Phosphodiesterase-1 Inhibition against Alpha-Synuclein Toxicity Revealed by Compound Screening in LUHMES Cells. *Sci. Rep.* **2017**, *7*, 11469. [CrossRef]
44. Shaffner, K. Alpha-Synuclein Overexpression Induces Epigenomic Dysregulation of Glutamate Signaling and Locomotor Pathways | Human Molecular Genetics | Oxford Academic. Available online: https://academic.oup.com/hmg/advance-article/doi/10.1093/hmg/ddac104/6585888?login=true (accessed on 6 July 2022).
45. Lv, Y.-C.; Gao, A.-B.; Yang, J.; Zhong, L.-Y.; Jia, B.; Ouyang, S.-H.; Gui, L.; Peng, T.-H.; Sun, S.; Cayabyab, F.S. Long-Term Adenosine A1 Receptor Activation-Induced Sortilin Expression Promotes α-Synuclein Upregulation in Dopaminergic Neurons. *Neural. Regen. Res.* **2019**, *15*, 712–723. [CrossRef]
46. Zhou, X.; Huang, N.; Hou, X.; Zhu, L.; Xie, Y.; Ba, Z.; Luo, Y. Icaritin Attenuates 6-OHDA-Induced MN9D Cell Damage by Inhibiting Oxidative Stress. *PeerJ.* **2022**, *10*, e13256. [CrossRef]
47. Qiu, W.-Q.; Ai, W.; Zhu, F.-D.; Zhang, Y.; Guo, M.-S.; Law, B.Y.-K.; Wu, J.-M.; Wong, V.K.-W.; Tang, Y.; Yu, L.; et al. Polygala Saponins Inhibit NLRP3 Inflammasome-Mediated Neuroinflammation via SHP-2-Mediated Mitophagy. *Free Radic. Biol. Med.* **2022**, *179*, 76–94. [CrossRef]
48. Yang, X.; Ma, H.; Yv, Q.; Ye, F.; He, Z.; Chen, S.; Keram, A.; Li, W.; Zhu, M. Alpha-Synuclein/MPP+ Mediated Activation of NLRP3 Inflammasome through Microtubule-Driven Mitochondrial Perinuclear Transport. *Biochem. Biophys. Res. Commun.* **2022**, *594*, 161–167. [CrossRef]
49. Li, N.; Stewart, T.; Sheng, L.; Shi, M.; Cilento, E.M.; Wu, Y.; Hong, J.-S.; Zhang, J. Immunoregulation of Microglial Polarization: An Unrecognized Physiological Function of α-Synuclein. *J. Neuroinflammation* **2020**, *17*, 272. [CrossRef]
50. Tu, H.; Yuan, B.; Hou, X.; Zhang, X.; Pei, C.; Ma, Y.; Yang, Y.; Fan, Y.; Qin, Z.; Liu, C.; et al. A-synuclein Suppresses Microglial Autophagy and Promotes Neurodegeneration in a Mouse Model of Parkinson's Disease. *Aging Cell* **2021**, *20*, e13522. [CrossRef]
51. Hoffmann, A.; Ettle, B.; Bruno, A.; Kulinich, A.; Hoffmann, A.-C.; von Wittgenstein, J.; Winkler, J.; Xiang, W.; Schlachetzki, J.C.M. Alpha-Synuclein Activates BV2 Microglia Dependent on Its Aggregation State. *Biochem. Biophys. Res. Commun.* **2016**, *479*, 881–886. [CrossRef]
52. Jovanovic, P.; Wang, Y.; Vit, J.-P.; Novinbakht, E.; Morones, N.; Hogg, E.; Tagliati, M.; Riera, C.E. Sustained Chemogenetic Activation of Locus Coeruleus Norepinephrine Neurons Promotes Dopaminergic Neuron Survival in Synucleinopathy. *PLoS ONE* **2022**, *17*, e0263074. [CrossRef]
53. Szegő, E.M.; Van den Haute, C.; Höfs, L.; Baekelandt, V.; Van der Perren, A.; Falkenburger, B.H. Rab7 Reduces α-Synuclein Toxicity in Rats and Primary Neurons. *Exp. Neurol.* **2022**, *347*, 113900. [CrossRef]
54. Nicholatos, J.W.; Tran, D.; Liu, Y.; Hirst, W.D.; Weihofen, A. Lysophosphatidylcholine Acyltransferase 1 Promotes Pathology and Toxicity in Two Distinct Cell-Based Alpha-Synuclein Models. *Neurosci. Lett.* **2022**, *772*, 136491. [CrossRef]
55. Brzozowski, C.F.; Hijaz, B.A.; Singh, V.; Gcwensa, N.Z.; Kelly, K.; Boyden, E.S.; West, A.B.; Sarkar, D.; Volpicelli-Daley, L.A. Inhibition of LRRK2 Kinase Activity Promotes Anterograde Axonal Transport and Presynaptic Targeting of α-Synuclein. *Acta Neuropathol. Commun.* **2021**, *9*, 180. [CrossRef]
56. Hlushchuk, I.; Barut, J.; Airavaara, M.; Luk, K.; Domanskyi, A.; Chmielarz, P. Cell Culture Media, Unlike the Presence of Insulin, Affect α-Synuclein Aggregation in Dopaminergic Neurons. *Biomolecules* **2022**, *12*, 563. [CrossRef]
57. Hideshima, M.; Kimura, Y.; Aguirre, C.; Kakuda, K.; Takeuchi, T.; Choong, C.-J.; Doi, J.; Nabekura, K.; Yamaguchi, K.; Nakajima, K.; et al. Two-Step Screening Method to Identify α-Synuclein Aggregation Inhibitors for Parkinson's Disease. *Sci. Rep.* **2022**, *12*, 351. [CrossRef]
58. Szegő, É.M.; Boß, F.; Komnig, D.; Gärtner, C.; Höfs, L.; Shaykhalishahi, H.; Wördehoff, M.M.; Saridaki, T.; Schulz, J.B.; Hoyer, W.; et al. A β-Wrapin Targeting the N-Terminus of α-Synuclein Monomers Reduces Fibril-Induced Aggregation in Neurons. *Front. Neurosci.* **2021**, *15*, 696440. [CrossRef]
59. Li, Y.; Glotfelty, E.J.; Karlsson, T.; Fortuno, L.V.; Harvey, B.K.; Greig, N.H. The Metabolite GLP-1 (9-36) Is Neuroprotective and Anti-Inflammatory in Cellular Models of Neurodegeneration. *J. Neurochem.* **2021**, *159*, 867–886. [CrossRef]
60. Choi, D.C.; Yoo, M.; Kabaria, S.; Junn, E. MicroRNA-7 Facilitates the Degradation of Alpha-Synuclein and Its Aggregates by Promoting Autophagy. *Neurosci. Lett.* **2018**, *678*, 118–123. [CrossRef]

61. Wiatrak, B.; Kubis-Kubiak, A.; Piwowar, A.; Barg, E. PC12 Cell Line: Cell Types, Coating of Culture Vessels, Differentiation and Other Culture Conditions. *Cells* **2020**, *9*, E958. [CrossRef]

62. Stefanis, L.; Larsen, K.E.; Rideout, H.J.; Sulzer, D.; Greene, L.A. Expression of A53T Mutant but Not Wild-Type Alpha-Synuclein in PC12 Cells Induces Alterations of the Ubiquitin-Dependent Degradation System, Loss of Dopamine Release, and Autophagic Cell Death. *J. Neurosci.* **2001**, *21*, 9549–9560. [CrossRef] [PubMed]

63. Larsen, K.E.; Schmitz, Y.; Troyer, M.D.; Mosharov, E.; Dietrich, P.; Quazi, A.Z.; Savalle, M.; Nemani, V.; Chaudhry, F.A.; Edwards, R.H.; et al. α-Synuclein Overexpression in PC12 and Chromaffin Cells Impairs Catecholamine Release by Interfering with a Late Step in Exocytosis. *J. Neurosci.* **2006**, *26*, 11915–11922. [CrossRef] [PubMed]

64. Lotharius, J.; Barg, S.; Wiekop, P.; Lundberg, C.; Raymon, H.K.; Brundin, P. Effect of Mutant Alpha-Synuclein on Dopamine Homeostasis in a New Human Mesencephalic Cell Line. *J. Biol. Chem.* **2002**, *277*, 38884–38894. [CrossRef] [PubMed]

65. Tönges, L.; Szegö, É.M.; Hause, P.; Saal, K.-A.; Tatenhorst, L.; Koch, J.C.; d'Hedouville, Z.; Dambeck, V.; Kügler, S.; Dohm, C.P.; et al. Alpha-Synuclein Mutations Impair Axonal Regeneration in Models of Parkinson's Disease. *Front. Aging Neurosci.* **2014**, *6*, 239. [CrossRef] [PubMed]

66. Li, L.; Nadanaciva, S.; Berger, Z.; Shen, W.; Paumier, K.; Schwartz, J.; Mou, K.; Loos, P.; Milici, A.J.; Dunlop, J.; et al. Human A53T α-Synuclein Causes Reversible Deficits in Mitochondrial Function and Dynamics in Primary Mouse Cortical Neurons. *PLoS ONE* **2013**, *8*, e85815. [CrossRef]

67. Engelender, S.; Kaminsky, Z.; Guo, X.; Sharp, A.H.; Amaravi, R.K.; Kleiderlein, J.J.; Margolis, R.L.; Troncoso, J.C.; Lanahan, A.A.; Worley, P.F.; et al. Synphilin-1 Associates with α-Synuclein and Promotes the Formation of Cytosolic Inclusions. *Nat. Genet.* **1999**, *22*, 110–114. [CrossRef]

68. Tanaka, M.; Kim, Y.M.; Lee, G.; Junn, E.; Iwatsubo, T.; Mouradian, M.M. Aggresomes Formed by α-Synuclein and Synphilin-1 Are Cytoprotective. *J. Biol. Chem.* **2004**, *279*, 4625–4631. [CrossRef]

69. Zhang, S.; Eitan, E.; Wu, T.-Y.; Mattson, M.P. Intercellular Transfer of Pathogenic α-Synuclein by Extracellular Vesicles Is Induced by the Lipid Peroxidation Product 4-Hydroxynonenal. *Neurobiol. Aging* **2018**, *61*, 52–65. [CrossRef]

70. Petrucci, S.; Ginevrino, M.; Valente, E.M. Phenotypic Spectrum of Alpha-Synuclein Mutations: New Insights from Patients and Cellular Models. *Park. Relat. Disord.* **2016**, *22*, S16–S20. [CrossRef]

71. Pacelli, C.; Giguère, N.; Bourque, M.-J.; Lévesque, M.; Slack, R.S.; Trudeau, L.-É. Elevated Mitochondrial Bioenergetics and Axonal Arborization Size Are Key Contributors to the Vulnerability of Dopamine Neurons. *Curr. Biol.* **2015**, *25*, 2349–2360. [CrossRef]

72. Liu, H.-S.; Jan, M.-S.; Chou, C.-K.; Chen, P.-H.; Ke, N.-J. Is Green Fluorescent Protein Toxic to the Living Cells? *Biochem. Biophys. Res. Commun.* **1999**, *260*, 712–717. [CrossRef]

73. Detrait, E.R.; Bowers, W.J.; Halterman, M.W.; Giuliano, R.E.; Bennice, L.; Federoff, H.J.; Richfield, E.K. Reporter Gene Transfer Induces Apoptosis in Primary Cortical Neurons. *Mol. Ther.* **2002**, *5*, 723–730. [CrossRef]

74. Howard, D.B.; Powers, K.; Wang, Y.; Harvey, B.K. Tropism and Toxicity of Adeno-Associated Viral Vector Serotypes 1, 2, 5, 6, 7, 8, and 9 in Rat Neurons and Glia in Vitro. *Virology* **2008**, *372*, 24–34. [CrossRef]

75. Xicoy, H.; Wieringa, B.; Martens, G.J.M. The SH-SY5Y Cell Line in Parkinson's Disease Research: A Systematic Review. *Mol. Neurodegener.* **2017**, *12*, 10. [CrossRef]

76. Luk, K.C.; Song, C.; O'Brien, P.; Stieber, A.; Branch, J.R.; Brunden, K.R.; Trojanowski, J.Q.; Lee, V.M.-Y. Exogenous α-Synuclein Fibrils Seed the Formation of Lewy Body-like Intracellular Inclusions in Cultured Cells. *Proc. Natl. Acad. Sci. USA* **2009**, *106*, 20051–20056. [CrossRef]

77. Wu, Q.; Takano, H.; Riddle, D.M.; Trojanowski, J.Q.; Coulter, D.A.; Lee, V.M.-Y. α-Synuclein (ASyn) Preformed Fibrils Induce Endogenous ASyn Aggregation, Compromise Synaptic Activity and Enhance Synapse Loss in Cultured Excitatory Hippocampal Neurons. *J. Neurosci.* **2019**, *39*, 5080–5094. [CrossRef]

78. Volpicelli-Daley, L.A.; Luk, K.C.; Lee, V.M.-Y. Addition of Exogenous α-Synuclein Preformed Fibrils to Primary Neuronal Cultures to Seed Recruitment of Endogenous α-Synuclein to Lewy Body and Lewy Neurite–like Aggregates. *Nat. Protoc.* **2014**, *9*, 2135–2146. [CrossRef]

79. Luk, K.C.; Kehm, V.M.; Zhang, B.; O'Brien, P.; Trojanowski, J.Q.; Lee, V.M.Y. Intracerebral Inoculation of Pathological α-Synuclein Initiates a Rapidly Progressive Neurodegenerative α-Synucleinopathy in Mice. *J. Exp. Med.* **2012**, *209*, 975–986. [CrossRef]

80. Luk, K.C.; Kehm, V.; Carroll, J.; Zhang, B.; O'Brien, P.; Trojanowski, J.Q.; Lee, V.M.-Y. Pathological α-Synuclein Transmission Initiates Parkinson-like Neurodegeneration in Nontransgenic Mice. *Science* **2012**, *338*, 949–953. [CrossRef]

81. Masuda-Suzukake, M.; Nonaka, T.; Hosokawa, M.; Oikawa, T.; Arai, T.; Akiyama, H.; Mann, D.M.A.; Hasegawa, M. Prion-like Spreading of Pathological α-Synuclein in Brain. *Brain* **2013**, *136*, 1128–1138. [CrossRef]

82. Cascella, R.; Chen, S.W.; Bigi, A.; Camino, J.D.; Xu, C.K.; Dobson, C.M.; Chiti, F.; Cremades, N.; Cecchi, C. The Release of Toxic Oligomers from α-Synuclein Fibrils Induces Dysfunction in Neuronal Cells. *Nat. Commun.* **2021**, *12*, 1814. [CrossRef] [PubMed]

83. Polinski, N.K.; Volpicelli-Daley, L.A.; Sortwell, C.E.; Luk, K.C.; Cremades, N.; Gottler, L.M.; Froula, J.; Duffy, M.F.; Lee, V.M.Y.; Martinez, T.N.; et al. Best Practices for Generating and Using Alpha-Synuclein Pre-Formed Fibrils to Model Parkinson's Disease in Rodents. *J. Parkinsons Dis.* **2018**, *8*, 303–322. [CrossRef] [PubMed]

84. Abdelmotilib, H.; Maltbie, T.; Delic, V.; Liu, Z.; Hu, X.; Fraser, K.B.; Moehle, M.S.; Stoyka, L.; Anabtawi, N.; Krendelchtchikova, V.; et al. α-Synuclein Fibril-Induced Inclusion Spread in Rats and Mice Correlates with Dopaminergic Neurodegeneration. *Neurobiol. Dis.* **2017**, *105*, 84–98. [CrossRef] [PubMed]

85. Er, S.; Hlushchuk, I.; Airavaara, M.; Chmielarz, P.; Domanskyi, A. Studying Pre-Formed Fibril Induced α-Synuclein Accumulation in Primary Embryonic Mouse Midbrain Dopamine Neurons. *J. Vis. Exp.* **2020**, *162*, e61118. [CrossRef]

86. Braak, H.; Del Tredici, K.; Rüb, U.; de Vos, R.A.I.; Jansen Steur, E.N.H.; Braak, E. Staging of Brain Pathology Related to Sporadic Parkinson's Disease. *Neurobiol. Aging* **2003**, *24*, 197–211. [CrossRef]

87. McCann, H.; Cartwright, H.; Halliday, G.M. Neuropathology of α-Synuclein Propagation and Braak Hypothesis. *Mov. Disord.* **2016**, *31*, 152–160. [CrossRef] [PubMed]

88. Holmes, B.B.; DeVos, S.L.; Kfoury, N.; Li, M.; Jacks, R.; Yanamandra, K.; Ouidja, M.O.; Brodsky, F.M.; Marasa, J.; Bagchi, D.P.; et al. Heparan Sulfate Proteoglycans Mediate Internalization and Propagation of Specific Proteopathic Seeds. *Proc. Natl. Acad. Sci. USA* **2013**, *110*, E3138–E3147. [CrossRef]

89. Ihse, E.; Yamakado, H.; van Wijk, X.M.; Lawrence, R.; Esko, J.D.; Masliah, E. Cellular Internalization of Alpha-Synuclein Aggregates by Cell Surface Heparan Sulfate Depends on Aggregate Conformation and Cell Type. *Sci. Rep.* **2017**, *7*, 9008. [CrossRef]

90. Abounit, S.; Bousset, L.; Loria, F.; Zhu, S.; de Chaumont, F.; Pieri, L.; Olivo-Marin, J.-C.; Melki, R.; Zurzolo, C. Tunneling Nanotubes Spread Fibrillar α-Synuclein by Intercellular Trafficking of Lysosomes. *EMBO J.* **2016**, *35*, 2120–2138. [CrossRef]

91. Grudina, C.; Kouroupi, G.; Nonaka, T.; Hasegawa, M.; Matsas, R.; Zurzolo, C. Human NPCs Can Degrade α-Syn Fibrils and Transfer Them Preferentially in a Cell Contact-Dependent Manner Possibly through TNT-like Structures. *Neurobiol. Dis.* **2019**, *132*, 104609. [CrossRef]

92. Senol, A.D.; Samarani, M.; Syan, S.; Guardia, C.M.; Nonaka, T.; Liv, N.; Latour-Lambert, P.; Hasegawa, M.; Klumperman, J.; Bonifacino, J.S.; et al. α-Synuclein Fibrils Subvert Lysosome Structure and Function for the Propagation of Protein Misfolding between Cells through Tunneling Nanotubes. *PLOS Biol.* **2021**, *19*, e3001287. [CrossRef]

93. Danzer, K.M.; Ruf, W.P.; Putcha, P.; Joyner, D.; Hashimoto, T.; Glabe, C.; Hyman, B.T.; McLean, P.J. Heat-Shock Protein 70 Modulates Toxic Extracellular α-Synuclein Oligomers and Rescues Trans-Synaptic Toxicity. *FASEB J. Off. Publ. Fed. Am. Soc. Exp. Biol.* **2011**, *25*, 326–336. [CrossRef]

94. Freundt, E.C.; Maynard, N.; Clancy, E.K.; Roy, S.; Bousset, L.; Sourigues, Y.; Covert, M.; Melki, R.; Kirkegaard, K.; Brahic, M. Neuron-to-Neuron Transmission of α-Synuclein Fibrils through Axonal Transport. *Ann. Neurol.* **2012**, *72*, 517–524. [CrossRef]

95. Brahic, M.; Bousset, L.; Bieri, G.; Melki, R.; Gitler, A.D. Axonal Transport and Secretion of Fibrillar Forms of α-Synuclein, Aβ42 Peptide and HTTExon 1. *Acta Neuropathol.* **2016**, *131*, 539–548. [CrossRef]

96. Halliday, G.M.; Holton, J.L.; Revesz, T.; Dickson, D.W. Neuropathology Underlying Clinical Variability in Patients with Synucleinopathies. *Acta Neuropathol.* **2011**, *122*, 187–204. [CrossRef]

97. Karpowicz, R.J.; Haney, C.M.; Mihaila, T.S.; Sandler, R.M.; Petersson, E.J.; Lee, V.M.-Y. Selective Imaging of Internalized Proteopathic α-Synuclein Seeds in Primary Neurons Reveals Mechanistic Insight into Transmission of Synucleinopathies. *J. Biol. Chem.* **2017**, *292*, 13482–13497. [CrossRef]

98. Shahmoradian, S.H.; Lewis, A.J.; Genoud, C.; Hench, J.; Moors, T.E.; Navarro, P.P.; Castaño-Díez, D.; Schweighauser, G.; Graff-Meyer, A.; Goldie, K.N.; et al. Lewy Pathology in Parkinson's Disease Consists of Crowded Organelles and Lipid Membranes. *Nat. Neurosci.* **2019**, *22*, 1099–1109. [CrossRef]

99. Mahul-Mellier, A.-L.; Burtscher, J.; Maharjan, N.; Weerens, L.; Croisier, M.; Kuttler, F.; Leleu, M.; Knott, G.W.; Lashuel, H.A. The Process of Lewy Body Formation, Rather than Simply α-Synuclein Fibrillization, Is One of the Major Drivers of Neurodegeneration. *PNAS* **2020**, *117*, 4971–4982. [CrossRef]

100. Trinkaus, V.A.; Riera-Tur, I.; Martínez-Sánchez, A.; Bäuerlein, F.J.B.; Guo, Q.; Arzberger, T.; Baumeister, W.; Dudanova, I.; Hipp, M.S.; Hartl, F.U.; et al. In Situ Architecture of Neuronal α-Synuclein Inclusions. *Nat. Commun.* **2021**, *12*, 2110. [CrossRef]

101. Fujioka, S.; Ogaki, K.; Tacik, P.; Uitti, R.J.; Ross, O.A.; Wszolek, Z.K. Update on Novel Familial Forms of Parkinson's Disease and Multiple System Atrophy. *Park. Relat. Disord.* **2014**, *20*, S29–S34. [CrossRef]

102. Oliveira, L.M.A.; Falomir-Lockhart, L.J.; Botelho, M.G.; Lin, K.-H.; Wales, P.; Koch, J.C.; Gerhardt, E.; Taschenberger, H.; Outeiro, T.F.; Lingor, P.; et al. Elevated α-Synuclein Caused by SNCA Gene Triplication Impairs Neuronal Differentiation and Maturation in Parkinson's Patient-Derived Induced Pluripotent Stem Cells. *Cell Death Dis.* **2015**, *6*, e1994. [CrossRef]

103. Lin, L.; Göke, J.; Cukuroglu, E.; Dranias, M.R.; VanDongen, A.M.J.; Stanton, L.W. Molecular Features Underlying Neurodegeneration Identified through In Vitro Modeling of Genetically Diverse Parkinson's Disease Patients. *Cell Rep.* **2016**, *15*, 2411–2426. [CrossRef]

104. Tagliafierro, L.; Zamora, M.E.; Chiba-Falek, O. Multiplication of the SNCA Locus Exacerbates Neuronal Nuclear Aging. *Hum. Mol. Genet.* **2019**, *28*, 407–421. [CrossRef]

105. Pozo Devoto, V.M.; Dimopoulos, N.; Alloatti, M.; Pardi, M.B.; Saez, T.M.; Otero, M.G.; Cromberg, L.E.; Marín-Burgin, A.; Scassa, M.E.; Stokin, G.B.; et al. A Synuclein Control of Mitochondrial Homeostasis in Human-Derived Neurons Is Disrupted by Mutations Associated with Parkinson's Disease. *Sci. Rep.* **2017**, *7*, 5042. [CrossRef]

106. Prots, I.; Grosch, J.; Brazdis, R.-M.; Simmnacher, K.; Veber, V.; Havlicek, S.; Hannappel, C.; Krach, F.; Krumbiegel, M.; Schütz, O.; et al. α-Synuclein Oligomers Induce Early Axonal Dysfunction in Human IPSC-Based Models of Synucleinopathies. *Proc. Natl. Acad. Sci. USA* **2018**, *115*, 7813–7818. [CrossRef]

107. Ryan, B.J.; Hoek, S.; Fon, E.A.; Wade-Martins, R. Mitochondrial Dysfunction and Mitophagy in Parkinson's: From Familial to Sporadic Disease. *Trends Biochem. Sci.* **2015**, *40*, 200–210. [CrossRef]

108. Rocha, E.M.; De Miranda, B.; Sanders, L.H. Alpha-Synuclein: Pathology, Mitochondrial Dysfunction and Neuroinflammation in Parkinson's Disease. *Neurobiol. Dis.* **2018**, *109*, 249–257. [CrossRef]

109. Heger, L.M.; Wise, R.M.; Hees, J.T.; Harbauer, A.B.; Burbulla, L.F. Mitochondrial Phenotypes in Parkinson's Diseases-A Focus on Human IPSC-Derived Dopaminergic Neurons. *Cells* **2021**, *10*, 3436. [CrossRef]

110. Grassi, D.; Diaz-Perez, N.; Volpicelli-Daley, L.A.; Lasmézas, C.I. Pα-Syn* Mitotoxicity Is Linked to MAPK Activation and Involves Tau Phosphorylation and Aggregation at the Mitochondria. *Neurobiol. Dis.* **2019**, *124*, 248–262. [CrossRef]

111. Seebauer, L.; Schneider, Y.; Drobny, A.; Plötz, S.; Koudelka, T.; Tholey, A.; Prots, I.; Winner, B.; Zunke, F.; Winkler, J.; et al. Interaction of Alpha Synuclein and Microtubule Organization Is Linked to Impaired Neuritic Integrity in Parkinson's Patient-Derived Neuronal Cells. *Int. J. Mol. Sci.* **2022**, *23*, 1812. [CrossRef]

112. Akrioti, E.; Karamitros, T.; Gkaravelas, P.; Kouroupi, G.; Matsas, R.; Taoufik, E. Early Signs of Molecular Defects in IPSC-Derived Neural Stems Cells from Patients with Familial Parkinson's Disease. *Biomolecules* **2022**, *12*, 876. [CrossRef] [PubMed]

113. Hallacli, E.; Kayatekin, C.; Nazeen, S.; Wang, X.H.; Sheinkopf, Z.; Sathyakumar, S.; Sarkar, S.; Jiang, X.; Dong, X.; Di Maio, R.; et al. The Parkinson's Disease Protein Alpha-Synuclein Is a Modulator of Processing Bodies and MRNA Stability. *Cell* **2022**, *185*, 2035–2056.e33. [CrossRef] [PubMed]

114. Prieto Huarcaya, S.; Drobny, A.; Marques, A.R.A.; Di Spiezio, A.; Dobert, J.P.; Balta, D.; Werner, C.; Rizo, T.; Gallwitz, L.; Bub, S.; et al. Recombinant Pro-CTSD (Cathepsin D) Enhances SNCA/α-Synuclein Degradation in α-Synucleinopathy Models. *Autophagy* **2022**, *18*, 1127–1151. [CrossRef] [PubMed]

115. Stojkovska, I.; Wani, W.Y.; Zunke, F.; Belur, N.R.; Pavlenko, E.A.; Mwenda, N.; Sharma, K.; Francelle, L.; Mazzulli, J.R. Rescue of α-Synuclein Aggregation in Parkinson's Patient Neurons by Synergistic Enhancement of ER Proteostasis and Protein Trafficking. *Neuron* **2022**, *110*, 436–451.e11. [CrossRef] [PubMed]

116. Ferrer-Lorente, R.; Lozano-Cruz, T.; Fernández-Carasa, I.; Miłowska, K.; de la Mata, F.J.; Bryszewska, M.; Consiglio, A.; Ortega, P.; Gómez, R.; Raya, A. Cationic Carbosilane Dendrimers Prevent Abnormal α-Synuclein Accumulation in Parkinson's Disease Patient-Specific Dopamine Neurons. *Biomacromolecules* **2021**, *22*, 4582–4591. [CrossRef] [PubMed]

117. Vajhøj, C.; Schmid, B.; Alik, A.; Melki, R.; Fog, K.; Holst, B.; Stummann, T.C. Establishment of a Human Induced Pluripotent Stem Cell Neuronal Model for Identification of Modulators of A53T α-Synuclein Levels and Aggregation. *PLoS ONE* **2021**, *16*, e0261536. [CrossRef]

118. Gribaudo, S.; Tixador, P.; Bousset, L.; Fenyi, A.; Lino, P.; Melki, R.; Peyrin, J.-M.; Perrier, A.L. Propagation of α-Synuclein Strains within Human Reconstructed Neuronal Network. *Stem Cell Rep.* **2019**, *12*, 230–244. [CrossRef]

119. Sonninen, T.-M.; Hämäläinen, R.H.; Koskuvi, M.; Oksanen, M.; Shakirzyanova, A.; Wojciechowski, S.; Puttonen, K.; Naumenko, N.; Goldsteins, G.; Laham-Karam, N.; et al. Metabolic Alterations in Parkinson's Disease Astrocytes. *Sci. Rep.* **2020**, *10*, 14474. [CrossRef]

120. Tsunemi, T.; Ishiguro, Y.; Yoroisaka, A.; Valdez, C.; Miyamoto, K.; Ishikawa, K.; Saiki, S.; Akamatsu, W.; Hattori, N.; Krainc, D. Astrocytes Protect Human Dopaminergic Neurons from α-Synuclein Accumulation and Propagation. *J. Neurosci.* **2020**, *40*, 8618–8628. [CrossRef]

121. Di Domenico, A.; Carola, G.; Calatayud, C.; Pons-Espinal, M.; Muñoz, J.P.; Richaud-Patin, Y.; Fernandez-Carasa, I.; Gut, M.; Faella, A.; Parameswaran, J.; et al. Patient-Specific IPSC-Derived Astrocytes Contribute to Non-Cell-Autonomous Neurodegeneration in Parkinson's Disease. *Stem Cell Rep.* **2019**, *12*, 213–229. [CrossRef]

122. Rostami, J.; Mothes, T.; Kolahdouzan, M.; Eriksson, O.; Moslem, M.; Bergström, J.; Ingelsson, M.; O'Callaghan, P.; Healy, L.M.; Falk, A.; et al. Crosstalk between Astrocytes and Microglia Results in Increased Degradation of α-Synuclein and Amyloid-β Aggregates. *J. Neuroinflammation* **2021**, *18*, 124. [CrossRef]

123. Russ, K.; Teku, G.; Bousset, L.; Redeker, V.; Piel, S.; Savchenko, E.; Pomeshchik, Y.; Savistchenko, J.; Stummann, T.C.; Azevedo, C.; et al. TNF-α and α-Synuclein Fibrils Differently Regulate Human Astrocyte Immune Reactivity and Impair Mitochondrial Respiration. *Cell Rep.* **2021**, *34*, 108895. [CrossRef]

124. Filippini, A.; Mutti, V.; Faustini, G.; Longhena, F.; Ramazzina, I.; Rizzi, F.; Kaganovich, A.; Roosen, D.A.; Landeck, N.; Duffy, M.; et al. Extracellular Clusterin Limits the Uptake of A-synuclein Fibrils by Murine and Human Astrocytes. *Glia* **2021**, *69*, 681–696. [CrossRef]

125. Pediaditakis, I.; Kodella, K.R.; Manatakis, D.V.; Le, C.Y.; Hinojosa, C.D.; Tien-Street, W.; Manolakos, E.S.; Vekrellis, K.; Hamilton, G.A.; Ewart, L.; et al. Modeling Alpha-Synuclein Pathology in a Human Brain-Chip to Assess Blood-Brain Barrier Disruption. *Nat. Commun.* **2021**, *12*, 5907. [CrossRef]

126. Azevedo, C.; Teku, G.; Pomeshchik, Y.; Reyes, J.F.; Chumarina, M.; Russ, K.; Savchenko, E.; Hammarberg, A.; Lamas, N.J.; Collin, A.; et al. Parkinson's Disease and Multiple System Atrophy Patient IPSC-Derived Oligodendrocytes Exhibit Alpha-Synuclein–Induced Changes in Maturation and Immune Reactive Properties. *Proc. Natl. Acad. Sci. USA* **2022**, *119*, e2111405119. [CrossRef]

127. Mohamed, N.-V.; Sirois, J.; Ramamurthy, J.; Mathur, M.; Lépine, P.; Deneault, E.; Maussion, G.; Nicouleau, M.; Chen, C.X.-Q.; Abdian, N.; et al. Midbrain Organoids with an SNCA Gene Triplication Model Key Features of Synucleinopathy. *Brain Commun.* **2021**, *3*, fcab223. [CrossRef]

128. Brown, S.J.; Boussaad, I.; Jarazo, J.; Fitzgerald, J.C.; Antony, P.; Keatinge, M.; Blechman, J.; Schwamborn, J.C.; Krüger, R.; Placzek, M.; et al. PINK1 Deficiency Impairs Adult Neurogenesis of Dopaminergic Neurons. *Sci Rep.* **2021**, *11*, 6617. [CrossRef]

129. Kano, M.; Takanashi, M.; Oyama, G.; Yoritaka, A.; Hatano, T.; Shiba-Fukushima, K.; Nagai, M.; Nishiyama, K.; Hasegawa, K.; Inoshita, T.; et al. Reduced Astrocytic Reactivity in Human Brains and Midbrain Organoids with PRKN Mutations. *NPJ Park. Dis.* **2020**, *6*, 33. [CrossRef]

130. Zhu, W.; Tao, M.; Hong, Y.; Wu, S.; Chu, C.; Zheng, Z.; Han, X.; Zhu, Q.; Xu, M.; Ewing, A.G.; et al. Dysfunction of Vesicular Storage in Young-Onset Parkinson's Patient-Derived Dopaminergic Neurons and Organoids Revealed by Single Cell Electrochemical Cytometry. *Chem. Sci.* **2022**, *13*, 6217–6223. [CrossRef]

131. Kim, H.; Park, H.J.; Choi, H.; Chang, Y.; Park, H.; Shin, J.; Kim, J.; Lengner, C.J.; Lee, Y.K.; Kim, J. Modeling G2019S-LRRK2 Sporadic Parkinson's Disease in 3D Midbrain Organoids. *Stem Cell Rep.* **2019**, *12*, 518–531. [CrossRef]

132. Zhao, J.; Lu, W.; Ren, Y.; Fu, Y.; Martens, Y.A.; Shue, F.; Davis, M.D.; Wang, X.; Chen, K.; Li, F.; et al. Apolipoprotein E Regulates Lipid Metabolism and α-Synuclein Pathology in Human IPSC-Derived Cerebral Organoids. *Acta Neuropathol.* **2021**, *142*, 807–825. [CrossRef]

133. Rodrigues, P.V.; de Godoy, J.V.P.; Bosque, B.P.; Amorim Neto, D.P.; Tostes, K.; Palameta, S.; Garcia-Rosa, S.; Tonoli, C.C.C.; de Carvalho, H.F.; de Castro Fonseca, M. Transcellular Propagation of Fibrillar α-Synuclein from Enteroendocrine to Neuronal Cells Requires Cell-to-Cell Contact and Is Rab35-Dependent. *Sci. Rep.* **2022**, *12*, 4168. [CrossRef]

134. Jo, J.; Yang, L.; Tran, H.-D.; Yu, W.; Sun, A.X.; Chang, Y.Y.; Jung, B.C.; Lee, S.-J.; Saw, T.Y.; Xiao, B.; et al. Lewy Body–like Inclusions in Human Midbrain Organoids Carrying Glucocerebrosidase and α-Synuclein Mutations. *Ann. Neurol.* **2021**, *90*, 490–505. [CrossRef]

135. Wulansari, N.; Darsono, W.H.W.; Woo, H.-J.; Chang, M.-Y.; Kim, J.; Bae, E.-J.; Sun, W.; Lee, J.-H.; Cho, I.-J.; Shin, H.; et al. Neurodevelopmental Defects and Neurodegenerative Phenotypes in Human Brain Organoids Carrying Parkinson's Disease-Linked DNAJC6 Mutations. *Sci. Adv.* **2021**, *7*, eabb1540. [CrossRef]

136. Bousset, L.; Pieri, L.; Ruiz-Arlandis, G.; Gath, J.; Jensen, P.H.; Habenstein, B.; Madiona, K.; Olieric, V.; Böckmann, A.; Meier, B.H.; et al. Structural and Functional Characterization of Two Alpha-Synuclein Strains. *Nat. Commun.* **2013**, *4*, 2575. [CrossRef]

137. Kuznetsov, I.A.; Kuznetsov, A.V. Can the Lack of Fibrillar Form of Alpha-Synuclein in Lewy Bodies Be Explained by Its Catalytic Activity? *Math. Biosci.* **2022**, *344*, 108754. [CrossRef]

138. Angelova, P.R.; Ludtmann, M.H.R.; Horrocks, M.H.; Negoda, A.; Cremades, N.; Klenerman, D.; Dobson, C.M.; Wood, N.W.; Pavlov, E.V.; Gandhi, S.; et al. Ca2+ Is a Key Factor in α-Synuclein-Induced Neurotoxicity. *J. Cell Sci.* **2016**, *129*, 1792–1801. [CrossRef]

139. Paillusson, S.; Gomez-Suaga, P.; Stoica, R.; Little, D.; Gissen, P.; Devine, M.J.; Noble, W.; Hanger, D.P.; Miller, C.C.J. α-Synuclein Binds to the ER–Mitochondria Tethering Protein VAPB to Disrupt Ca2+ Homeostasis and Mitochondrial ATP Production. *Acta Neuropathol.* **2017**, *134*, 129–149. [CrossRef]

140. Tanudjojo, B.; Shaikh, S.S.; Fenyi, A.; Bousset, L.; Agarwal, D.; Marsh, J.; Zois, C.; Heman-Ackah, S.; Fischer, R.; Sims, D.; et al. Phenotypic Manifestation of α-Synuclein Strains Derived from Parkinson's Disease and Multiple System Atrophy in Human Dopaminergic Neurons. *Nat. Commun.* **2021**, *12*, 3817. [CrossRef]

141. Yehezkel, S.; Rebibo-Sabbah, A.; Segev, Y.; Tzukerman, M.; Shaked, R.; Huber, I.; Gepstein, L.; Skorecki, K.; Selig, S. Reprogramming of Telomeric Regions during the Generation of Human Induced Pluripotent Stem Cells and Subsequent Differentiation into Fibroblast-like Derivatives. *Epigenetics* **2011**, *6*, 63–75. [CrossRef]

142. Rohani, L.; Johnson, A.A.; Arnold, A.; Stolzing, A. The Aging Signature: A Hallmark of Induced Pluripotent Stem Cells? *Aging Cell* **2014**, *13*, 2–7. [CrossRef] [PubMed]

143. Tansey, M.G.; Wallings, R.L.; Houser, M.C.; Herrick, M.K.; Keating, C.E.; Joers, V. Inflammation and Immune Dysfunction in Parkinson Disease. *Nat. Rev. Immunol.* **2022**, 1–17. [CrossRef] [PubMed]

144. Blasi, E.; Barluzzi, R.; Bocchini, V.; Mazzolla, R.; Bistoni, F. Immortalization of Murine Microglial Cells by a V-Raf/v-Myc Carrying Retrovirus. *J. Neuroimmunol.* **1990**, *27*, 229–237. [CrossRef]

145. Grozdanov, V.; Bousset, L.; Hoffmeister, M.; Bliederhaeuser, C.; Meier, C.; Madiona, K.; Pieri, L.; Kiechle, M.; McLean, P.J.; Kassubek, J.; et al. Increased Immune Activation by Pathologic α-Synuclein in Parkinson's Disease. *Ann. Neurol.* **2019**, *86*, 593–606. [CrossRef] [PubMed]

146. Bliederhaeuser, C.; Grozdanov, V.; Speidel, A.; Zondler, L.; Ruf, W.P.; Bayer, H.; Kiechle, M.; Feiler, M.S.; Freischmidt, A.; Brenner, D.; et al. Age-Dependent Defects of Alpha-Synuclein Oligomer Uptake in Microglia and Monocytes. *Acta Neuropathol.* **2016**, *131*, 379–391. [CrossRef] [PubMed]

147. Trudler, D.; Nazor, K.L.; Eisele, Y.S.; Grabauskas, T.; Dolatabadi, N.; Parker, J.; Sultan, A.; Zhong, Z.; Goodwin, M.S.; Levites, Y.; et al. Soluble α-Synuclein–Antibody Complexes Activate the NLRP3 Inflammasome in HiPSC-Derived Microglia. *Proc. Natl. Acad. Sci. USA* **2021**, *118*, e2025847118. [CrossRef] [PubMed]

148. Booth, H.D.E.; Hirst, W.D.; Wade-Martins, R. The Role of Astrocyte Dysfunction in Parkinson's Disease Pathogenesis. *Trends Neurosci.* **2017**, *40*, 358–370. [CrossRef]

149. Sofroniew, M.V.; Vinters, H.V. Astrocytes: Biology and Pathology. *Acta Neuropathol.* **2010**, *119*, 7–35. [CrossRef]

150. Phatnani, H.; Maniatis, T. Astrocytes in Neurodegenerative Disease. *Cold Spring Harb. Perspect Biol.* **2015**, *7*, a020628. [CrossRef]

151. Klegeris, A.; Giasson, B.I.; Zhang, H.; Maguire, J.; Pelech, S.; McGeer, P.L. Alpha-Synuclein and Its Disease-Causing Mutants Induce ICAM-1 and IL-6 in Human Astrocytes and Astrocytoma Cells. *FASEB J.* **2006**, *20*, 2000–2008. [CrossRef]

152. Rostami, J.; Holmqvist, S.; Lindström, V.; Sigvardson, J.; Westermark, G.T.; Ingelsson, M.; Bergström, J.; Roybon, L.; Erlandsson, A. Human Astrocytes Transfer Aggregated Alpha-Synuclein via Tunneling Nanotubes. *J. Neurosci.* **2017**, *37*, 11835–11853. [CrossRef]

153. Daneman, R.; Prat, A. The Blood–Brain Barrier. *Cold Spring Harb. Perspect Biol.* **2015**, *7*, a020412. [CrossRef]

154. Sweeney, M.D.; Zhao, Z.; Montagne, A.; Nelson, A.R.; Zlokovic, B.V. Blood-Brain Barrier: From Physiology to Disease and Back. *Physiol. Rev.* **2019**, *99*, 21–78. [CrossRef]

155. Sui, Y.-T.; Bullock, K.M.; Erickson, M.A.; Zhang, J.; Banks, W.A. Alpha Synuclein Is Transported into and out of the Brain by the Blood–Brain Barrier. *Peptides* **2014**, *62*, 197–202. [CrossRef]

156. Matsumoto, J.; Stewart, T.; Sheng, L.; Li, N.; Bullock, K.; Song, N.; Shi, M.; Banks, W.A.; Zhang, J. Transmission of α-Synuclein-Containing Erythrocyte-Derived Extracellular Vesicles across the Blood-Brain Barrier via Adsorptive Mediated Transcytosis: Another Mechanism for Initiation and Progression of Parkinson's Disease? *Acta Neuropathol. Commun.* **2017**, *5*, 71. [CrossRef]

157. Elabi, O.; Gaceb, A.; Carlsson, R.; Padel, T.; Soylu-Kucharz, R.; Cortijo, I.; Li, W.; Li, J.-Y.; Paul, G. Human α-Synuclein Overexpression in a Mouse Model of Parkinson's Disease Leads to Vascular Pathology, Blood Brain Barrier Leakage and Pericyte Activation. *Sci. Rep.* **2021**, *11*, 1120. [CrossRef]

158. Lan, G.; Wang, P.; Chan, R.B.; Liu, Z.; Yu, Z.; Liu, X.; Yang, Y.; Zhang, J. Astrocytic VEGFA: An Essential Mediator in Blood–Brain-Barrier Disruption in Parkinson's Disease. *Glia* **2022**, *70*, 337–353. [CrossRef]

159. Dohgu, S.; Takata, F.; Matsumoto, J.; Kimura, I.; Yamauchi, A.; Kataoka, Y. Monomeric α-Synuclein Induces Blood–Brain Barrier Dysfunction through Activated Brain Pericytes Releasing Inflammatory Mediators in Vitro. *Microvasc. Res.* **2019**, *124*, 61–66. [CrossRef]

160. Kuan, W.-L.; Bennett, N.; He, X.; Skepper, J.N.; Martynyuk, N.; Wijeyekoon, R.; Moghe, P.V.; Williams-Gray, C.H.; Barker, R.A. α-Synuclein Pre-Formed Fibrils Impair Tight Junction Protein Expression without Affecting Cerebral Endothelial Cell Function. *Exp. Neurol.* **2016**, *285*, 72–81. [CrossRef]

161. Richter-Landsberg, C.; Gorath, M.; Trojanowski, J.Q.; Lee, V.M. Alpha-Synuclein Is Developmentally Expressed in Cultured Rat Brain Oligodendrocytes. *J. Neurosci. Res.* **2000**, *62*, 9–14. [CrossRef]

162. Djelloul, M.; Holmqvist, S.; Boza-Serrano, A.; Azevedo, C.; Yeung, M.S.; Goldwurm, S.; Frisén, J.; Deierborg, T.; Roybon, L. Alpha-Synuclein Expression in the Oligodendrocyte Lineage: An In Vitro and In Vivo Study Using Rodent and Human Models. *Stem Cell Rep.* **2015**, *5*, 174–184. [CrossRef]

163. Ahmed, Z.; Asi, Y.T.; Lees, A.J.; Revesz, T.; Holton, J.L. Identification and Quantification of Oligodendrocyte Precursor Cells in Multiple System Atrophy, Progressive Supranuclear Palsy and Parkinson's Disease. *Brain Pathol.* **2012**, *23*, 263–273. [CrossRef]

164. May, V.E.L.; Ettle, B.; Poehler, A.-M.; Nuber, S.; Ubhi, K.; Rockenstein, E.; Winner, B.; Wegner, M.; Masliah, E.; Winkler, J. α-Synuclein Impairs Oligodendrocyte Progenitor Maturation in Multiple System Atrophy. *Neurobiol. Aging* **2014**, *35*, 2357–2368. [CrossRef]

165. Ettle, B.; Reiprich, S.; Deusser, J.; Schlachetzki, J.C.M.; Xiang, W.; Prots, I.; Masliah, E.; Winner, B.; Wegner, M.; Winkler, J. Intracellular Alpha-Synuclein Affects Early Maturation of Primary Oligodendrocyte Progenitor Cells. *Mol. Cell Neurosci.* **2014**, *0*, 68–78. [CrossRef]

166. Sekiya, H.; Kowa, H.; Koga, H.; Takata, M.; Satake, W.; Futamura, N.; Funakawa, I.; Jinnai, K.; Takahashi, M.; Kondo, T.; et al. Wide Distribution of Alpha-Synuclein Oligomers in Multiple System Atrophy Brain Detected by Proximity Ligation. *Acta Neuropathol.* **2019**, *137*, 455–466. [CrossRef]

167. Camp, J.G.; Badsha, F.; Florio, M.; Kanton, S.; Gerber, T.; Wilsch-Bräuninger, M.; Lewitus, E.; Sykes, A.; Hevers, W.; Lancaster, M.; et al. Human Cerebral Organoids Recapitulate Gene Expression Programs of Fetal Neocortex Development. *Proc. Natl. Acad. Sci. USA* **2015**, *112*, 15672–15677. [CrossRef]

168. Lancaster, M.A.; Renner, M.; Martin, C.-A.; Wenzel, D.; Bicknell, L.S.; Hurles, M.E.; Homfray, T.; Penninger, J.M.; Jackson, A.P.; Knoblich, J.A. Cerebral Organoids Model Human Brain Development and Microcephaly. *Nature* **2013**, *501*, 373–379. [CrossRef]

169. Luo, C.; Lancaster, M.A.; Castanon, R.; Nery, J.R.; Knoblich, J.A.; Ecker, J.R. Cerebral Organoids Recapitulate Epigenomic Signatures of the Human Fetal Brain. *Cell Rep.* **2016**, *17*, 3369–3384. [CrossRef] [PubMed]

170. Jo, J.; Xiao, Y.; Sun, A.X.; Cukuroglu, E.; Tran, H.-D.; Göke, J.; Tan, Z.Y.; Saw, T.Y.; Tan, C.-P.; Lokman, H.; et al. Midbrain-like Organoids from Human Pluripotent Stem Cells Contain Functional Dopaminergic and Neuromelanin-Producing Neurons. *Cell Stem Cell* **2016**, *19*, 248–257. [CrossRef] [PubMed]

171. Renner, H.; Grabos, M.; Becker, K.J.; Kagermeier, T.E.; Wu, J.; Otto, M.; Peischard, S.; Zeuschner, D.; TsyTsyura, Y.; Disse, P.; et al. A Fully Automated High-Throughput Workflow for 3D-Based Chemical Screening in Human Midbrain Organoids. *eLife* **2020**, *9*, e52904. [CrossRef] [PubMed]

172. Renner, H.; Grabos, M.; Schöler, H.R.; Bruder, J.M. Generation and Maintenance of Homogeneous Human Midbrain Organoids. *Bio Protoc.* **2021**, *11*, e4049. [CrossRef]

173. Kierdorf, K.; Erny, D.; Goldmann, T.; Sander, V.; Schulz, C.; Perdiguero, E.G.; Wieghofer, P.; Heinrich, A.; Riemke, P.; Hölscher, C.; et al. Microglia Emerge from Erythromyeloid Precursors via Pu.1- and Irf8-Dependent Pathways. *Nat. Neurosci.* **2013**, *16*, 273–280. [CrossRef]

174. Gomez Perdiguero, E.; Klapproth, K.; Schulz, C.; Busch, K.; Azzoni, E.; Crozet, L.; Garner, H.; Trouillet, C.; de Bruijn, M.F.; Geissmann, F.; et al. Tissue-Resident Macrophages Originate from Yolk-Sac-Derived Erythro-Myeloid Progenitors. *Nature* **2015**, *518*, 547–551. [CrossRef]

175. Hoeffel, G.; Chen, J.; Lavin, Y.; Low, D.; Almeida, F.F.; See, P.; Beaudin, A.E.; Lum, J.; Low, I.; Forsberg, E.C.; et al. C-Myb(+) Erythro-Myeloid Progenitor-Derived Fetal Monocytes Give Rise to Adult Tissue-Resident Macrophages. *Immunity* **2015**, *42*, 665–678. [CrossRef]

176. Chambers, S.M.; Fasano, C.A.; Papapetrou, E.P.; Tomishima, M.; Sadelain, M.; Studer, L. Highly Efficient Neural Conversion of Human ES and IPS Cells by Dual Inhibition of SMAD Signaling. *Nat. Biotechnol.* **2009**, *27*, 275–280. [CrossRef]

177. Ormel, P.R.; Vieira de Sá, R.; van Bodegraven, E.J.; Karst, H.; Harschnitz, O.; Sneeboer, M.A.M.; Johansen, L.E.; van Dijk, R.E.; Scheefhals, N.; Berdenis van Berlekom, A.; et al. Microglia Innately Develop within Cerebral Organoids. *Nat. Commun.* **2018**, *9*, 4167. [CrossRef]

178. Abud, E.M.; Ramirez, R.N.; Martinez, E.S.; Healy, L.M.; Nguyen, C.H.H.; Newman, S.A.; Yeromin, A.V.; Scarfone, V.M.; Marsh, S.E.; Fimbres, C.; et al. IPSC-Derived Human Microglia-like Cells to Study Neurological Diseases. *Neuron* **2017**, *94*, 278–293.e9. [CrossRef]

179. Fagerlund, I.; Dougalis, A.; Shakirzyanova, A.; Gómez-Budia, M.; Pelkonen, A.; Konttinen, H.; Ohtonen, S.; Fazaludeen, M.F.; Koskuvi, M.; Kuusisto, J.; et al. Microglia-like Cells Promote Neuronal Functions in Cerebral Organoids. *Cells* **2021**, *11*, 124. [CrossRef]

180. Xu, R.; Boreland, A.J.; Li, X.; Erickson, C.; Jin, M.; Atkins, C.; Pang, Z.P.; Daniels, B.P.; Jiang, P. Developing Human Pluripotent Stem Cell-Based Cerebral Organoids with a Controllable Microglia Ratio for Modeling Brain Development and Pathology. *Stem Cell Rep.* **2021**, *16*, 1923–1937. [CrossRef]

181. Sabate-Soler, S.; Nickels, S.L.; Saraiva, C.; Berger, E.; Dubonyte, U.; Barmpa, K.; Lan, Y.J.; Kouno, T.; Jarazo, J.; Robertson, G.; et al. Microglia Integration into Human Midbrain Organoids Leads to Increased Neuronal Maturation and Functionality. *Glia* **2022**, *70*, 1267–1288. [CrossRef]

182. Mansour, A.A.; Gonçalves, J.T.; Bloyd, C.W.; Li, H.; Fernandes, S.; Quang, D.; Johnston, S.; Parylak, S.L.; Jin, X.; Gage, F.H. An in Vivo Model of Functional and Vascularized Human Brain Organoids. *Nat. Biotechnol.* **2018**, *36*, 432–441. [CrossRef] [PubMed]

183. Pham, M.T.; Pollock, K.M.; Rose, M.D.; Cary, W.A.; Stewart, H.R.; Zhou, P.; Nolta, J.A.; Waldau, B. Generation of Human Vascularized Brain Organoids. *Neuroreport* **2018**, *29*, 588–593. [CrossRef] [PubMed]

184. Cakir, B.; Xiang, Y.; Tanaka, Y.; Kural, M.H.; Parent, M.; Kang, Y.-J.; Chapeton, K.; Patterson, B.; Yuan, Y.; He, C.-S.; et al. Development of Human Brain Organoids with Functional Vascular-like System. *Nat. Methods* **2019**, *16*, 1169–1175. [CrossRef]

185. Shi, Y.; Sun, L.; Wang, M.; Liu, J.; Zhong, S.; Li, R.; Li, P.; Guo, L.; Fang, A.; Chen, R.; et al. Vascularized Human Cortical Organoids (VOrganoids) Model Cortical Development in Vivo. *PLoS Biol.* **2020**, *18*, e3000705. [CrossRef] [PubMed]

186. Ahn, Y.; An, J.-H.; Yang, H.-J.; Lee, D.G.; Kim, J.; Koh, H.; Park, Y.-H.; Song, B.-S.; Sim, B.-W.; Lee, H.J.; et al. Human Blood Vessel Organoids Penetrate Human Cerebral Organoids and Form a Vessel-Like System. *Cells* **2021**, *10*, 2036. [CrossRef] [PubMed]

187. Kook, M.G.; Lee, S.-E.; Shin, N.; Kong, D.; Kim, D.-H.; Kim, M.-S.; Kang, H.K.; Choi, S.W.; Kang, K.-S. Generation of Cortical Brain Organoid with Vascularization by Assembling with Vascular Spheroid. *Int. J. Stem Cells* **2022**, *15*, 85–94. [CrossRef] [PubMed]

188. Sun, X.-Y.; Ju, X.-C.; Li, Y.; Zeng, P.-M.; Wu, J.; Zhou, Y.-Y.; Shen, L.-B.; Dong, J.; Chen, Y.-J.; Luo, Z.-G. Generation of Vascularized Brain Organoids to Study Neurovascular Interactions. *eLife* **2022**, *11*, e76707. [CrossRef]

189. Kwak, T.H.; Kang, J.H.; Hali, S.; Kim, J.; Kim, K.-P.; Park, C.; Lee, J.-H.; Ryu, H.K.; Na, J.E.; Jo, J.; et al. Generation of Homogeneous Midbrain Organoids with in Vivo-like Cellular Composition Facilitates Neurotoxin-Based Parkinson's Disease Modeling. *Stem Cells* **2020**, *38*, 727–740. [CrossRef]

190. Dickson, D.W.; Heckman, M.G.; Murray, M.E.; Soto, A.I.; Walton, R.L.; Diehl, N.N.; van Gerpen, J.A.; Uitti, R.J.; Wszolek, Z.K.; Ertekin-Taner, N.; et al. APOE E4 Is Associated with Severity of Lewy Body Pathology Independent of Alzheimer Pathology. *Neurology* **2018**, *91*, e1182–e1195. [CrossRef]

191. Su, C.-J.; Feng, Y.; Liu, T.-T.; Liu, X.; Bao, J.-J.; Shi, A.-M.; Hu, D.-M.; Liu, T.; Yu, Y.-L. Thioredoxin-Interacting Protein Induced α-Synuclein Accumulation via Inhibition of Autophagic Flux: Implications for Parkinson's Disease. *CNS Neurosci. Ther.* **2017**, *23*, 717–723. [CrossRef]

192. Nogueira, G.O.; Garcez, P.P.; Bardy, C.; Cunningham, M.O.; Sebollela, A. Modeling the Human Brain With Ex Vivo Slices and in Vitro Organoids for Translational Neuroscience. *Front. Neurosci.* **2022**, *16*, 838594. [CrossRef]

193. Emin, D.; Zhang, Y.P.; Lobanova, E.; Miller, A.; Li, X.; Xia, Z.; Dakin, H.; Sideris, D.I.; Lam, J.Y.L.; Ranasinghe, R.T.; et al. Small Soluble α-Synuclein Aggregates Are the Toxic Species in Parkinson's Disease. *Nat. Commun.* **2022**, *13*, 5512. [CrossRef]

194. Schweighauser, M.; Shi, Y.; Tarutani, A.; Kametani, F.; Murzin, A.G.; Ghetti, B.; Matsubara, T.; Tomita, T.; Ando, T.; Hasegawa, K.; et al. Structures of α-Synuclein Filaments from Multiple System Atrophy. *Nature* **2020**, *558*, 464–469. [CrossRef]

195. Koga, S.; Sekiya, H.; Kondru, N.; Ross, O.A.; Dickson, D.W. Neuropathology and Molecular Diagnosis of Synucleinopathies. *Mol. Neurodegener.* **2021**, *16*, 83. [CrossRef]

196. Berg, D.; Postuma, R.B.; Bloem, B.; Chan, P.; Dubois, B.; Gasser, T.; Goetz, C.G.; Halliday, G.M.; Hardy, J.; Lang, A.E.; et al. Time to Redefine PD? Introductory Statement of the MDS Task Force on the Definition of Parkinson's Disease. *Mov. Disord.* **2014**, *29*, 454–462. [CrossRef]

197. Parkkinen, L.; O'Sullivan, S.S.; Collins, C.; Petrie, A.; Holton, J.L.; Revesz, T.; Lees, A.J. Disentangling the Relationship between Lewy Bodies and Nigral Neuronal Loss in Parkinson's Disease. *J. Park. Dis.* **2011**, *1*, 277–286. [CrossRef]

 *biomedicines*

*Review*

# Synuclein Proteins in MPTP-Induced Death of Substantia Nigra Pars Compacta Dopaminergic Neurons

Valeria V. Goloborshcheva [1,*], Valerian G. Kucheryanu [1], Natalia A. Voronina [1], Ekaterina V. Teterina [2], Aleksey A. Ustyugov [2] and Sergei G. Morozov [1]

[1] Institute of General Pathology and Pathophysiology, 125315 Moscow, Russia
[2] Institute of Physiologically Active Compounds, Russian Academy of Sciences, 142432 Chernogolovka, Russia
* Correspondence: educadobelleza@gmail.com; Tel.: +7-(909)-644-92-31

**Abstract:** Parkinson's disease (PD) is one of the key neurodegenerative disorders caused by a dopamine deficiency in the striatum due to the death of dopaminergic (DA) neurons of the substantia nigra pars compacta. The initially discovered A53T mutation in the alpha-synuclein gene was linked to the formation of cytotoxic aggregates: Lewy bodies in the DA neurons of PD patients. Further research has contributed to the discovery of beta- and gamma-synucleins, which presumably compensate for the functional loss of either member of the synuclein family. Here, we review research from 1-methyl-4-phenyl-1,2,3,6-tetrahydropyridine (MPTP) toxicity models and various synuclein-knockout animals. We conclude that the differences in the sensitivity of the synuclein-knockout animals compared with the MPTP neurotoxin are due to the ontogenetic selection of early neurons followed by a compensatory effect of beta-synuclein, which optimizes dopamine capture in the synapses. Triple-knockout synuclein studies have confirmed the higher sensitivity of DA neurons to the toxic effects of MPTP. Nonetheless, beta-synuclein could modulate the alpha-synuclein function, preventing its aggregation and loss of function. Overall, the use of knockout animals has helped to solve the riddle of synuclein functions, and these proteins could be promising molecular targets for the development of therapies that are aimed at optimizing the synaptic function of dopaminergic neurons.

**Keywords:** synucleins; dopaminergic neuron; MPTP; knockout mice; Parkinson's disease

**Citation:** Goloborshcheva, V.V.; Kucheryanu, V.G.; Voronina, N.A.; Teterina, E.V.; Ustyugov, A.A.; Morozov, S.G. Synuclein Proteins in MPTP-Induced Death of Substantia Nigra Pars Compacta Dopaminergic Neurons. *Biomedicines* **2022**, *10*, 2278. https://doi.org/10.3390/biomedicines10092278

Academic Editor: Carmela Matrone

Received: 30 July 2022
Accepted: 6 September 2022
Published: 14 September 2022

**Publisher's Note:** MDPI stays neutral with regard to jurisdictional claims in published maps and institutional affiliations.

## 1. Introduction

The synuclein family consists of three highly homologous genes encoding proteins similar in structure: alpha-, beta-, and gamma-synuclein. Among the three representatives of the synuclein family, alpha-synuclein is the best-studied and the volume of scientific research devoted to its functions significantly exceeds the much-needed attention to the other two members altogether [1]. Despite extensive international studies of the synuclein family of proteins, their physiological functions as well as their pathophysiological role in synuclein-associated neurodegenerative diseases have not been fully resolved [2]. The question remains open whether the formation of Lewy bodies is the primary cause of Parkinson's disease (PD) or whether it is a by-product of the activation of intracellular defense mechanisms against the ongoing debilitating neurodegenerative process.

In order to understand these fundamental questions, modern experimental science is developing new hybrid forms of parkinsonism in laboratory model animal systems. A toxic PD model that was initiated by single or multiple treatments of the neurotoxin 1-methyl-4-phenyl-1,2,3,6-tetrahydropyridine (MPTP) was actively used in knockout animals lacking one or more synucleins as well as in mice overexpressing a mutant form of the human protein [3–7]. In this review, we focus on current findings on the potential role of the synuclein family of proteins during the MPTP-induced death of substantia nigra pars compacta (SNpc) dopaminergic neurons (DA neurons) of the midbrain.

## 2. Synuclein Structure and Functions

Synucleins are a family of small soluble proteins that have at least five amino acid repeats located in the N-terminal region, resulting in an alpha-helical conformation with the C-terminal region remaining unstructured [8–10]. In contrast to alpha-synuclein, beta-synuclein does not contain the internal hydrophobic region corresponding with the non-beta-amyloid component (NAC) peptide, which makes alpha-synuclein capable of forming aggregates [11]. The alpha-synuclein protein was first detected in the Torpedo California electric scat in 1988 [12], but was later identified as a precursor protein in the amyloid plaques of Alzheimer's disease patients [13]. Somewhat later, beta-synuclein was isolated from the presynaptic endings of rat and bovine brains [14,15]; gamma-synuclein was found in breast cancer metastases [16], but was further isolated from the mouse trigeminal nerve [17].

All synucleins are actively expressed in nervous system tissues. High expression levels in the neocortex, hippocampus, striatum, and cerebellum are typical for alpha- and beta-synuclein [18], but, in addition to the CNS, these proteins can also be found in blood cells, astrocytes, skeletal muscles, and the liver [11,19,20]. The first two proteins are highly represented in many structures of the brain and their levels in the spinal cord and peripheral nervous system are relatively low; the opposite is found for gamma-synuclein, with a high expression level in the motor neurons of the spinal cord and medulla oblongata, neurons of the sympathetic and parasympathetic peripheral nervous system, tumor entities, and retinal ganglion cells [9,21].

Despite independent roles in the cell, synucleins are highly homologous and have similar functions, often compensating for the dysfunction between each other. Synucleins are important for the synaptic transmission and circulation of synaptic vesicles [22–27]. Alpha-synuclein modulates the release of neurotransmitters from presynaptic terminals by binding and clustering synaptic vesicles and chaperoning the soluble N-ethylmaleimide-sensitive factor attachment protein receptor (SNARE) complex assembly by binding to the protein synaptobrevin-2 (VAMP2) [28,29] whereas beta-synuclein and gamma-synuclein modulate the synaptic vesicular binding of alpha-synuclein and thus reduce the synaptic physiological activity of alpha-synuclein [30,31] (Table 1). Moreover, in vitro and in vivo experiments have revealed that all three members of the synuclein family have chaperone activity [32–34].

**Table 1.** Physiological functions of synuclein proteins.

| Functions | α-syn | β-syn | γ-syn | Ref. |
|---|---|---|---|---|
| Neurotransmission | ✔ | ✔ | ✔ | [22–27] |
| Chaperoning | ✔ | ✔ | ✔ | [32–34] |
| SNARE assembly | ✔ | Maintenance | Maintenance | [28–31] |
| DAT transporter delivery to the presynapse | ✔ | ✔ | ✔ | [11,31,35] |
| Regulation of DAT transporter activity | ✔ | NA | Maintenance | [36–38] |
| Regulation of dopamine homeostasis | ✔ | ? | NA | [27,39–44] |
| Potentiation of vesicular dopamine uptake | NA | ✔ | NA | [4] |
| Lipid structure or morphology changes | ✔ | ✔ | ✔ | [11,31,45] |
| Regulation of lipid metabolism | NA | NA | ✔ | [46] |
| Anti-apoptosis | ✔ | ✔ | NA | [47,48] |
| Mitochondrial regulation | ? | NA | NA | [49–52] |

**Table 1.** *Cont.*

| Functions | α-syn | β-syn | γ-syn | Ref. |
|---|---|---|---|---|
| Regulation of cellular metal homeostasis | NA | ✔ | NA | [11,53,54] |
| Regulation of the autophagic–lysosomal pathway | NA | ? | ✔ | [55–57] |
| Interaction with proteasomes | ✔ | ✔ | ✔ | [58–60] |
| Cytoskeleton stabilization | ? | NA | ✔ | [61–63] |
| Regulation of the growth of neurons in SNpc | ✔ | NA | NA | [64] |
| Regeneration of damaged neurons | ? | ? | ? | [65,66] |

✔: involved; NA: not available; ?: hypothesis.

Alpha-, beta-, and gamma-synucleins can bind to the dopamine transporter (DAT) and modulate its delivery to the synaptic membrane, thereby affecting dopamine neurotransmitter reuptake [11,31,35]. In turn, it has been shown through protein–protein interactions that alpha-synuclein can affect DAT activity and this effect is regulated by the gamma-synuclein concentration [36–38]. Alpha-synuclein is involved in maintaining the required level of dopamine (DA) and if its function cannot be performed due to a mutation, a vesicle degradation occurs [27]. The mechanism of this effect has multiple roots: alpha-synuclein regulates synaptic DA homeostasis [39], affects the expression of DA synthesis member genes (such as GTP-cyclohydrolase, tyrosine hydroxylase (TH), and aromatic acid decarboxylase) [40], modulates synaptic DA reuptake by binding to DAT [41], and inhibits DA release in response to repeated excitation [42,43]. Previously, there was no evidence for an interaction between beta-synuclein and TH, but it has been suggested that it may functionally overlap with alpha-synuclein [44]. Moreover, a recent study convincingly demonstrated that beta-synuclein potentiates vesicular dopamine uptake, presumably by the assembly of the TH/AADC/VMAT-2 protein complex, which is probably not functionally compensated by alpha- or gamma-synuclein [4].

Synucleins are also lipid-binding proteins capable of inducing membrane curvature and turning large vesicles into highly curved formations [11,31,45]. Moreover, gamma-synuclein regulates lipid metabolism in adipocytes and the lack of this protein has a significant impact on the energy metabolism of the whole organism [46]. In addition, alpha- and beta-synucleins prevent cell autolysis. For example, beta-synuclein possesses p53-dependent anti-apoptosis properties at low physiological concentrations, inhibiting caspase-3 activation by binding to Akt [47,48].

A number of studies have found that alpha-synuclein is able to bind to the mitochondria and even penetrate them through VDAC channels (the outer membrane metabolic channel), thus probably targeting the mitochondrial respiratory chain complexes in the inner membrane [49–51], but the physiological significance of this interaction remains unclear. A difference in the lipid composition of the mitochondrial membrane is a regulatory link in the affinity with the alpha-synuclein–VDAC interaction [52].

Beta-synuclein binds to metals to regulate cellular metal homeostasis, particularly chelated copper ions, which can produce free radicals and promote the formation of cytotoxic alpha-synuclein oligomers [11,53,54]. There is also a suggestion that beta-synuclein can affect the autophagic–lysosomal pathway, removing damaged or toxic protein molecules and even aggregates [55,56]. In turn, gamma-synuclein optimizes the autophagy process, which protects colon cancer cells from endoplasmic reticulum stress [57].

The ubiquitin–proteasome system that provides controlled protein degradation is extremely important for the removal of toxic oligomers and soluble protofibrillar structures formed by proteins prone to aggregation, including synucleins. All three members of the synuclein family are able to interact with proteasomes but with different efficiencies. For alpha-synuclein, the interaction depends on the degree of its aggregation [58,59].

Monomeric beta-synuclein also has a low inhibitory effect on 20S and 26S proteasome complexes, but monomeric gamma-synuclein inhibits ubiquitin-independent proteolysis much more effectively. Interestingly, beta-synuclein acts as a negative regulator of alpha-synuclein in these processes [60].

Gamma-synuclein is involved in the stabilization of the cell cytoskeleton [61]. Although alpha-synuclein is capable of interacting with a few components of the cytoskeleton—in particular, with tubulin—the putative effects of alpha-synuclein on its polymerization are not clear [62]. In the lysates of cancer cells, gamma-synuclein was found both in the cytosolic fraction and in the cytoskeleton fraction and the role of gamma-synuclein in stabilizing the neurofilament network in neurons was also revealed [63].

Interestingly, several studies have shown a modulating role of alpha-synuclein in the formation of populations of the SNpc DA neurons of the midbrain. Alpha-synuclein takes part in the maturation of SNpc DA neurons whilst the development of the adjacent similar anatomical structure, the ventral tegmental area (VTA), proceeds independently [64]. In turn, one of the possible roles of synucleins is considered to be participation in the regeneration of damaged neural tissues. It was found that the concentration of alpha- and beta-synucleins (gamma-synuclein-less) was significantly increased around damaged neural endings [65,66]. Hence, the link between synucleins and neurodegeneration can be explained not only by pathological aggregation and its induced toxicity, but also by the loss of normal function. Disruptions in the structure, intracellular localization, and compartmentalization of the synuclein family of proteins result in pathological conditions called synucleinopathies.

## 3. Parkinson's Disease Is a Form of Synucleinopathy

Parkinsonian syndrome (or parkinsonism) is a neurological condition with a multifactorial etiology caused by a disorder in the extrapyramidal system of the brain. Parkinsonism is clinically characterized by a triad of signs (bradykinesia, rigidity, and tremor) and it has additional motor and non-motor pathological manifestations. The debut of the disease usually occurs between the ages of 65 and 70, with less than 5% of cases in patients younger than 45 [67,68].

According to worldwide statistics, the prevalence of parkinsonism in the general population ranges from 100 to 200 cases per 100,000 people, with an annual increase of 15 cases per 100,000 people [69]. In reality, these figures are underestimated due to the low detection rate at the initial stages of the disease and difficulties in the differential diagnosis of various extrapyramidal pathologies burdened with a PD-like set of symptoms.

Parkinson's disease (PD) is the most common form of parkinsonian syndrome and it is etiopathogenetically designated as primary or idiopathic parkinsonism. However, there are other clinical forms of neurodegenerative diseases to be considered. These include progressive supranuclear palsy (Steele–Richardson–Olszewski syndrome), Huntington's chorea, and corticobasal degeneration (CBD) as well as secondary drug-induced toxic parkinsonism and many others [70,71]. These diseases can be differentiated on the basis of key clinical features as well as a clear understanding of the pathogenetic mechanisms underlying PD, which is crucial for the diagnosis, treatment, and prognosis of the neurodegenerative process in the extrapyramidal system.

Pathophysiologically, PD is characterized by the degeneration of dopaminergic neurons in the substantia nigra of the midbrain due to the cytotoxic aggregation and formation of cytoplasmic inclusions—Lewy bodies (LBs)—resulting in a dopamine deficiency in the striatum and in other associated structures of the basal ganglia [72–74]. LBs contain aggregated forms of the alpha-synuclein protein, which is also present in other neurodegenerative disorders, including multiple system atrophy, dementia with Lewy bodies [75,76], Hallervorden–Spatz disease, and many others that are collectively referred to as "synucleinopathies" [77]. Although a small percentage of patients with PD have a monogenic form of the disease (LRRK2, parkin, etc.), in most cases the disorder is sporadic with an unknown etiology. Normally, alpha-synuclein is present in several states, such as monomeric, dimeric, oligomeric, and fibrillar forms. However, alpha-synuclein oligomers exert the most toxic effects on DA neurons [78]. An increased concentration of alpha-synuclein oligomers was found in the substantia nigra [79,80], cerebrospinal fluid [81], and blood [82] of PD patients. The intranasal administration of oligomeric forms of alpha-synuclein to C57BL/6J mice caused PD-like symptoms [83]. These data all suggest that the oligomeric form of alpha-synuclein has a pathogenetic significance in the development of PD. However, the exact mechanisms of the involvement of alpha-synuclein oligomers in the death of nigrostriatal dopaminergic neurons are currently unknown.

A joint injection of MPTP and alpha-synuclein oligomers into the striatum of C57BL/6J mice resulted in the activation of astrocytes and microglia in the substantia nigra and increased the loss of nigral TH+ neurons and the development of motor deficits in animals to a greater extent than MPTP-only treatments. These results indicate that alpha-synuclein oligomerization induces a neurotoxic effect on DA neurons in SN [6]. Activated microglia secrete proinflammatory cytokines IL-1$\beta$, IL-6, IL-10, interferon gamma (IFN-$\gamma$), and tumor necrosis factor-$\alpha$ (TNF-$\alpha$). These secretions activate the nuclear transcription factor NF-kB, triggering core apoptosis and inducible NO synthase (iNOS), leading NO and other ROS and cyclooxygenase-2 (COX2) to increase the formation of prostaglandin E2. The presence of these pathogenic factors eventually causes the death of SNpc DA neurons [84]. Thus, it is crucial to use various models of parkinsonism—including laboratory animals such as transgenic mice with an overexpression of a mutant form of human alpha-synuclein (A53T; A30P), toxin-induced models (6-hydroxydopamine (6-OHDA), MPTP, and reserpine), and knockout mice lines with a depletion of Parkin/Park genes (Pink-1, DJ-1, and synuclein family proteins)—in order to fully understand the mechanisms of PD pathogenesis.

### 3.1. Toxic Animal Modeling of Parkinsonism Using MPTP

The toxic modeling of parkinsonism with MPTP was proposed at the end of the 20th century. Dr. Langston discovered clinical PD symptoms in addicts of "synthetic heroin", which contained MPTP as one of its byproducts [85]. The discovery of MPTP-induced parkinsonian syndrome provoked a number of scientific studies worldwide that were aimed at determining the pathophysiological mechanisms underlying parkinsonism and it raised the disciplines of neurochemistry and neurobiology to a new level. Thus, MPTP was found to cause the extensive selective death of dopaminergic neurons in the substantia nigra [86]. The results of biochemical studies and an analysis of the cytoarchitectonics of SNpc revealed a decrease in the dopamine content in the striatum and a decrease in the number of nigrostriatal DA neurons in various MPTP-treated animals, including monkeys [87], dogs [88], cats [89], mice [64], and even frogs [90]. A local neurodegeneration caused by a single injection of MPTP at relatively low doses (5–10 mg/kg for dogs and cats; 30 mg/kg for mice) resulted in symptoms (hypokinesia, muscle rigidity, and tremor) that were typical for idiopathic parkinsonism. Yet, not all laboratory animals are sensitive to MPTP. For example, rats, rabbits, and guinea pigs required relatively high doses of MPTP (50–70 mg/kg) in order to manifest the neurological signs of an extrapyramidal system dysfunction, which leads to the development of parkinsonism [91].

MPTP is a lipophilic compound that freely crosses the blood–brain barrier and is metabolized by MAO-B in the glial cells to 1-methyl-4-phenylpyridine in an ionic form (MPP+), which is a highly toxic final metabolite [85,92]. DA neurons in the SNpc then selectively capture MPP+ from the intercellular space using the membrane transporter DAT due to its structural similarities to the dopamine molecule [93]. MPP+ accumulates in the mitochondria where it inhibits complex I of the electron transport chain, leading to the inhibition of cellular respiration [94,95], decreased ATP production [96,97], oxidative stress [98,99], the activation of the caspase cascade [100], and, ultimately, cell death.

The MPTP-toxic model of parkinsonism induced in C57BL/6J mice is widely accepted as the primary system to study the pathogenetic mechanisms that underlie extrapyramidal system disorders and that contribute to PD as well as to develop prospective neuroprotection strategies. Over the past decades, numerous protocols have been created to model toxic parkinsonism. These protocols are grouped based on the speed and severity of the clinical signs into three main categories: "acute administration" (several MPTP doses in one day); "subchronic administration" (usually 1–2 doses a day for a 5-day period); and "chronic" administration (multiple injections for 1 month or more) [101,102].

As indicated earlier, synuclein family proteins are actively involved in the processes of dopamine neurotransmission in the presynaptic endings of SNpc DA neurons. The saturation of the presynaptic endings of DA neurons with the toxic end-metabolite of MPTP—1-methyl-4-phenylpyridine in an ionic form (MPP+), which has a high affinity with the plasma membrane transporter DAT—is presumably directly related to the activity of synuclein family proteins (Figure 1). Thus, the selective pathological effect of MPP+ is based on the ability of neurons to reuptake the neurotransmitter from the synaptic cleft in order to replenish the intracellular stores and form new vesicles [103] where synucleins could play a special role.

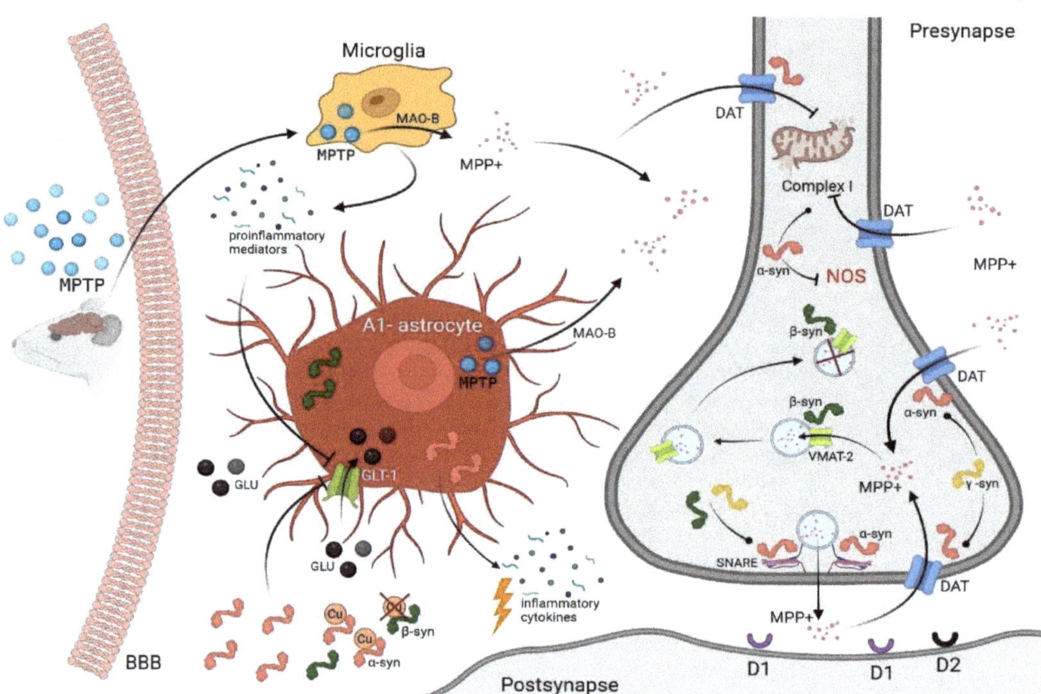


**Figure 1.** The role of synucleins in the mechanisms of SNpc DA neurons during MPTP-induced parkinsonism. Key regulatory factors include the regulatory activity of all synucleins toward the presynaptic membrane of the dopamine transporter (DAT); increased DAT/VMAT-2 ratio and SNARE assembly due to the presence of alpha-synuclein and support from other members of the synuclein family; the inability of beta-synuclein in the presence of alpha- and gamma-synucleins to potentiate VMAT-2-dependent MPP+ capture to further sequester these molecules; the involvement of alpha-synuclein in the neuroinflammatory response; and glutamate toxicity induced by glial cells. These, as well as other unexplored effects of alpha-synuclein binding and penetration into damaged mitochondria, may have a special effect on the MPTP-induced death of DA neurons. Created with BioRender.com (accessed on 2 November 2021).

### 3.2. Synucleins and MPTP Toxicity

Dopamine is the most important signaling neurotransmitter that regulates the motor function of the entire extrapyramidal system, which is responsible for the superstructure of movements [104]. MPP+ is structurally similar to dopamine and it competes for binding sites on the presynaptic membrane of DA neurons. In toxic conditions, such as parkinsonism, DA neurons are particularly sensitive and vulnerable to the pathological effects of MPP+, which entails a series of dramatic events leading to the complete degeneration of the nigrostriatal pathway because DA neuron bodies lie in the substantia nigra of the midbrain with axons extending to the dorsal striatum. On the other hand, it is not quite clear what role synuclein family proteins play in these processes as the main representative of the family, alpha-synuclein, acts as a pathological marker of PD.

The first and subsequent studies on the effects of MPTP toxicity in alpha-synuclein-deficient animals showed surprising results: acute and chronic neurotoxin administration protocols did not have the desired effect on the death of the DA neurons of the SNpc despite lower cell counts [105–108] (Table 2). Moreover, several in vitro studies demonstrated that an overexpression of human alpha-synuclein was associated with enhanced cell death after MPP+ exposure [109,110]. MPTP administration to mice with a selective inactivation of alpha-synuclein in a few cases resulted in a dopamine deficiency and the manifestation of early clinical symptoms of a dopaminergic system dysfunction typical of the early stages of PD [7], which indirectly indicated the activation of the compensatory mechanisms of DA/MPP+ neurotransmission. It is worth emphasizing that phenotypically alpha-synuclein-knockout mice do not differ from wild-type animals [111,112]. However, decreased levels of striatal dopamine in a few lines [113,114] resulted in a reduced availability of DAT on neuronal surfaces [107] and the early debut of Parkinson-like symptoms in aging mice [114–116]. Although neurons manage to compensate for a lack of alpha-synuclein, this takes a toll on the restructuring of the defense systems, which, under certain conditions, can lead to the development of pathological processes, primarily in those cellular compartments where alpha-synuclein normally functions.

**Table 2.** Main phenotypic changes in synuclein-knockout animals before and after MPTP injections.

| Effect | MPTP * | α-syn KO | β-syn KO | γ-syn KO | αβγ-syn KO |
|---|---|---|---|---|---|
| Clinical manifestation | − | ≈ | ≈ | ≈ | ≈ |
| | + | ✔ | NA | NA | ✘ |
| Striatal dopamine | − | ≈ | ▼ | ≈ | ▼ |
| | + | ▼ | ▼ | NA | ▼ |
| DAT expression | − | ▼ | NA | NA | NA |
| | + | NA | NA | NA | NA |
| SNpc neurons | − | ▼ | ≈ | ▼ | ≈ |
| | + | resistant | ▼ | resistant | ▼ |

≈: similar to wild-type animals; ✔: presence; ✘: absence; ▼: decrease; NA: not available; *: subchronic MPTP administration.

In turn, animals with a gamma-synuclein deficiency showed a similar response to MPTP-induced dopaminergic neurodegeneration. Here, the main feature was also the resistance of SNpc DA neurons to the toxic effect of MPTP [5,113,116]. Notably, a comparative immunoblotting analysis of the synuclein levels in the midbrain of gamma- and alpha-synuclein-knockout vs. wild-type mice showed increased levels of beta-synuclein [5,117]. This phenomenon led to a further strategy to investigate the role of synucleins in the development of MPTP resistance.

Recent studies have convincingly demonstrated that beta-synuclein is involved in optimizing the capture of dopamine and probably that of structurally similar molecules via VMAT-2 (vesicular monoamine transporter-2) [4]. Moreover, there was a loss of resistance of the DA neurons in the SNpc to MPP+, which is a toxic metabolite of MPTP, in beta-synuclein knockouts. A similar effect was observed in triple-knockout mice (triple synuclein-deficient mice), where the initial population of DA neurons in the SNpc was similar to wild-type mice [4]. In cases of alpha- and/or gamma-synuclein deficiency there was a 2.8-fold increase in the VMAT-2 density per vesicle [107], probably due to the increased presence of beta-synuclein at the presynaptic end, which was consistent with other studies [5]. However, DA neurons in the SNpc are known to be particularly susceptible to MPP+ because they have a higher DAT/VMAT-2 ratio than other brain neurons [11]. Thus, a reduced DAT transporter in the presynapse, combined with an increased VMAT-2 density in the vesicles, changed the VMAT-2/DAT ratio, leading to the utilization of toxic MPP+ molecules. Taken together, these results suggest a direct involvement of beta-synuclein in the developmental processes of the resistance of SNpc DA neurons to neurotoxins rather than the absence of alpha- or gamma-synucleins *per se*.

The potential neuroprotective properties of beta-synuclein also include the regulation of cellular apoptosis. Serine threonine kinase (Akt) is an enzyme that inhibits apoptosis by phosphorylating the Mdm2 protein that binds to p53 in the nucleus. In an experiment by Hashimoto et al., it was shown that a beta-synuclein overexpression in a rat neuroblastoma B103 cell line resulted in the resistance of these cells to the toxic action of rotenone, which, in a similar manner to MPTP, inhibits mitochondrial respiratory chain complex I. However, an Akt inhibition in this cell line resulted in the loss of neuronal resistance to neurotoxin exposure [118].

The specific damaging effect of MPTP on catecholaminergic neurons is also associated with the activation of toxic A-astrocytes, which, under the influence of proinflammatory mediators, inhibit the glutamate capture via GLT-1 and induce the production of inflammatory cytokines, leading to neuroinflammation [119]. Moreover, a disruption of the Nrf2 system in astrocytes leads to a decrease in the number of antioxidant molecules, resulting in oxidative stress. Damaged DA neurons secrete oligomeric alpha-synuclein in PD. The transfer of alpha-synuclein from neurons to astrocytes, with the subsequent accumulation and deposition in astrocytes, leads to the formation of proinflammatory cytokines and the disruption of the glutamate capture via GLT-1 [119]. Such a scenario is possible in the case of a long-term protocol of chronic MPTP administration, for which the presence of amyloid-like inclusions in both the astrocytes and DA neurons in the SNpc has been noted [120,121].

An abnormal aggregation of alpha-synuclein can increase the degree of glutamate excitotoxicity. Alpha-synuclein accumulation in astrocytes affects the glutamate transport, causing increased extracellular glutamate concentrations and excitotoxicity, further aggravating the damage to the dopaminergic neurons [122]. These data emphasize that alpha-synuclein increases the glutamate release. The concentration of alpha-synuclein itself depends on the release of activity-dependent presynaptic glutamate from the endings of the forebrain neurons [123]. In addition, the overexpression of alpha-synuclein increases the phosphorylation of N-methyl-D-aspartate (NMDA) receptors, thereby increasing the formation of NR1 and NR2B subunits and the sensitivity of NMDA receptors to developing glutamate excitotoxicity [124]. Increased levels of glutamate in the intercellular space activates glutamate NMDA receptors, leading to a calcium overload and the death of DA neurons [125–127]. Alpha-synuclein can also enhance glutamate excitotoxicity by accelerating α-amino-3-hydroxy-5-methyl-4-isoxazolepropionic acid (AMPA) receptor signaling [128].

The formation of reactive oxygen species (ROS) is directly involved in the pathogenesis of MPTP-induced parkinsonism [103]. It is unclear how synuclein family proteins are related to these events. It has been established that an alpha-synuclein deficiency leads to the inhibition of nitric oxide synthase (NOS), which forms another powerful oxidant, peroxynitrite ($ONOO^-$), by interacting with ROS [107]. Thus, NOS activation is an important step in MPTP-induced toxicity and it can be inhibited by a targeted inactivation of alpha-synuclein. Therefore, this targeted inactivation could be a promising direction for the development of a PD therapy.

Finally, there is an assumption that alpha-synuclein specifically interacts with the mitochondria by blocking the toxic effect of neurotoxins, which have an established pathogenic action on DA neurons, leading to the development of PD [129,130]. However, this protective function of synucleins does not extend to all cells; in particular, not to differentiated

DA neurons. This may imply that the cytoprotective properties of alpha-synuclein are aimed at optimizing the mitochondrial function and directly depend on the stage of cell differentiation; i.e., are linked to aging [27]. This is indirectly confirmed by studies of the role of alpha-synuclein in the maturation of SNpc DA neurons in the early postnatal developmental period [64].

## 4. Concluding Remarks and Future Directions

All proteins of the synuclein family are distributed throughout the nervous system, predominantly performing the optimization and systematization functions of various processes. Based on all the studies summarized in this review, we conclude that the differences in the sensitivity of synuclein-knockout animals compared with MPTP neurotoxin models are due to and result from the ontogenetic selection of early neurons followed by a compensatory effect of beta-synuclein, which optimizes the DA capture in the synapses. This is supported by MPTP toxicity data from synuclein-free animals with the inactivation of all three members. Compared with single alpha- or gamma-synuclein knockouts, the sensitivity of DA neurons to the toxic effects of MPTP is higher in triple-knockout animals and almost identical to the levels shown in wild-type controls, suggesting that beta-synuclein could modulate the alpha-synuclein function, preventing its aggregation and a loss of function. Thus, synucleins can be considered to be promising molecular targets for the development of therapies that are aimed at optimizing the synaptic function of dopaminergic neurons. Knockout mice lacking any of the three synuclein members could be used as a promising tool to study the mechanisms of the neurodegenerative processes of synucleinopathies such as PD.

**Author Contributions:** V.V.G. drafted the work; V.G.K. helped in writing the manuscript and approved the submitted version; N.A.V. and E.V.T. reviewed the literature; A.A.U. contributed to the conception or design of the work, and S.G.M. substantively revised it and approved the submitted version. All authors have read and agreed to the published version of the manuscript.

**Funding:** The work is supported by grant agreement No. 075-15-2020-795 and state contract No. 13.1902.21.0027 of 29.09.2020 unique project ID: RF-190220X0027.

**Institutional Review Board Statement:** Not applicable.

**Informed Consent Statement:** Not applicable.

**Data Availability Statement:** Not applicable.

**Acknowledgments:** We would like to thank Claire McQuerry for valuable comments and for proofreading the manuscript.

**Conflicts of Interest:** The authors declare no conflict of interest.

## Abbreviations

| | |
|---|---|
| PD | Parkinson's disease |
| DA | Dopamine |
| NF-kB | Nuclear factor kappa-light-chain-enhancer of activated B cells |
| DA neurons | Dopaminergic neurons |
| MPTP | 1-methyl-4-phenyl-1,2,3,6-tetrahydropyridine |
| MPP+ | 1-methyl-4-phenylpyridine in ionic form |
| SNpc | Substantia nigra pars compacta |
| NAC | Non-beta-amyloid component |
| CNS | Central nervous system |
| SNARE | Soluble N-ethylmaleimide-sensitive factor attachment protein receptor |
| VAMP-2 | Protein synaptobrevin-2 |
| VMAT-2 | Vesicular monoamine transporter-2 |
| DAT | Dopamine transporter |

| TH | Tyrosine hydroxylase |
| VDAC | Voltage-dependent anion channels |
| VTA | Ventral tegmental area |
| CBD | Corticobasal degeneration |
| IL-1$\beta$ | Interleukin-1$\beta$ |
| IL-6 | Interleukin-6 |
| IL-10 | Interleukin-10 |
| IFN-$\gamma$ | Interferon gamma |
| TNF-$\alpha$ | Tumor necrosis factor-$\alpha$ |
| MAO-B | Monoamine oxidase B |
| iNOS | Inducible NO synthase |
| NO | Nitric oxide |
| ONOO | Peroxynitrite |
| COX2 | Cyclooxygenase-2 enzyme |
| NMDA | N-methyl-D-aspartate |
| AMPA | $\alpha$-Amino-3-hydroxy-5-methyl-4-isoxazolepropionic acid |
| ROS | Reactive oxygen species |

## References


1. Bras, J.; Gibbons, E.; Guerreiro, R. Genetics of synucleins in neurodegenerative diseases. *Acta Neuropathol.* **2021**, *141*, 471–490. [CrossRef]
2. Mahoney-Sanchez, L.; Bouchaoui, H.; Ayton, S.; Devos, D.; Duce, J.A.; Devedjian, J.C. Ferroptosis and its potential role in the physiopathology of Parkinson's Disease. *Prog. Neurobiol.* **2021**, *196*, 101890. [CrossRef]
3. Schluter, O.M.; Fornai, F.; Alessandri, M.G.; Takamori, S.; Geppert, M.; Jahn, R.; Sudhof, T.C. Role of alpha-synuclein in 1-methyl-4-phenyl-1,2,3,6-tetrahydropyridine-induced parkinsonism in mice. *Neuroscience* **2003**, *118*, 985–1002. [CrossRef]
4. Ninkina, N.; Millership, S.J.; Peters, O.M.; Connor-Robson, N.; Chaprov, K.; Kopylov, A.T.; Montoya, A.; Kramer, H.; Withers, D.J.; Buchman, V.L. Beta-synuclein potentiates synaptic vesicle dopamine uptake and rescues dopaminergic neurons from MPTP-induced death in the absence of other synucleins. *J. Biol. Chem.* **2021**, *297*, 101375. [CrossRef]
5. Robertson, D.C.; Schmidt, O.; Ninkina, N.; Jones, P.A.; Sharkey, J.; Buchman, V.L. Developmental loss and resistance to MPTP toxicity of dopaminergic neurones in substantia nigra pars compacta of gamma-synuclein, alpha-synuclein and double alpha/gamma-synuclein null mutant mice. *J. Neurochem.* **2004**, *89*, 1126–1136. [CrossRef]
6. Merghani, M.M.; Ardah, M.T.; Al Shamsi, M.; Kitada, T.; Haque, M.E. Dose-related biphasic effect of the Parkinson's disease neurotoxin MPTP, on the spread, accumulation, and toxicity of alpha-synuclein. *Neurotoxicology* **2021**, *84*, 41–52. [CrossRef]
7. Chaprov, K.D.; Teterina, E.V.; Roman, A.Y.; Ivanova, T.A.; Goloborshcheva, V.V.; Kucheryanu, V.G.; Morozov, S.G.; Lysikova, E.A.; Lytkina, O.A.; Koroleva, I.V.; et al. Comparative Analysis of MPTP Neurotoxicity in Mice with a Constitutive Knockout of the alpha-Synuclein Gene. *Mol. Biol.* **2021**, *55*, 152–163. [CrossRef]
8. Patel, D.; Bordoni, B. *Physiology, Synuclein*; StatPearls Publishing: Treasure Island, FL, USA, 2022.
9. Burre, J.; Sharma, M.; Sudhof, T.C. Cell Biology and Pathophysiology of alpha-Synuclein. *Cold Spring Harb. Perspect Med.* **2018**, *8*, a024091. [CrossRef]
10. Carija, A.; Pinheiro, F.; Pujols, J.; Bras, I.C.; Lazaro, D.F.; Santambrogio, C.; Grandori, R.; Outeiro, T.F.; Navarro, S.; Ventura, S. Biasing the native alpha-synuclein conformational ensemble towards compact states abolishes aggregation and neurotoxicity. *Redox Biol.* **2019**, *22*, 101135. [CrossRef]
11. Hayashi, J.; Carver, J.A. beta-Synuclein: An Enigmatic Protein with Diverse Functionality. *Biomolecules* **2022**, *12*, 142. [CrossRef]
12. Maroteaux, L.; Campanelli, J.T.; Scheller, R.H. Synuclein: A neuron-specific protein localized to the nucleus and presynaptic nerve terminal. *J. Neurosci.* **1988**, *8*, 2804–2815. [CrossRef] [PubMed]
13. Ueda, K.; Fukushima, H.; Masliah, E.; Xia, Y.; Iwai, A.; Yoshimoto, M.; Otero, D.A.; Kondo, J.; Ihara, Y.; Saitoh, T. Molecular cloning of cDNA encoding an unrecognized component of amyloid in Alzheimer disease. *Proc. Natl. Acad. Sci. USA* **1993**, *90*, 11282–11286. [CrossRef]
14. Nakajo, S.; Omata, K.; Aiuchi, T.; Shibayama, T.; Okahashi, I.; Ochiai, H.; Nakai, Y.; Nakaya, K.; Nakamura, Y. Purification and characterization of a novel brain-specific 14-kDa protein. *J. Neurochem.* **1990**, *55*, 2031–2038. [CrossRef]
15. Tobe, T.; Nakajo, S.; Tanaka, A.; Mitoya, A.; Omata, K.; Nakaya, K.; Tomita, M.; Nakamura, Y. Cloning and characterization of the cDNA encoding a novel brain-specific 14-kDa protein. *J. Neurochem.* **1992**, *59*, 1624–1629. [CrossRef]
16. Ji, H.; Liu, Y.E.; Jia, T.; Wang, M.; Liu, J.; Xiao, G.; Joseph, B.K.; Rosen, C.; Shi, Y.E. Identification of a breast cancer-specific gene, BCSG1, by direct differential cDNA sequencing. *Cancer Res.* **1997**, *57*, 759–764.
17. Buchman, V.L.; Adu, J.; Pinon, L.G.; Ninkina, N.N.; Davies, A.M. Persyn, a member of the synuclein family, influences neurofilament network integrity. *Nat. Neurosci.* **1998**, *1*, 101–103. [CrossRef]
18. Iwai, A.; Masliah, E.; Yoshimoto, M.; Ge, N.; Flanagan, L.; de Silva, H.A.; Kittel, A.; Saitoh, T. The precursor protein of non-A beta component of Alzheimer's disease amyloid is a presynaptic protein of the central nervous system. *Neuron* **1995**, *14*, 467–475. [CrossRef]

19. Jakes, R.; Spillantini, M.G.; Goedert, M. Identification of two distinct synucleins from human brain. *FEBS Lett.* **1994**, *345*, 27–32. [CrossRef]

20. Tanji, K.; Mori, F.; Nakajo, S.; Imaizumi, T.; Yoshida, H.; Hirabayashi, T.; Yoshimoto, M.; Satoh, K.; Takahashi, H.; Wakabayashi, K. Expression of beta-synuclein in normal human astrocytes. *Neuroreport* **2001**, *12*, 2845–2848. [CrossRef]

21. Le, T.; Winham, C.L.; Andromidas, F.; Silver, A.C.; Jellison, E.R.; Levesque, A.A.; Koob, A.O. Chimera RNA interference knockdown of gamma-synuclein in human cortical astrocytes results in mitotic catastrophe. *Neural. Regen. Res.* **2020**, *15*, 1894–1902. [CrossRef]

22. Butler, B.; Sambo, D.; Khoshbouei, H. Alpha-synuclein modulates dopamine neurotransmission. *J. Chem. Neuroanat.* **2017**, *83–84*, 41–49. [CrossRef] [PubMed]

23. Bernal-Conde, L.D.; Ramos-Acevedo, R.; Reyes-Hernandez, M.A.; Balbuena-Olvera, A.J.; Morales-Moreno, I.D.; Arguero-Sanchez, R.; Schule, B.; Guerra-Crespo, M. Alpha-Synuclein Physiology and Pathology: A Perspective on Cellular Structures and Organelles. *Front. Neurosci.* **2019**, *13*, 1399. [CrossRef] [PubMed]

24. Lazarevic, V.; Yang, Y.; Paslawski, W.; Svenningsson, P. alpha-Synuclein induced cholesterol lowering increases tonic and reduces depolarization-evoked synaptic vesicle recycling and glutamate release. *NPJ Parkinsons Dis.* **2022**, *8*, 71. [CrossRef] [PubMed]

25. Kachappilly, N.; Srivastava, J.; Swain, B.P.; Thakur, P. Interaction of alpha-synuclein with lipids. *Methods Cell Biol.* **2022**, *169*, 43–66. [CrossRef]

26. Uceda, A.B.; Frau, J.; Vilanova, B.; Adrover, M. Glycation of alpha-synuclein hampers its binding to synaptic-like vesicles and its driving effect on their fusion. *Cell Mol. Life Sci.* **2022**, *79*, 342. [CrossRef]

27. Sulzer, D.; Edwards, R.H. The physiological role of alpha-synuclein and its relationship to Parkinson's Disease. *J. Neurochem.* **2019**, *150*, 475–486. [CrossRef]

28. Burre, J.; Sharma, M.; Tsetsenis, T.; Buchman, V.; Etherton, M.R.; Sudhof, T.C. Alpha-synuclein promotes SNARE-complex assembly in vivo and in vitro. *Science* **2010**, *329*, 1663–1667. [CrossRef]

29. Guo, J.T.; Chen, A.Q.; Kong, Q.; Zhu, H.; Ma, C.M.; Qin, C. Inhibition of vesicular monoamine transporter-2 activity in alpha-synuclein stably transfected SH-SY5Y cells. *Cell Mol. Neurobiol.* **2008**, *28*, 35–47. [CrossRef]

30. Carnazza, K.E.; Komer, L.E.; Xie, Y.X.; Pineda, A.; Briano, J.A.; Gao, V.; Na, Y.; Ramlall, T.; Buchman, V.L.; Eliezer, D.; et al. Synaptic vesicle binding of alpha-synuclein is modulated by beta- and gamma-synucleins. *Cell Rep.* **2022**, *39*, 110675. [CrossRef]

31. Carnazza, K.E.; Komer, L.E.; Pineda, A.; Na, Y.; Ramlall, T.; Buchman, V.L.; Eliezer, D.; Sharma, M.; Burre, J. Beta- and gamma-synucleins modulate synaptic vesicle-binding of alpha-synuclein. *bioRxiv* **2020**. [CrossRef]

32. Scheibe, C.; Karreman, C.; Schildknecht, S.; Leist, M.; Hauser, K. Synuclein Family Members Prevent Membrane Damage by Counteracting alpha-Synuclein Aggregation. *Biomolecules* **2021**, *11*, 1067. [CrossRef] [PubMed]

33. Yates, D. Processing alpha-synuclein interactions. *Nat. Rev. Neurosci.* **2022**, *23*, 456–457. [CrossRef] [PubMed]

34. Liu, C.; Zhao, Y.; Xi, H.; Jiang, J.; Yu, Y.; Dong, W. The Membrane Interaction of Alpha-Synuclein. *Front. Cell Neurosci.* **2021**, *15*, 633727. [CrossRef] [PubMed]

35. Jeannotte, A.M.; McCarthy, J.G.; Sidhu, A. Desipramine induced changes in the norepinephrine transporter, alpha- and gamma-synuclein in the hippocampus, amygdala and striatum. *Neurosci. Lett.* **2009**, *467*, 86–89. [CrossRef]

36. Bu, M.; Farrer, M.J.; Khoshbouei, H. Dynamic control of the dopamine transporter in neurotransmission and homeostasis. *NPJ Parkinsons Dis.* **2021**, *7*, 22. [CrossRef]

37. Threlfell, S.; Mohammadi, A.S.; Ryan, B.J.; Connor-Robson, N.; Platt, N.J.; Anand, R.; Serres, F.; Sharp, T.; Bengoa-Vergniory, N.; Wade-Martins, R.; et al. Striatal Dopamine Transporter Function Is Facilitated by Converging Biology of alpha-Synuclein and Cholesterol. *Front. Cell Neurosci.* **2021**, *15*, 658244. [CrossRef]

38. Longhena, F.; Faustini, G.; Missale, C.; Pizzi, M.; Bellucci, A. Dopamine Transporter/alpha-Synuclein Complexes Are Altered in the Post Mortem Caudate Putamen of Parkinson's Disease: An In Situ Proximity Ligation Assay Study. *Int. J. Mol. Sci.* **2018**, *19*, 1611. [CrossRef]

39. Lotharius, J.; Barg, S.; Wiekop, P.; Lundberg, C.; Raymon, H.K.; Brundin, P. Effect of mutant alpha-synuclein on dopamine homeostasis in a new human mesencephalic cell line. *J. Biol. Chem.* **2002**, *277*, 38884–38894. [CrossRef]

40. Ghamgosha, M.; Latifi, A.M.; Meftahi, G.H.; Mohammadi, A. Cellular, Molecular and Non-Pharm.ogical Therapeutic Advances for the Treatment of Parkinson's Disease: Separating Hope from Hype. *Curr. Gene. Ther.* **2018**, *18*, 206–224. [CrossRef]

41. Wersinger, C.; Sidhu, A. Attenuation of dopamine transporter activity by alpha-synuclein. *Neurosci. Lett.* **2003**, *340*, 189–192. [CrossRef]

42. Bridi, J.C.; Hirth, F. Mechanisms of alpha-Synuclein Induced Synaptopathy in Parkinson's Disease. *Front. Neurosci.* **2018**, *12*, 80. [CrossRef] [PubMed]

43. Yavich, L.; Tanila, H.; Vepsalainen, S.; Jakala, P. Role of alpha-synuclein in presynaptic dopamine recruitment. *J. Neurosci.* **2004**, *24*, 11165–11170. [CrossRef] [PubMed]

44. Surguchov, A. Molecular and cellular biology of synucleins. *Int. Rev. Cell Mol. Biol.* **2008**, *270*, 225–317. [CrossRef]

45. Fanning, S.; Selkoe, D.; Dettmer, U. Parkinson's disease: Proteinopathy or lipidopathy? *NPJ Parkinsons Dis* **2020**, *6*, 3. [CrossRef]

46. Millership, S.; Ninkina, N.; Rochford, J.J.; Buchman, V.L. gamma-synuclein is a novel player in the control of body lipid metabolism. *Adipocyte* **2013**, *2*, 276–280. [CrossRef]

47. da Costa, C.A.; Masliah, E.; Checler, F. Beta-synuclein displays an antiapoptotic p53-dependent phenotype and protects neurons from 6-hydroxydopamine-induced caspase 3 activation: Cross-talk with alpha-synuclein and implication for Parkinson's disease. *J. Biol. Chem.* **2003**, *278*, 37330–37335. [CrossRef]

48. Brockhaus, K.; Bohm, M.R.R.; Melkonyan, H.; Thanos, S. Age-related Beta-synuclein Alters the p53/Mdm2 Pathway and Induces the Apoptosis of Brain Microvascular Endothelial Cells In Vitro. *Cell Transpl.* **2018**, *27*, 796–813. [CrossRef]

49. Martinez, J.H.; Fuentes, F.; Vanasco, V.; Alvarez, S.; Alaimo, A.; Cassina, A.; Coluccio Leskow, F.; Velazquez, F. Alpha-synuclein mitochondrial interaction leads to irreversible translocation and complex I impairment. *Arch. Biochem. Biophys.* **2018**, *651*, 1–12. [CrossRef]

50. Wang, X.; Becker, K.; Levine, N.; Zhang, M.; Lieberman, A.P.; Moore, D.J.; Ma, J. Pathogenic alpha-synuclein aggregates preferentially bind to mitochondria and affect cellular respiration. *Acta Neuropathol. Commun.* **2019**, *7*, 41. [CrossRef]

51. Rostovtseva, T.K.; Gurnev, P.A.; Protchenko, O.; Hoogerheide, D.P.; Yap, T.L.; Philpott, C.C.; Lee, J.C.; Bezrukov, S.M. alpha-Synuclein Shows High Affinity Interaction with Voltage-dependent Anion Channel, Suggesting Mechanisms of Mitochondrial Regulation and Toxicity in Parkinson Disease. *J. Biol. Chem.* **2015**, *290*, 18467–18477. [CrossRef]

52. Hoogerheide, D.P.; Rostovtseva, T.K.; Bezrukov, S.M. Exploring lipid-dependent conformations of membrane-bound alpha-synuclein with the VDAC nanopore. *Biochim. Biophys. Acta Biomembr.* **2021**, *1863*, 183643. [CrossRef] [PubMed]

53. McHugh, P.C.; Wright, J.A.; Brown, D.R. Transcriptional regulation of the beta-synuclein 5'-promoter metal response element by metal transcription factor-1. *PLoS ONE* **2011**, *6*, e17354. [CrossRef] [PubMed]

54. Rodriguez, E.E.; Rios, A.; Trujano-Ortiz, L.G.; Villegas, A.; Castaneda-Hernandez, G.; Fernandez, C.O.; Gonzalez, F.J.; Quintanar, L. Comparing the copper binding features of alpha and beta synucleins. *J. Inorg. Biochem.* **2022**, *229*, 111715. [CrossRef] [PubMed]

55. Finkbeiner, S. The Autophagy Lysosomal Pathway and Neurodegeneration. *Cold Spring Harb. Perspect Biol.* **2020**, *12*, a033993. [CrossRef] [PubMed]

56. Popova, B.; Kleinknecht, A.; Arendarski, P.; Mischke, J.; Wang, D.; Braus, G.H. Sumoylation Protects Against beta-Synuclein Toxicity in Yeast. *Front. Mol. Neurosci.* **2018**, *11*, 94. [CrossRef]

57. Ye, Q.; Peng, Y.; Huang, F.; Chen, J.; Xu, Y.; Li, Y.; Liu, S.; Huang, L. gamma-Synuclein is Closely Involved in Autophagy that Protects Colon Cancer Cell from Endoplasmic Reticulum Stress. *Anticancer Agents Med. Chem.* **2021**, *21*, 2385–2396. [CrossRef]

58. Madsen, D.A.; Schmidt, S.I.; Blaabjerg, M.; Meyer, M. Interaction between Parkin and alpha-Synuclein in PARK2-Mediated Parkinson's Disease. *Cells* **2021**, *10*, 283. [CrossRef]

59. Tanaka, Y.; Engelender, S.; Igarashi, S.; Rao, R.K.; Wanner, T.; Tanzi, R.E.; Sawa, A.; Dawson, V.L.; Dawson, T.M.; Ross, C.A. Inducible expression of mutant alpha-synuclein decreases proteasome activity and increases sensitivity to mitochondria-dependent apoptosis. *Hum. Mol. Genet.* **2001**, *10*, 919–926. [CrossRef]

60. Barba, L.; Paolini Paoletti, F.; Bellomo, G.; Gaetani, L.; Halbgebauer, S.; Oeckl, P.; Otto, M.; Parnetti, L. Alpha and Beta Synucleins: From Pathophysiology to Clinical Application as Biomarkers. *Mov. Disord.* **2022**, *37*, 669–683. [CrossRef]

61. Hanson, K.A.; Kim, S.H.; Wassarman, D.A.; Tibbetts, R.S. Ubiquilin modifies TDP-43 toxicity in a Drosophila model of amyotrophic lateral sclerosis (ALS). *J. Biol. Chem.* **2010**, *285*, 11068–11072. [CrossRef]

62. Zhou, R.M.; Huang, Y.X.; Li, X.L.; Chen, C.; Shi, Q.; Wang, G.R.; Tian, C.; Wang, Z.Y.; Jing, Y.Y.; Gao, C.; et al. Molecular interaction of alpha-synuclein with tubulin influences on the polymerization of microtubule in vitro and structure of microtubule in cells. *Mol. Biol. Rep.* **2010**, *37*, 3183–3192. [CrossRef] [PubMed]

63. Liu, Y.; Tapia, M.L.; Yeh, J.; He, R.C.; Pomerleu, D.; Lee, R.K. Differential Gamma-Synuclein Expression in Acute and Chronic Retinal Ganglion Cell Death in the Retina and Optic Nerve. *Mol. Neurobiol.* **2020**, *57*, 698–709. [CrossRef] [PubMed]

64. Tarasova, T.V.; Lytkina, O.A.; Goloborshcheva, V.V.; Skuratovskaya, L.N.; Antohin, A.I.; Ovchinnikov, R.K.; Kukharsky, M.S. Genetic inactivation of alpha-synuclein affects embryonic development of dopaminergic neurons of the substantia nigra, but not the ventral tegmental area, in mouse brain. *PeerJ* **2018**, *6*, e4779. [CrossRef] [PubMed]

65. Norris, E.H.; Giasson, B.I.; Lee, V.M. Alpha-synuclein: Normal function and role in neurodegenerative diseases. *Curr. Top. Dev. Biol.* **2004**, *60*, 17–54. [CrossRef]

66. Quilty, M.C.; Gai, W.P.; Pountney, D.L.; West, A.K.; Vickers, J.C. Localization of alpha-, beta-, and gamma-synuclein during neuronal development and alterations associated with the neuronal response to axonal trauma. *Exp. Neurol.* **2003**, *182*, 195–207. [CrossRef]

67. Beitz, J.M. Parkinson's disease: A review. *Front. Biosci.* **2014**, *6*, 65–74. [CrossRef]

68. Hayes, M.T. Parkinson's Disease and Parkinsonism. *Am. J. Med.* **2019**, *132*, 802–807. [CrossRef]

69. Tysnes, O.B.; Storstein, A. Epidemiology of Parkinson's disease. *J. Neural. Transm.* **2017**, *124*, 901–905. [CrossRef]

70. Cardoso, F.; Jankovic, J. Movement disorders. *Neurol. Clin.* **1993**, *11*, 625–638. [CrossRef]

71. Keener, A.M.; Bordelon, Y.M. Parkinsonism. *Semin. Neurol.* **2016**, *36*, 330–334. [CrossRef]

72. Kalia, L.V.; Lang, A.E. Parkinson disease in 2015: Evolving basic, pathological and clinical concepts in PD. *Nat. Rev. Neurol.* **2016**, *12*, 65–66. [CrossRef] [PubMed]

73. Reich, S.G.; Savitt, J.M. Parkinson's Disease. *Med. Clin. N. Am.* **2019**, *103*, 337–350. [CrossRef] [PubMed]

74. Riederer, P.; Berg, D.; Casadei, N.; Cheng, F.; Classen, J.; Dresel, C.; Jost, W.; Kruger, R.; Muller, T.; Reichmann, H.; et al. alpha-Synuclein in Parkinson's disease: Causal or bystander? *J. Neural. Transm.* **2019**, *126*, 815–840. [CrossRef]

75. Polymeropoulos, M.H.; Lavedan, C.; Leroy, E.; Ide, S.E.; Dehejia, A.; Dutra, A.; Pike, B.; Root, H.; Rubenstein, J.; Boyer, R.; et al. Mutation in the alpha-synuclein gene identified in families with Parkinson's disease. *Science* **1997**, *276*, 2045–2047. [CrossRef] [PubMed]

76. Spillantini, M.G.; Schmidt, M.L.; Lee, V.M.; Trojanowski, J.Q.; Jakes, R.; Goedert, M. Alpha-synuclein in Lewy bodies. *Nature* **1997**, *388*, 839–840. [CrossRef]

77. Goedert, M.; Jakes, R.; Spillantini, M.G. The Synucleinopathies: Twenty Years On. *J. Parkinsons Dis.* **2017**, *7*, S51–S69. [CrossRef]

78. Gadhe, L.; Sakunthala, A.; Mukherjee, S.; Gahlot, N.; Bera, R.; Sawner, A.S.; Kadu, P.; Maji, S.K. Intermediates of alpha-synuclein aggregation: Implications in Parkinson's disease pathogenesis. *Biophys. Chem.* **2022**, *281*, 106736. [CrossRef]

79. Sharon, R.; Bar-Joseph, I.; Frosch, M.P.; Walsh, D.M.; Hamilton, J.A.; Selkoe, D.J. The formation of highly soluble oligomers of alpha-synuclein is regulated by fatty acids and enhanced in Parkinson's disease. *Neuron* **2003**, *37*, 583–595. [CrossRef]

80. Paleologou, K.E.; Kragh, C.L.; Mann, D.M.; Salem, S.A.; Al-Shami, R.; Allsop, D.; Hassan, A.H.; Jensen, P.H.; El-Agnaf, O.M. Detection of elevated levels of soluble alpha-synuclein oligomers in post-mortem brain extracts from patients with dementia with Lewy bodies. *Brain* **2009**, *132*, 1093–1101. [CrossRef]

81. Park, M.J.; Cheon, S.M.; Bae, H.R.; Kim, S.H.; Kim, J.W. Elevated levels of alpha-synuclein oligomer in the cerebrospinal fluid of drug-naive patients with Parkinson's disease. *J. Clin. Neurol.* **2011**, *7*, 215–222. [CrossRef]

82. El-Agnaf, O.M.; Salem, S.A.; Paleologou, K.E.; Curran, M.D.; Gibson, M.J.; Court, J.A.; Schlossmacher, M.G.; Allsop, D. Detection of oligomeric forms of alpha-synuclein protein in human plasma as a potential biomarker for Parkinson's disease. *FASEB J* **2006**, *20*, 419–425. [CrossRef] [PubMed]

83. Gruden, M.A.; Davidova, T.V.; Yanamandra, K.; Kucheryanu, V.G.; Morozova-Roche, L.A.; Sherstnev, V.V.; Sewell, R.D. Nasal inoculation with alpha-synuclein aggregates evokes rigidity, locomotor deficits and immunity to such misfolded species as well as dopamine. *Behav. Brain Res.* **2013**, *243*, 205–212. [CrossRef] [PubMed]

84. Pajares, M.; Rojo, A.I.; Manda, G.; Bosca, L.; Cuadrado, A. Inflammation in Parkinson's Disease: Mechanisms and Therapeutic Implications. *Cells* **2020**, *9*, 1687. [CrossRef] [PubMed]

85. Langston, J.W. The MPTP Story. *J. Parkinsons Dis.* **2017**, *7*, S11–S19. [CrossRef]

86. Vivacqua, G.; Biagioni, F.; Busceti, C.L.; Ferrucci, M.; Madonna, M.; Ryskalin, L.; Yu, S.; D'Este, L.; Fornai, F. Motor Neurons Pathology After Chronic Exposure to MPTP in Mice. *Neurotox. Res.* **2020**, *37*, 298–313. [CrossRef]

87. Fox, S.H.; Brotchie, J.M. The MPTP-lesioned non-human primate models of Parkinson's disease. Past, present, and future. *Prog. Brain Res.* **2010**, *184*, 133–157. [CrossRef]

88. Johannessen, J.N.; Sobotka, T.J.; Weise, V.K.; Markey, S.P. Prolonged alterations in canine striatal dopamine metabolism following subtoxic doses of 1-methyl-4-phenyl-1,2,3,6-tetrahydropyridine (MPTP) and 4'-amino-MPTP are linked to the persistence of pyridinium metabolites. *J. Neurochem.* **1991**, *57*, 981–990. [CrossRef]

89. Aznavour, N.; Cendres-Bozzi, C.; Lemoine, L.; Buda, C.; Sastre, J.P.; Mincheva, Z.; Zimmer, L.; Lin, J.S. MPTP animal model of Parkinsonism: Dopamine cell death or only tyrosine hydroxylase impairment? A study using PET imaging, autoradiography, and immunohistochemistry in the cat. *CNS Neurosci. Ther.* **2012**, *18*, 934–941. [CrossRef]

90. Sokolowski, A.L.; Larsson, B.S.; Lindquist, N.G. Distribution of 1-(3H)-methyl-4-phenyl-1,2,3,6-tetrahydropyridine (3H-MPTP) in the frog: Uptake in neuromelanin. *Pharm. Toxicol.* **1990**, *66*, 252–258. [CrossRef]

91. Ya, V.M. Experimental reproduction of catecholamine deficiency states and the problem of parkinsonism. *Neurophysiology* **1990**, *22*, 401–414. (In Russian)

92. Sablin, S.O.; Krueger, M.J.; Bachurin, S.O.; Solyakov, L.S.; Efange, S.M.; Singer, T.P. Oxidation products arising from the action of monoamine oxidase B on 1-methyl-4-benzyl-1,2,3,6-tetrahydropyridine, a nonneurotoxic analogue of 1-methyl-4-phenyl-1,2,3,6-tetrahydropyridine. *J. Neurochem.* **1994**, *62*, 2012–2016. [CrossRef] [PubMed]

93. Mayer, R.A.; Kindt, M.V.; Heikkila, R.E. Prevention of the nigrostriatal toxicity of 1-methyl-4-phenyl-1,2,3,6-tetrahydropyridine by inhibitors of 3,4-dihydroxyphenylethylamine transport. *J. Neurochem.* **1986**, *47*, 1073–1079. [CrossRef] [PubMed]

94. Mat Taib, C.N.; Mustapha, M. MPTP-induced mouse model of Parkinson's disease: A promising direction of therapeutic strategies. *Bosn. J. Basic Med. Sci.* **2020**, *21*, 422–433. [CrossRef] [PubMed]

95. Wu, W.J.; Lu, C.W.; Wang, S.E.; Lin, C.L.; Su, L.Y.; Wu, C.H. MPTP toxicity causes vocal, auditory, orientation and movement defects in the echolocation bat. *Neuroreport* **2021**, *32*, 125–134. [CrossRef]

96. Haga, H.; Matsuo, K.; Yabuki, Y.; Zhang, C.; Han, F.; Fukunaga, K. Enhancement of ATP production ameliorates motor and cognitive impairments in a mouse model of MPTP-induced Parkinson's disease. *Neurochem. Int.* **2019**, *129*, 104492. [CrossRef]

97. Vyas, I.; Heikkila, R.E.; Nicklas, W.J. Studies on the neurotoxicity of 1-methyl-4-phenyl-1,2,3,6-tetrahydropyridine: Inhibition of NAD-linked substrate oxidation by its metabolite, 1-methyl-4-phenylpyridinium. *J. Neurochem.* **1986**, *46*, 1501–1507. [CrossRef]

98. Prasad, E.M.; Hung, S.Y. Behavioral Tests in Neurotoxin-Induced Animal Models of Parkinson's Disease. *Antioxidants* **2020**, *9*, 1007. [CrossRef]

99. Rossetti, Z.L.; Sotgiu, A.; Sharp, D.E.; Hadjiconstantinou, M.; Neff, N.H. 1-Methyl-4-phenyl-1,2,3,6-tetrahydropyridine (MPTP) and free radicals in vitro. *Biochem. Pharm.* **1988**, *37*, 4573–4574. [CrossRef]

100. Przedborski, S.; Chen, Q.; Vila, M.; Giasson, B.I.; Djaldatti, R.; Vukosavic, S.; Souza, J.M.; Jackson-Lewis, V.; Lee, V.M.; Ischiropoulos, H. Oxidative post-translational modifications of alpha-synuclein in the 1-methyl-4-phenyl-1,2,3,6-tetrahydropyridine (MPTP) mouse model of Parkinson's disease. *J. Neurochem.* **2001**, *76*, 637–640. [CrossRef]

101. Baranyi, M.; Porceddu, P.F.; Goloncser, F.; Kulcsar, S.; Otrokocsi, L.; Kittel, A.; Pinna, A.; Frau, L.; Huleatt, P.B.; Khoo, M.L.; et al. Novel (Hetero)arylalkenyl propargylamine compounds are protective in toxin-induced models of Parkinson's disease. *Mol. Neurodegener* **2016**, *11*, 6. [CrossRef]

102. Wada, M.; Ang, M.J.; Weerasinghe-Mudiyanselage, P.D.E.; Kim, S.H.; Kim, J.C.; Shin, T.; Moon, C. Behavioral characterization in MPTP/p mouse model of Parkinson's disease. *J. Integr. Neurosci.* **2021**, *20*, 307–320. [CrossRef] [PubMed]

103. Goloborshcheva, V.; Voronina, N.; Ovchinnikov, R.; Kucheryanu, V.; Morozov, S. MPTP-induced Parkinsonism in genetically modified mice. *Pathogenesis* **2021**, *19*, 12–23. (In Russian) [CrossRef]

104. Speranza, L.; di Porzio, U.; Viggiano, D.; de Donato, A.; Volpicelli, F. Dopamine: The Neuromodulator of Long-Term Synaptic Plasticity, Reward and Movement Control. *Cells* **2021**, *10*, 735. [CrossRef] [PubMed]

105. Goloborshcheva, V.V.N.; Ovchinnikov, R.; Kucheryanu, V.; Morozov, S. Morphometric analysis of dopaminergic neurons (substantia nigra) in the brain of MPTP treated alpha synuclein knockout mice. *Pathogenesis* **2021**, *19*, 32–37. (In Russian)

106. Klivenyi, P.; Siwek, D.; Gardian, G.; Yang, L.; Starkov, A.; Cleren, C.; Ferrante, R.J.; Kowall, N.W.; Abeliovich, A.; Beal, M.F. Mice lacking alpha-synuclein are resistant to mitochondrial toxins. *Neurobiol. Dis.* **2006**, *21*, 541–548. [CrossRef]

107. Zharikov, A.; Bai, Q.; De Miranda, B.R.; Van Laar, A.; Greenamyre, J.T.; Burton, E.A. Long-term RNAi knockdown of alpha-synuclein in the adult rat substantia nigra without neurodegeneration. *Neurobiol. Dis.* **2019**, *125*, 146–153. [CrossRef]

108. Goloborshcheva, V.V.; Chaprov, K.D.; Teterina, E.V.; Ovchinnikov, R.; Buchman, V.L. Reduced complement of dopaminergic neurons in the substantia nigra pars compacta of mice with a constitutive "low footprint" genetic knockout of alpha-synuclein. *Mol. Brain* **2020**, *13*, 75. [CrossRef]

109. Lehmensiek, V.; Tan, E.M.; Schwarz, J.; Storch, A. Expression of mutant alpha-synucleins enhances dopamine transporter-mediated MPP+ toxicity in vitro. *Neuroreport* **2002**, *13*, 1279–1283. [CrossRef]

110. Van Laar, V.S.; Chen, J.; Zharikov, A.D.; Bai, Q.; Di Maio, R.; Dukes, A.A.; Hastings, T.G.; Watkins, S.C.; Greenamyre, J.T.; St Croix, C.M.; et al. alpha-Synuclein amplifies cytoplasmic peroxide flux and oxidative stress provoked by mitochondrial inhibitors in CNS dopaminergic neurons in vivo. *Redox. Biol.* **2020**, *37*, 101695. [CrossRef]

111. Cabin, D.E.; Shimazu, K.; Murphy, D.; Cole, N.B.; Gottschalk, W.; McIlwain, K.L.; Orrison, B.; Chen, A.; Ellis, C.E.; Paylor, R.; et al. Synaptic vesicle depletion correlates with attenuated synaptic responses to prolonged repetitive stimulation in mice lacking alpha-synuclein. *J. Neurosci.* **2002**, *22*, 8797–8807. [CrossRef]

112. Ninkina, N.; Tarasova, T.V.; Chaprov, K.D.; Roman, A.Y.; Kukharsky, M.S.; Kolik, L.G.; Ovchinnikov, R.; Ustyugov, A.A.; Durnev, A.D.; Buchman, V.L. Alterations in the nigrostriatal system following conditional inactivation of alpha-synuclein in neurons of adult and aging mice. *Neurobiol. Aging.* **2020**, *91*, 76–87. [CrossRef] [PubMed]

113. Ninkina, N.N.; Tarasova, T.V.; Chaprov, K.D.; Goloborshcheva, V.V.; Bachurin, S.O.; Buchman, V.L. Synuclein Deficiency Decreases the Efficiency of Dopamine Uptake by Synaptic Vesicles. *Dokl. Biochem. Biophys.* **2019**, *486*, 168–170. [CrossRef] [PubMed]

114. Connor-Robson, N.; Peters, O.M.; Millership, S.; Ninkina, N.; Buchman, V.L. Combinational losses of synucleins reveal their differential requirements for compensating age-dependent alterations in motor behavior and dopamine metabolism. *Neurobiol. Aging.* **2016**, *46*, 107–112. [CrossRef] [PubMed]

115. Al-Wandi, A.; Ninkina, N.; Millership, S.; Williamson, S.J.; Jones, P.A.; Buchman, V.L. Absence of alpha-synuclein affects dopamine metabolism and synaptic markers in the striatum of aging mice. *Neurobiol. Aging.* **2010**, *31*, 796–804. [CrossRef]

116. Pavia-Collado, R.; Rodriguez-Aller, R.; Alarcon-Aris, D.; Miquel-Rio, L.; Ruiz-Bronchal, E.; Paz, V.; Campa, L.; Galofre, M.; Sgambato, V.; Bortolozzi, A. Up and Down gamma-Synuclein Transcription in Dopamine Neurons Translates into Changes in Dopamine Neurotransmission and Behavioral Performance in Mice. *Int. J. Mol. Sci.* **2022**, *23*, 1807. [CrossRef]

117. Thomas, B.; Mandir, A.S.; West, N.; Liu, Y.; Andrabi, S.A.; Stirling, W.; Dawson, V.L.; Dawson, T.M.; Lee, M.K. Resistance to MPTP-neurotoxicity in alpha-synuclein knockout mice is complemented by human alpha-synuclein and associated with increased beta-synuclein and Akt activation. *PLoS ONE* **2011**, *6*, e16706. [CrossRef]

118. Hashimoto, M.; Bar-On, P.; Ho, G.; Takenouchi, T.; Rockenstein, E.; Crews, L.; Masliah, E. Beta-synuclein regulates Akt activity in neuronal cells. A possible mechanism for neuroprotection in Parkinson's disease. *J. Biol. Chem.* **2004**, *279*, 23622–23629. [CrossRef]

119. Miyazaki, I.; Asanuma, M. Neuron-Astrocyte Interactions in Parkinson's Disease. *Cells* **2020**, *9*, 2623. [CrossRef]

120. Gibrat, C.; Saint-Pierre, M.; Bousquet, M.; Levesque, D.; Rouillard, C.; Cicchetti, F. Differences between subacute and chronic MPTP mice models: Investigation of dopaminergic neuronal degeneration and alpha-synuclein inclusions. *J. Neurochem.* **2009**, *109*, 1469–1482. [CrossRef]

121. Fornai, F.; Schluter, O.M.; Lenzi, P.; Gesi, M.; Ruffoli, R.; Ferrucci, M.; Lazzeri, G.; Busceti, C.L.; Pontarelli, F.; Battaglia, G.; et al. Parkinson-like syndrome induced by continuous MPTP infusion: Convergent roles of the ubiquitin-proteasome system and alpha-synuclein. *Proc. Natl. Acad. Sci. USA* **2005**, *102*, 3413–3418. [CrossRef]

122. Gu, X.L.; Long, C.X.; Sun, L.; Xie, C.; Lin, X.; Cai, H. Astrocytic expression of Parkinson's disease-related A53T alpha-synuclein causes neurodegeneration in mice. *Mol. Brain* **2010**, *3*, 12. [CrossRef] [PubMed]

123. Sarafian, T.A.; Littlejohn, K.; Yuan, S.; Fernandez, C.; Cilluffo, M.; Koo, B.K.; Whitelegge, J.P.; Watson, J.B. Stimulation of synaptoneurosome glutamate release by monomeric and fibrillated alpha-synuclein. *J. Neurosci. Res.* **2017**, *95*, 1871–1887. [CrossRef] [PubMed]

124. Yang, J.; Hertz, E.; Zhang, X.; Leinartaite, L.; Lundius, E.G.; Li, J.; Svenningsson, P. Overexpression of alpha-synuclein simultaneously increases glutamate NMDA receptor phosphorylation and reduces glucocerebrosidase activity. *Neurosci. Lett.* **2016**, *611*, 51–58. [CrossRef] [PubMed]

125. Voronina, N.A.; Lisina, O.Y.; Krasilnikova, I.A.; Kucheryanu, V.G.; Kapitsa, I.G.; Voronina, T.A.; Surin, A.M. Influence of hemantane on changes in Ca$^{2+}$ and Na$^+$ caused by activation of NMDA channels in cultured rat brain neurons. *Neurochem. J.* **2021**, *15*, 8–17. [CrossRef]
126. Bashkatova, V. Metabotropic glutamate receptors and nitric oxide in dopaminergic neurotoxicity. *World J. Psychiatry* **2021**, *11*, 830–840. [CrossRef]
127. Fairless, R.; Bading, H.; Diem, R. Pathophysiological Ionotropic Glutamate Signalling in Neuroinflammatory Disease as a Therapeutic Target. *Front. Neurosci.* **2021**, *15*, 741280. [CrossRef]
128. Huls, S.; Hogen, T.; Vassallo, N.; Danzer, K.M.; Hengerer, B.; Giese, A.; Herms, J. AMPA-receptor-mediated excitatory synaptic transmission is enhanced by iron-induced alpha-synuclein oligomers. *J. Neurochem.* **2011**, *117*, 868–878. [CrossRef]
129. Jensen, P.J.; Alter, B.J.; O'Malley, K.L. Alpha-synuclein protects naive but not dbcAMP-treated dopaminergic cell types from 1-methyl-4-phenylpyridinium toxicity. *J. Neurochem.* **2003**, *86*, 196–209. [CrossRef]
130. Rosencrans, W.M.; Aguilella, V.M.; Rostovtseva, T.K.; Bezrukov, S.M. alpha-Synuclein emerges as a potent regulator of VDAC-facilitated calcium transport. *Cell Calcium* **2021**, *95*, 102355. [CrossRef]

*Commentary*

# Surgical Management of Synucleinopathies

Sai Sriram, Kevin Root, Kevin Chacko, Aashay Patel and Brandon Lucke-Wold *

Department of Neurosurgery, University of Florida, Gainesville, FL 32608, USA
* Correspondence: brandon.lucke-wold@neurosurgery.ufl.edu

**Abstract:** Synucleinopathies represent a diverse set of pathologies with significant morbidity and mortality. In this review, we highlight the surgical management of three synucleinopathies: Parkinson's disease (PD), dementia with Lewy bodies (DLB), and multiple system atrophy (MSA). After examining underlying molecular mechanisms and the medical management of these diseases, we explore the role of deep brain stimulation (DBS) in the treatment of synuclein pathophysiology. Further, we examine the utility of focused ultrasound (FUS) in the treatment of synucleinopathies such as PD, including its role in blood–brain barrier (BBB) opening for the delivery of novel drug therapeutics and gene therapy vectors. We also discuss other recent advances in the surgical management of MSA and DLB. Together, we give a diverse overview of current techniques in the neurosurgical management of these pathologies.

**Keywords:** synucleinopathies; Parkinson's disease; multiple system atrophy; dementia with Lewy bodies; deep brain stimulation; focused ultrasound; gene therapy; surgical techniques

**Citation:** Sriram, S.; Root, K.; Chacko, K.; Patel, A.; Lucke-Wold, B. Surgical Management of Synucleinopathies. *Biomedicines* 2022, 10, 2657. https://doi.org/10.3390/biomedicines10102657

Academic Editor: Natalia Ninkina

Received: 20 September 2022
Accepted: 19 October 2022
Published: 21 October 2022

**Publisher's Note:** MDPI stays neutral with regard to jurisdictional claims in published maps and institutional affiliations.



## 1. Introduction

The three members of the α-synucleinopathy family—Parkinson's disease (PD), multiple system atrophy (MSA), and dementia with Lewy bodies (DLB)—are some of the most common and costly neurodegenerative diseases affecting neurosurgical patients [1]. Parkinson's disease alone affects more than one million individuals in the United States [1–3]. In addition to the untold emotional burden, these afflictions place an economic strain upon the nation currently exceeding USD 50 billion per annum [4]. Considering Parkinson's disease alone is projected to double in incidence rate by 2030 due to an aging population, it is essential to establish optimal management protocols for α-synucleinopathies [5]. In this regard, we aim to highlight current medical standards of practice as well as the efficacy of available neurosurgical interventions. Specifically, we will examine the role of deep brain stimulation (DBS), focused ultrasound (FUS), and recent advancements in the treatment of synucleinopathies.

### 1.1. Parkinson's Disease

With a current incidence rate of 1–2 per 1000 individuals, Parkinson's disease (PD) is the second most common neurodegenerative disease following Alzheimer's disease [5,6]. The prevalence of PD significantly rises with age, increasing more than 5-fold from the sixth to ninth decades of life [7]. Furthermore, this prevalence is projected to rise drastically by 2030 due to an aging population, underscoring the necessity of understanding and advancing its management. Decades of investigation have given way to a current consensus that attributes PD to a combination of environmental and genetic factors [8]. The pathogenesis of PD is multifactorial, implicating oxidative stress, altered protein handling, and environmental mitochondrial toxins [8,9]. Of note to this review, mutations in the SNCA gene and mitochondrial dysfunction have been shown to result in α-synuclein accumulation [10,11]. In select brain regions in animal models, accumulations of α-synuclein have been identified to inhibit complex I of the mitochondria [12–14]. Additionally, α-synuclein toxicity is well supported to play a significant role in the pathogenesis of PD, although more evidence regarding the underlying molecular mechanism is needed [15,16]. The clinical progression of PD is also highly variable among patients [17]. However, the classic presentation is well documented to include tremors at rest, rigidity, bradykinesia, loss of postural reflexes, and shuffling gait [18]. Pharmacological management often focuses on symptomatic treatment, specifically targeting motor disturbances. Levodopa serves as a standard first-line treatment, functioning to supplement decreased dopamine [19]. Other dopamine modifying medications include NMDA receptor antagonists, muscarinic receptors, or dopamine receptor agonists. Degradation of dopamine is also targeted through catechol-O-methyltransferase inhibitors, and monoamine oxidase type B inhibitors [20]. Recently, pimavanserin, an atypical antipsychotic, has been described to reduce psychosis in PD, allowing for the treatment of some of the psychiatric comorbidities in PD1. Additional clinical symptoms of PD include disturbances in autonomic function, sleep disruptions, neuropsychiatric as well as sensory symptoms, and dementia, all of which are also managed pharmacologically with variable success [21–23]. Notably, dopaminergic medications, particularly at the higher doses required by patients with progressive disease, are associated with debilitating adverse effects. Unpredictable fluctuations of on and off states as well as levodopa induced dyskinesias represent significant barriers to effective treatment [24].

*1.2. Multiple System Atrophy and Dementia with Lewy Bodies*

In the other two diseases of the α-synucleinopathy family, Multiple system atrophy (MSA) and dementia with Lewy bodies (DLB), age is also the primary risk factor. In MSA, the annual incidence rate is 3 cases per 100,000 for a population older than 50 years [25]. Similarly, the incidence of DLB peaks in the sixth decade of life and has an incidence rate of 0.87 cases/1000 person-years in the general population [26]. Risk factors for MSA extend to environmental factors—including exposure to organic solvents, pesticides, metals, and monomers—and genetic factors—involving impaired variants of the enzyme encoded by COQ2 [27,28]. General risk factors for DLB include depression, anxiety, low caffeine intake, and stroke, as well as a genetic predisposition with specific APOE ε4 alleles [29]. MSA is a unique member of the α-synucleinopathy family, as α-synuclein deposits in oligodendrocytes in lieu of neurons [30]. Consequently, the disease pathology stems from oligodendrogliopathy with myelin disruption from α-synuclein positive glial cytoplasmic inclusions (GCI), which leads to axon degeneration and eventual neuron degeneration [28]. Though the exact mechanism of their action is yet unknown, α-synuclein GCIs are required for the diagnosis of MSA and their density correlates with disease severity [31,32]. Similarly, though the precise pathogenesis of DLB is unknown, it is theorized to involve several metabolic pathways that lead to dysfunctions in mitochondria, purine metabolism, protein synthesis, energy metabolism [33], and α-synuclein deposits in neuronal Lewy bodies [34]. In MSA, nonmotor disruptions (i.e., respiratory, autonomic, and urogenital symptoms) often manifest first; however, it is commonly diagnosed only after motor dysfunction occurs [35]. These motor disruptions are sporadic in frequency and parkinsonian in nature, including many of the hallmark symptoms such as rigidity, bradykinesia, and postural instability. Furthermore, early autonomic dysfunctions are classic presentation features of MSA [36]. While the same parkinsonism extends to DLB, visual hallucinations, variable mental status, and dementia are also common features [37]. Treatment options for MSA are limited. MSA is characterized by a poor response to dopaminergic therapy, with only a transient improvement noted in 40% of patients [38]. Alternatively, the disease is often managed by nonpharmacologic strategies such as a decreased salt intake for orthostatic hypotension [36]. Conversely, DLB is effectively managed by acetylcholinesterase inhibitors as a first line treatment and responds well to classic dopaminergic therapy [39].

## 2. Deep Brain Stimulation (DBS)

*2.1. DBS and α-Synuclein in PD*

Despite promising pharmacological treatment options, α-synucleinopathies can prove to be extremely difficult to manage medically. For one, prolonged levodopa treatment is associated with significant side effects in PD such as dyskinesias and motor fluctuations between the on and off state [40]. Unfortunately, due to the progressive nature of the disease, dose escalation is an eventuality in most patients [41]. Deep brain stimulation (DBS) offers a promising surgical option where a high frequency electrode is implanted into a target structure in the basal ganglia or thalamus such as the globus pallidus internus (GPi), subthalamic nucleus (STN) or ventral intermediate nucleus (VIM) [42]. High frequency electrical stimulation is then delivered from the DBS electrode to normalize the pathologically exaggerated basal ganglia bursting activity seen in PD [43]. There is a potential for significant stimulation induced side effects such as paresthesia, involuntary motor contractions, speech impairment, and mood changes. To minimize these side effects, parameters of stimulation such as its amplitude and directionality relative to the DBS electrode are controlled in programming sessions in the weeks following initial implantation. DBS therapy often leads to significant reduction in medication regimens and significant prolongation of "on" medication periods. In one meta-analysis of 22 studies, L-dopa regimens were reduced over 50%, with dyskinesias reduced around 70%, and 70% reduction in "off" periods [44].

Despite the ability of DBS therapy to significantly improve quality of life, it is widely believed that it cannot stop or reverse the effects of α-synuclein-mediated neurodegeneration in PD. In fact, disease progression with loss of dopaminergic neurons is thought to occur rapidly, within 4 years [45]. Nevertheless, several recent works have expanded knowledge concerning the effect of DBS on disease progression, though there remains considerable debate as to whether long term DBS stimulation confers a neuroprotective effect. In one study, rats were induced to overexpress α-synuclein using intranigral injections of an adeno-associated viral (AAV) vector and were subsequently implanted with STN DBS [46]. Limb use by the animals was observed at baseline as well as after electrode implantation and after electrode stimulation for a period of 26 days and was compared to mice that were implanted but did not receive stimulation. Of note, though there was impaired contralateral limb use following the α-syn vector injection in all rats, there was no difference in limb use between stimulation and nonstimulation groups. Neurodegenerative changes were assessed using tyrosine hydroxylase staining (a marker of dopaminergic neurons), which also did not differ between stimulation and nonstimulation groups. This result suggests that DBS stimulation does not protect against the impairment nor the neurodegenerative changes that accompany α-synuclein accumulation [46]. Despite this, another very similar study using an AAV induced α-synuclein overexpression rat model implanted with STN DBS reported different findings. In particular, results of this study suggest that motor performance after 3 weeks of stimulation was significantly improved compared to rats in the nonstimulation group, even with stimulation turned off during motor testing. Tyrosine hydroxylase was also significantly increased in the stimulation group [47].

In addition, beta oscillations are a significant pathologic alteration observed in PD and are linked to some of its symptoms [48]. Recent evidence suggests that STN DBS suppresses these pathological beta oscillations in an AAV induced α-synuclein overexpression rat model, even 2 weeks after stimulation [49]. Other studies have also demonstrated the possible neuroprotective effects of DBS. Specifically, brain derived neurotrophic factor (BDNF) has come under interest recently for its roles in plasticity, neurogenesis, and neuroprotection [50]. In one study of rats injected with preformed fibrils of α-synuclein, though STN DBS did not impact α-synuclein deposition or total BDNF, rats in the stimulation group displayed restoration of striatal BDNF when compared to those in the nonstimulation group [51]. Furthermore, DBS may exert a neuroprotective effect through the inhibition of neuroinflammatory cytokines and pathways [52] and the modulation of synaptic plasticity [53]. Taken together, the findings related to α-synuclein deposition in PD reiterate both sides of thought related to DBS-induced neuroprotection (Figure 1).

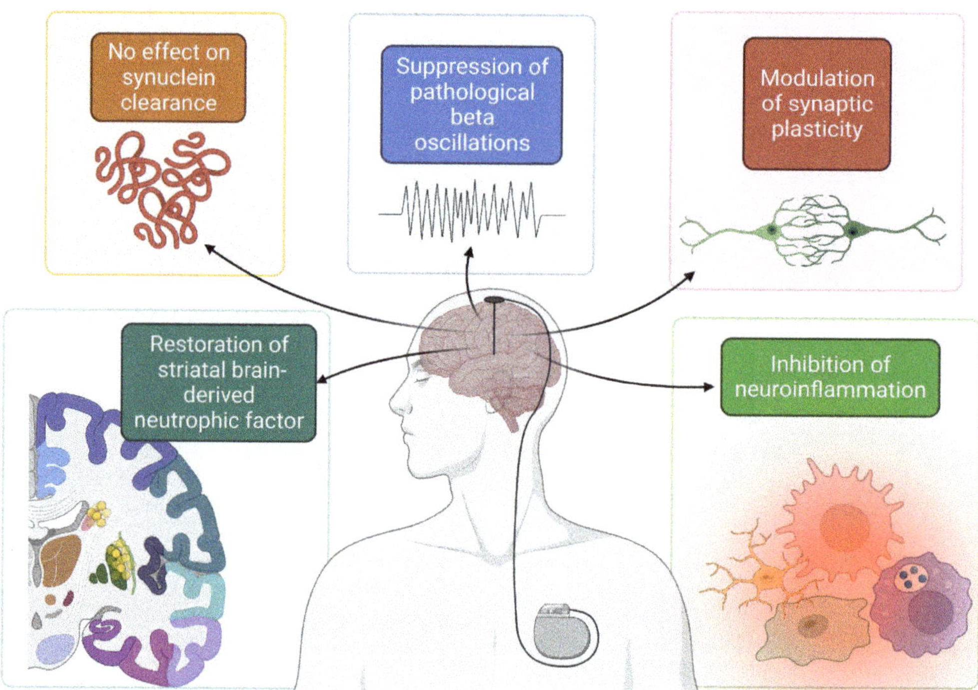


**Figure 1.** Summary of mechanisms by which DBS possibly confers a neuroprotective effect. Figure created with BioRender.com, accessed on 19 September 2022.

One central regulator of α-synuclein accumulation is autophagy. Rapamycin, for instance, stimulates autophagy and improves clearance of α-synuclein [54]. Notably, derailments in α-synuclein and autophagy in the perioperative environment can lead to pathology. In turn, an important consideration in the surgical treatment of synucleinopathies is the effect of anesthetic on α-synuclein, which was explored in a recent preclinical study. In this study, rats were anesthetized with propofol for 4 h and were assessed for neurobehavioral and cognitive deficits and hippocampal α-synuclein deposition. Along with elevated hippocampal α-synuclein deposition and impaired autophagy, 4 h of propofol induced worse performance on the Morris water maze test and shorter freezing times in the freezing conditioning test [55]. These results have important implications in the surgical management of PD with DBS, where electrode implantation being conducted awake or under general anesthetic is highly center dependent.

### 2.2. DBS in DLB and MSA

Despite the abundance of work related to DBS and PD, there is a relative paucity of evidence regarding the indications and benefits related to DBS use in the remaining synucleinopathies. The risk of cognitive decline is a central issue in DBS, making preoperative neuropsychiatric evaluation of paramount importance. In one study of 60 patients treated with STN DBS, executive functioning was significantly reduced when compared to a control group receiving standard medical therapy but no DBS [56]. Risk factors for cognitive decline following DBS included age and larger preoperative medication doses. Given that patients with DLB and MSA are already at significant risk for dysfunctional cognition, this possible adverse effect should be carefully considered in the context of each patient being evaluated for DBS therapy. Nevertheless, there has been some investigation into its use in DLB and MSA. In one randomized, double blind crossover study of 6 DLB patients implanted with bilateral nucleus basalis of Meynert (NBM) electrodes, there was no difference between stimulation and control conditions in a variety of cognitive tasks. Importantly, the procedure was well-tolerated and there was a positive impact on neuropsychiatric inventory (NPI) scores in that study [57]. However, in another more recent phase 1 study of six patients with bilateral NBM DBS, some cognitive decline occurred following electrode implantation [58]. MSA is also thought to be largely unresponsive to DBS and may be an underlying explanation for why neuromodulation fails in some patients with PD [59,60]. In one recent systematic review of 12 studies representing 22 patients with MSA, the majority were treated with bilateral STN electrodes ($n$ = 18) or bilateral GPi ($n$ = 3) [61]. At a median follow-up of 12 months, though subjective improvements in bradykinesia, gait, and rigidity were accompanied by a more than 12-point reduction in UPDRS-III score, a significant 23% of patients displayed neurobehavioral or neurocognitive side effects following DBS implantation. In MSA patients, who are particularly at risk for cognitive impairments [62], this represents a significant barrier to the safety and tolerability of DBS in MSA61. Furthermore, this systematic review also included 12 patients with DLB treated with bilateral NBM DBS. Again, no significant improvements in quality of life nor cognitive measurements were appreciated in this study [61]. Taken together, this suggests that the surgical indications for DBS therapy in MSA and DLB are still poorly understood and require further investigation.

### 3. Focused Ultrasound

Focused ultrasound (FUS) therapy is an actively studied alternative to current standard treatments involving open surgeries [63]. The primary benefit of FUS therapy is its ability to induce biological effects on deeper target tissue without damaging surrounding tissues [63]. FUS therapy is achieved using a piezoelectric ultrasound transducer to deliver a FUS beam and is guided using imaging modalities such as a traditional ultrasound or Magnetic Resonance Imaging (MRI) [64] to provide simultaneous monitoring of tissue effects [63]. The FUS beam is steered with precision by mechanically manipulating the transducer [64], and the spatial specificity of the beam and depth of its effects can be parametrized by varying the delivered sonication frequency and intensity [65]. At high intensity, the delivered FUS beam induces two main effects on target tissue: thermoablation and cavitation. Thermoablation results from tissue absorption of the beam energy, which rapidly increases the tissue temperature to irreversibly cytotoxic levels [66]. FUS mechanically induces cavitation, or the creation of gas cavities, in tissue by expanding and compressing the tissue as it travels through it [66]. These effects are purposely leveraged in clinical treatments to target varying tissues of interest as an alternative approach to surgery. Previous research demonstrates the use of FUS thermoablation to treat uterine fibroids [67], advanced stage renal malignancy [68], and primary bone tumors [69], and thus may offer a potential alternative to contemporary invasive surgeries. FUS has emerged as a new modality for treating movement disorders—such as essential tremor and PD—through noninvasive lesioning. Additionally, FUS therapy does not require hardware placement, such as an electrode, minimizing the risk of perioperative infection. However, FUS therapy faces limitations such as attenuation from overlying tissue, sensitivity to patient movement, possible treatment times of up to several hours, and the fact that lesioning is permanent and irreversible [64,70].

### 3.1. FUS and Synuclein

One emerging field of research utilizing FUS therapy involves leveraging cavitation to open the blood–brain barrier (BBB) [71]. The BBB historically has been a major limiting factor in drug delivery to brain parenchyma [71,72] as it is only permeable to lipid-soluble molecules smaller than 400 Da [72], which restricts pharmaceutical therapy options for neurological disorders. However, recent research has shown that FUS cavitation, in conjunction with localized microbubble injections, has the ability to noninvasively and reversibly open the BBB at specific targets in vivo, providing the possibility for localized drug therapies [71,73]. In a recent phase 1 trial in PD patients, the use of FUS in the parieto-occipito-temporal junction resulted in no serious adverse effects post-treatment but exhibited an opening of the BBB, demonstrating the feasibility of this approach to enhance drug delivery [73]. In another study, FUS cavitation was used to open the BBB in mice substantia nigra and striatal caudoputamen and was coupled with intravenous administration of the potentially neuroprotective and pro-dopaminergic neurturin neurotrophic factor (NTN). It was demonstrated that, in both targeted locations, NTN was successfully delivered into brain parenchyma with minimal diffusion to nontargeted areas [71]. Though NTN bioavailability was assessed, cognitive outcomes were not reported [71]. These drug delivery results mirror the effects observed in another study that utilized localized FUS to open BBB to deliver anti-α-syn antibodies in the left hippocampus, caudate putamen, and substantia nigra of PD-model mice [74]. Importantly, the α-synuclein load was decreased without impairing neuronal cell count [74]. Another recent study utilized FU to enhance delivery of copper nanoparticles (Cu-NPs) targeted to open TRPV1 channels [75]. The opening of TRPV1 channels is proposed to induce a $Ca^{2+}$-dependent signaling cascade, culminating in improved phagocytosis and elimination of a-syn. In this study, FU-mediated delivery of Cu-NPs ameliorated the histopathological alterations in tyrosine hydroxylase, glial fibrillary acidic protein, and α-syn. Importantly, motor, memory, and anxiety tests in mice initially worsened by α-syn aggregation were also improved with Cu-NPs [75]. Collectively, these studies demonstrate the ability of FUS cavitation to safely and effectively disrupt the BBB in vivo to facilitate targeted drug delivery into the brain parenchyma. Thus, with future research, FUS therapy coupled with localized drug delivery is an optimistic noninvasive therapeutic option that can be utilized in treating neurodegenerative diseases such as PD (Figure 2).

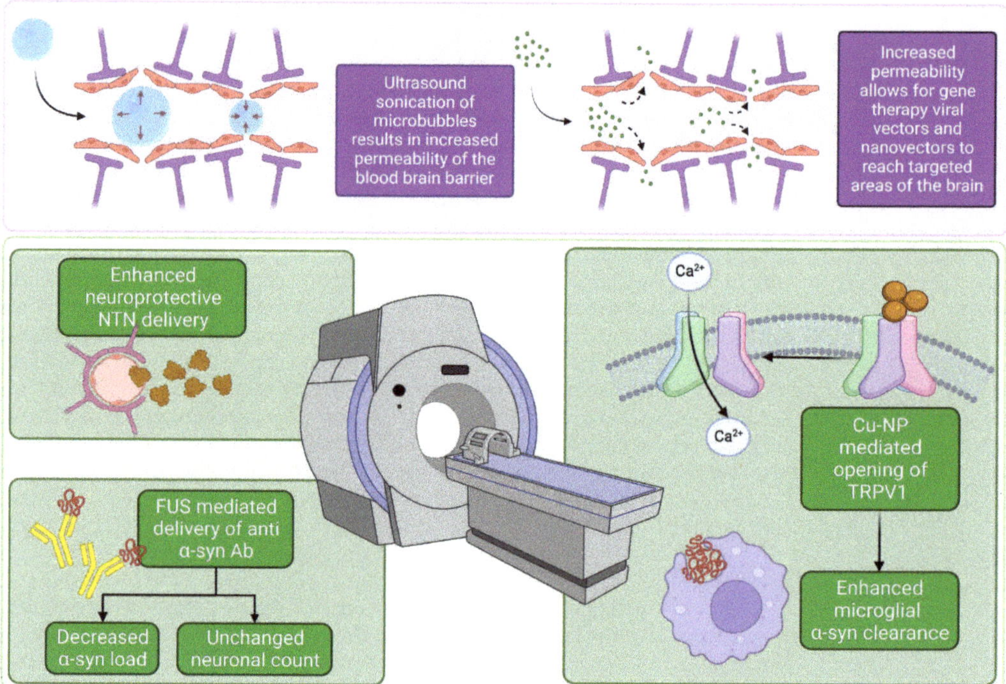


**Figure 2.** Summary of applications of focused ultrasound to synucleinopathies. Figure created with BioRender.com, accessed on 19 September 2022.

Beyond opening the BBB and facilitating drug delivery, FUS therapy offers potential symptomatic management options for treating PD with thermoablation. To date, there are three primary approaches of FUS therapy to treat PD symptoms: thalamotomy, subthalamotomy, and pallidotomy [76]. In one study unilaterally targeting the ventral intermediate nucleus (VIM) of the thalamus in PD patients with medication-resistant tremor, tremor was completely abolished immediately following the treatment [77]. The procedure was met with mild transient adverse effects such as headache [77]. Another recent study investigated the ability of FU therapy to abate parkinsonian symptoms by unilaterally targeting the subthalamic nucleus in PD patients with asymmetric parkinsonism [78]. Here, in a cohort of 40 patients treated with FUS or sham procedure of the STN, FUS was delivered until adequate control of tremor symptoms was achieved. UPDRS III score improved by almost 10 points at 4 months in the treatment group, which was significantly different from the 1.7-point improvement in the sham group. Specifically, FUS performed well in improving rigidity and tremor. Notably, dyskinesia, weakness, and abnormalities in gait and speech were common adverse effects. In line with previous reports, there were transient side effects related to treatment such as headache which resolved within a brief period following treatment. Importantly, there were no major cognitive or behavioral complications that developed as a result of thermoablation when measured at 4 months post-treatment [78]. FUS therapy has also been used for unilateral pallidotomy of the globus pallidus internus to improve symptoms in dyskinesia-dominant PD, again associated with transient side effects such as minor headache [79]. Overall, these different approaches utilizing FUS therapy demonstrate consistent findings, in that FUS therapy generally improves parkinsonian symptoms with only minor transient side effects. Furthermore, all of these approaches highlight the effectiveness of FUS therapy in noninvasively improving patient quality of life.

### 3.2. FUS Gene Therapy Approaches

Gene therapy is a therapeutic approach that aims to genetically modify cells through transcription and/or translation of transferred genetic material and/or integration into host genomes [80]. With PD, the goal of gene therapy is to treat disease symptoms and, ideally, to reverse disease progression. Gene therapy has been studied as a potential therapeutic for PD and other neurodegenerative diseases for years. Trophic factors, such as glial cell line derived neurotrophic factor (GDNF) and neurturin (NTN) have been explored as potential agents for PD gene therapy [81,82]. GDNF shows promise as a gene therapy agent due to its neurotrophic and neuroprotective effects. In primate models, the overexpression of neuroprotective agents such as GDNF has been demonstrated to decrease symptom severity and slow PD progression [83]. NTN is a structural and functional analogue of GDNF that has also demonstrated its ability to improve dopaminergic activity in animal models of PD [82]. Although preclinical trials provide ample evidence supporting GDNF and NTN gene therapy for the treatment of PD, clinical trials to date have not proven successful [84]. One of the key issues hindering the success of previous clinical trials has been low volumetric coverage of the gene therapy to targeted areas after direct injection [85]. In the CERE-120 clinical trial, which directly injected NTN into the putamen of study subjects, histological assessments show that the distribution of NTN was restricted [82]. Implementing FUS along with gene therapy has the potential to eliminate this issue that direct injection presents. Preclinical studies have used FUS to deliver vectors to specific animal models with encouraging results [86,87]. Xhima et al., for example, used recombinant adeno associated virus serotype 9 (AAV9) along with FUS to enhance delivery of an α-synuclein gene silencing short hairpin RNA sequence in mice overexpressing α-synuclein. FUS was targeted in the hippocampus, substantia nigra, olfactory bulb, and dorsal motor nucleus of the vagus. Decreased α-synuclein immunoreactivity was reported in these targets one month after FUS. Importantly, tyrosine hydroxylase (the rate-limiting enzyme in norepinephrine, epinephrine, and dopamine synthesis) and synaptophysin expression was not altered in targeted brain regions [86]. Mead et al. used a nanovector, particularly brain penetrating nanoparticles (BPN), along with FUS to promote GDNF transgene expression in target brain areas of rats. After only one treatment, there were therapeutically relevant levels of GDNF in targeted brain tissue. Furthermore, these therapeutic levels persisted for 10 weeks [88]. As previously mentioned, another key feature in the pathophysiology of PD is an elevated state of oxidative stress. The pro-oxidative environment established in PD potentiates neuronal stress, a pro-death environment, and progression of the disease [89]. In one study, FUS was used to deliver Nrf2, a nuclear factor which promotes downstream antioxidative elements, into the brains of PD rat models [90]. Though motor and behavioral outcomes were not measured in this study, Nrf2 expression was significantly elevated and led to a reduction in the pro-oxidative superoxide dismutase. Taken together, this suggests that FUS with viral vectors or nanovectors may show promise in developing gene therapy approaches for patients with PD. However, human evidence is lacking currently and requires more investigation.

### 4. Other Approaches in the Management of MSA and DLB

Though surgical interventions are poorly understood in the context of MSA and DLB, several promising frontiers of treatment have recently come to light. For example, in MSA, autonomic dysfunction remains a critical issue, characterized by progressive sympathetic failure, namely in the form of orthostatic hypotension [91]. This orthostatic hypotension can lead to a number of downstream sequelae, including cerebral hypoperfusion and increased risk for falls from resulting dizziness [91]. In fact, autonomic dysregulation is a predictor of worse outcomes in MSA and faster disease progression [92]. One recent study is poised to target this debilitating comorbidity of MSA [93]. This study describes the implantation of an epidural thoracic cord stimulator in a 48-year-old woman with progressive sympathetic dysfunction with resultant orthostatic hypotension. This stimulator is paired with an accelerometer which detects when the patient stands up and controls the delivery of stimulation to the thoracic cord. With the stimulator off, an 85 mmHg drop in systolic blood pressure was observed within 3 min of tilting the patient upright. This is compared to an 85 mmHg drop in systolic blood pressure occurring over 10 min after the stimulation was turned on [93]. Though this study lacks the power of a large clinical trial, it represents a promising treatment option for a very debilitating comorbidity of MSA.

In DLB, pathology stems in part from the deposition of extracellular synuclein, which leads to dysfunctional synaptic transmission and plasticity [33]. Transcranial direct current stimulation (tDCS) may play a role in modulating cortical excitability and is thought to induce changes in synaptic plasticity [94]. One recent double-blind clinical trial tested the efficacy of 10 days of tDCS sessions in improving the cognitive and psychiatric assessments of 11 DLB patients versus sham tDCS. Though there were no adverse effects from the treatment, no significant cognitive or psychiatric differences were found between the groups [87].

### 5. Conclusions

In this review, we have described the epidemiology, clinical presentation, molecular mechanisms, and standard of care for members of the synucleinopathy family of diseases: PD, MSA, and DLB. We examined the role of DBS in PD, including its potential for significant side effects, with special attention towards both how it is affected by and how it influences synuclein aggregation. Neuroinflammation, oxidative stress, and plasticity are central concepts in the pathogenesis of synucleinopathies. In PD, DBS may play a neuroprotective role, but these mechanisms are still poorly understood and require further investigation, as does the role of DBS in MSA and DLB. Furthermore, we examined the mechanism of FUS in opening the BBB and enhancing drug delivery, inducing thermoablation, and delivering gene therapy vectors. In summary, FUS represents a novel treatment paradigm and a promising area of future study. Finally, though they represent difficult pathological entities to treat, we further covered some novel treatment strategies for MSA and DLB.

**Author Contributions:** Writing—original draft preparation, S.S., K.R., K.C. and A.P.; writing—review and editing, B.L.-W. All authors have read and agreed to the published version of the manuscript.

**Funding:** This research received no external funding.

**Acknowledgments:** The authors have no competing interests to declare. Figures created with BioRender.com, accessed on 19 September 2022.

**Conflicts of Interest:** The authors declare no conflict of interest.

## References


1. Kianirad, Y.; Simuni, T. Pimavanserin, a novel antipsychotic for management of Parkinson's disease psychosis. *Expert Rev. Clin. Pharmacol.* **2017**, *10*, 1161–1168. [CrossRef] [PubMed]
2. Abrantes, A.M.; Friedman, J.H.; Brown, R.A.; Strong, D.R.; Desaulniers, J.; Ing, E.; Saritelli, J.; Riebe, D. Physical Activity and Neuropsychiatric Symptoms of Parkinson Disease. *J. Geriatr. Psychiatry Neurol.* **2012**, *25*, 138–145. [CrossRef] [PubMed]
3. Wickremaratchi, M.M.; Ben-Shlomo, Y.; Morris, H.R. The effect of onset age on the clinical features of Parkinson's disease. *Eur. J. Neurol.* **2009**, *16*, 450–456. [CrossRef] [PubMed]
4. Yang, W.; Hamilton, J.L.; Kopil, C.; Beck, J.C.; Tanner, C.M.; Albin, R.L.; Ray Dorsey, E.; Dahodwala, N.; Cintina, I.; Hogan, P.; et al. Current and projected future economic burden of Parkinson's disease in the U.S. *NPJ Park. Dis.* **2020**, *6*, 1–9. [CrossRef]
5. De Souza, R.M.; Moro, E.; Lang, A.E.; Schapira, A.H.V. Timing of Deep Brain Stimulation in Parkinson Disease: A Need for Reappraisal? *Ann. Neurol.* **2013**, *73*, 565–575. [CrossRef] [PubMed]
6. Tysnes, O.B.; Storstein, A. Epidemiology of Parkinson's disease. *J. Neural Transm.* **2017**, *124*, 901–905. [CrossRef]
7. Marras, C.; Beck, J.C.; Bower, J.H.; Roberts, E.; Ritz, B.; Ross, G.W.; Abbott, R.D.; Savica, R.; Van Den Eeden, S.K.; Willis, A.W.; et al. Prevalence of Parkinson's disease across North America. *NPJ Park. Dis.* **2018**, *4*, 1–7. [CrossRef]
8. Schapira, A.H.; Jenner, P. Etiology and pathogenesis of Parkinson's disease: Etiology and Pathogenesis. *Mov. Disord.* **2011**, *26*, 1049–1055. [CrossRef]
9. Schapira, A.H.; Agid, Y.; Barone, P.; Jenner, P.; Lemke, M.R.; Poewe, W.; Rascol, O.; Reichmann, H.; Tolosa, E. Perspectives on recent advances in the understanding and treatment of Parkinson's disease. *Eur. J. Neurol.* **2009**, *16*, 1090–1099. [CrossRef]
10. Singleton, A.B.; Farrer, M.; Johnson, J.; Singleton, A.; Hague, S.; Kachergus, J.; Hulihan, M.; Peuralinna, T.; Dutra, A.; Nussbaum, R.; et al. α-Synuclein Locus Triplication Causes Parkinson's Disease. *Science* **2003**, *302*, 841. [CrossRef]
11. Rocha, E.M.; De Miranda, B.; Sanders, L.H. Alpha-synuclein: Pathology, mitochondrial dysfunction and neuroinflammation in Parkinson's disease. *Neurobiol. Dis.* **2018**, *109*, 249–257. [CrossRef] [PubMed]
12. Li, W.W.; Yang, R.; Guo, J.C.; Ren, H.M.; Zha, X.L.; Cheng, J.S.; Cai, D.F. Localization of alpha-synuclein to mitochondria within midbrain of mice. *Neuroreport* **2007**, *18*, 1543–1546. [CrossRef] [PubMed]
13. Liu, G.; Zhang, C.; Yin, J.; Li, X.; Cheng, F.; Li, Y.; Yang, H.; Uéda, K.; Chan, P.; Yu, S. alpha-Synuclein is differentially expressed in mitochondria from different rat brain regions and dose-dependently down-regulates complex I activity. *Neurosci. Lett.* **2009**, *454*, 187–192. [CrossRef]
14. Martin, L.J.; Pan, Y.; Price, A.C.; Sterling, W.; Copeland, N.G.; Jenkins, N.A.; Price, D.L.; Lee, M.K. Parkinson's disease alpha-synuclein transgenic mice develop neuronal mitochondrial degeneration and cell death. *J. Neurosci. Off. J. Soc. Neurosci.* **2006**, *26*, 41–50. [CrossRef] [PubMed]
15. Choi, B.K.; Choi, M.G.; Kim, J.Y.; Yang, Y.; Lai, Y.; Kweon, D.H.; Lee, N.K.; Shin, Y.K. Large α-synuclein oligomers inhibit neuronal SNARE-mediated vesicle docking. *Proc. Natl. Acad. Sci. USA* **2013**, *110*, 4087–4092. [CrossRef]
16. Braak, H.; Sastre, M.; Del Tredici, K. Development of alpha-synuclein immunoreactive astrocytes in the forebrain parallels stages of intraneuronal pathology in sporadic Parkinson's disease. *Acta Neuropathol.* **2007**, *114*, 231–241. [CrossRef] [PubMed]
17. Poewe, W. The natural history of Parkinson's disease. *J. Neurol.* **2006**, *253* (Suppl. S7), VII2–VII6. [CrossRef]
18. Jankovic, J. Parkinson's disease: Clinical features and diagnosis. *J. Neurol. Neurosurg. Psychiatry* **2008**, *79*, 368–376. [CrossRef]
19. Sveinbjornsdottir, S. The clinical symptoms of Parkinson's disease. *J. Neurochem.* **2016**, *139*, 318–324. [CrossRef] [PubMed]
20. Jankovic, J.; Stacy, M. Medical management of levodopa-associated motor complications in patients with Parkinson's disease. *CNS Drugs* **2007**, *21*, 677–692. [CrossRef] [PubMed]
21. Jost, W.H.; Augustis, S. Severity of orthostatic hypotension in the course of Parkinson's disease: No correlation with the duration of the disease. *Park. Relat. Disord.* **2015**, *21*, 314–316. [CrossRef] [PubMed]
22. Riedel, O.; Klotsche, J.; Spottke, A.; Deuschl, G.; Förstl, H.; Henn, F.; Heuser, I.; Oertel, W.; Reichmann, H.; Riederer, P.; et al. Cognitive impairment in 873 patients with idiopathic Parkinson's disease. Results from the German Study on Epidemiology of Parkinson's Disease with Dementia (GEPAD). *J. Neurol.* **2008**, *255*, 255–264. [CrossRef] [PubMed]
23. Knie, B.; Mitra, M.T.; Logishetty, K.; Chaudhuri, K.R. Excessive Daytime Sleepiness in Patients with Parkinson's Disease. *CNS Drugs* **2011**, *25*, 203–212. [CrossRef] [PubMed]
24. Freitas, M.E.; Hess, C.W.; Fox, S.H. Motor Complications of Dopaminergic Medications in Parkinson's Disease. *Semin. Neurol.* **2017**, *37*, 147–157. [CrossRef] [PubMed]
25. Bower, J.H.; Maraganore, D.M.; McDonnell, S.K.; Rocca, W.A. Incidence of progressive supranuclear palsy and multiple system atrophy in Olmsted County, Minnesota, 1976 to 1990. *Neurology* **1997**, *49*, 1284–1288. [CrossRef] [PubMed]

26. Vann Jones, S.A.; O'Brien, J.T. The prevalence and incidence of dementia with Lewy bodies: A systematic review of population and clinical studies. *Psychol. Med.* **2014**, *44*, 673–683. [CrossRef]

27. Multiple-System Atrophy Research Collaboration. Mutations in COQ2 in familial and sporadic multiple-system atrophy. *N. Engl. J. Med.* **2013**, *369*, 233–244. [CrossRef]

28. Wenning, G.K.; Krismer, F. Multiple system atrophy. In *Handbook of Clinical Neurology*; Elsevier: Amsterdam, The Netherlands, 2013; Volume 117, pp. 229–241. [CrossRef]

29. Boot, B.P.; Orr, C.F.; Ahlskog, J.E.; Ferman, T.J.; Roberts, R.; Pankratz, V.S.; Dickson, D.W.; Parisi, J.; Aakre, J.A.; Geda, Y.E.; et al. Risk factors for dementia with Lewy bodies: A case-control study. *Neurology* **2013**, *81*, 833–840. [CrossRef]

30. Kim, W.S.; Kågedal, K.; Halliday, G.M. Alpha-synuclein biology in Lewy body diseases. *Alzheimers Res. Ther.* **2014**, *6*, 73. [CrossRef]

31. Gilman, S.; Wenning, G.K.; Low, P.A.; Brooks, D.J.; Mathias, C.J.; Trojanowski, J.Q.; Wood, N.W.; Colosimo, C.; Dürr, A.; Fowler, C.J.; et al. Second consensus statement on the diagnosis of multiple system atrophy. *Neurology* **2008**, *71*, 670–676. [CrossRef]

32. Ozawa, T.; Paviour, D.; Quinn, N.P.; Josephs, K.A.; Sangha, H.; Kilford, L.; Healy, D.G.; Wood, N.W.; Lees, A.J.; Holton, J.L.; et al. The spectrum of pathological involvement of the striatonigral and olivopontocerebellar systems in multiple system atrophy: Clinicopathological correlations. *Brain J. Neurol.* **2004**, *127 Pt 12*, 2657–2671. [CrossRef] [PubMed]

33. Garcia-Esparcia, P.; López-González, I.; Grau-Rivera, O.; García-Garrido, M.F.; Konetti, A.; Llorens, F.; Zafar, S.; Carmona, M.; del Rio, J.A.; Zerr, I.; et al. Dementia with Lewy Bodies: Molecular Pathology in the Frontal Cortex in Typical and Rapidly Progressive Forms. *Front. Neurol.* **2017**, *8*, 89. [CrossRef] [PubMed]

34. Halliday, G.M.; Holton, J.L.; Revesz, T.; Dickson, D.W. Neuropathology underlying clinical variability in patients with synucleinopathies. *Acta Neuropathol.* **2011**, *122*, 187–204. [CrossRef] [PubMed]

35. Jecmenica-Lukic, M.; Poewe, W.; Tolosa, E.; Wenning, G.K. Premotor signs and symptoms of multiple system atrophy. *Lancet Neurol.* **2012**, *11*, 361–368. [CrossRef]

36. Erkkinen, M.G.; Kim, M.O.; Geschwind, M.D. Clinical Neurology and Epidemiology of the Major Neurodegenerative Diseases. *Cold Spring Harb. Perspect. Biol.* **2018**, *10*, a033118. [CrossRef]

37. Burn, D.J.; Rowan, E.N.; Minett, T.; Sanders, J.; Myint, P.; Richardson, J.; Thomas, A.; Newby, J.; Reid, J.; O'Brien, J.T.; et al. Extrapyramidal features in Parkinson's disease with and without dementia and dementia with Lewy bodies: A cross-sectional comparative study. *Mov. Disord. Off. J. Mov. Disord. Soc.* **2003**, *18*, 884–889. [CrossRef]

38. Köllensperger, M.; Geser, F.; Ndayisaba, J.P.; Boesch, S.; Seppi, K.; Ostergaard, K.; Dupont, E.; Cardozo, A.; Tolosa, E.; Abele, M.; et al. Presentation, diagnosis, and management of multiple system atrophy in Europe: Final analysis of the European multiple system atrophy registry. *Mov. Disord. Off. J. Mov. Disord. Soc.* **2010**, *25*, 2604–2612. [CrossRef]

39. Hamilton, J.M.; Salmon, D.P.; Galasko, D.; Raman, R.; Emond, J.; Hansen, L.A.; Masliah Eliezer Thal, L.J. Visuospatial Deficits Predict Rate of Cognitive Decline in Autopsy-Verified Dementia with Lewy Bodies. *Neuropsychology* **2008**, *22*, 729–737. [CrossRef]

40. Abbott, A. Levodopa: The story so far. *Nature* **2010**, *466*, S6–S7. [CrossRef]

41. Brodell, D.W.; Stanford, N.T.; Jacobson, C.E.; Schmidt, P.; Okun, M.S. Carbidopa/levodopa dose elevation and safety concerns in Parkinson's patients: A cross-sectional and cohort design. *BMJ Open* **2012**, *2*, e001971. [CrossRef]

42. Honey, C.R.; Hamani, C.; Kalia, S.K.; Sankar, T.; Picillo, M.; Munhoz, R.P.; Fasano, A.; Panisset, M. Deep Brain Stimulation Target Selection for Parkinson's Disease. *Can. J. Neurol. Sci.* **2017**, *44*, 3–8. [CrossRef] [PubMed]

43. Tinkhauser, G.; Pogosyan, A.; Tan, H.; Herz, D.M.; Kühn, A.A.; Brown, P. Beta burst dynamics in Parkinson's disease OFF and ON dopaminergic medication. *Brain* **2017**, *140*, 2968–2981. [CrossRef] [PubMed]

44. Kleiner-Fisman, G.; Herzog, J.; Fisman, D.N.; Tamma, F.; Lyons, K.E.; Pahwa, R.; Lang, A.E.; Deuschl, G. Subthalamic nucleus deep brain stimulation: Summary and meta-analysis of outcomes. *Mov. Disord.* **2006**, *21* (Suppl. S14), S290–S304. [CrossRef] [PubMed]

45. Kordower, J.H.; Olanow, C.W.; Dodiya, H.B.; Chu, Y.; Beach, T.G.; Adler, C.H.; Halliday, G.M.; Bartus, R.T. Disease duration and the integrity of the nigrostriatal system in Parkinson's disease. *Brain* **2013**, *136*, 2419–2431. [CrossRef]

46. Fischer, D.L.; Manfredsson, F.P.; Kemp, C.J.; Cole-Strauss, A.; Lipton, J.W.; Duffy, M.F.; Polinski, N.K.; Steece-Collier, K.; Collier, T.J.; Gombash, S.E.; et al. Subthalamic Nucleus Deep Brain Stimulation Does Not Modify the Functional Deficits or Axonopathy Induced by Nigrostriatal α-Synuclein Overexpression. *Sci. Rep.* **2017**, *7*, 16356. [CrossRef]

47. Musacchio, T.; Rebenstorff, M.; Fluri, F.; Brotchie, J.M.; Volkmann, J.; Koprich, J.B.; Ip, C.W. Subthalamic nucleus deep brain stimulation is neuroprotective in the A53T α-synuclein Parkinson's disease rat model. *Ann. Neurol.* **2017**, *81*, 825–836. [CrossRef] [PubMed]

48. Little, S.; Brown, P. The functional role of beta oscillations in Parkinson's disease. *Park. Relat. Disord.* **2014**, *20*, S44–S48. [CrossRef]

49. Chen, J.Z.; Hofman, K.; Koprich, J.B.; Brotchie, J.M.; Volkmann, J.; Ip, C.W. P 15 Long-term subthalamic deep brain stimulation modulates pathological beta oscillations in the AAV-A53T-Synuclein Parkinson's disease rat model. *Clin. Neurophysiol.* **2022**, *137*, e23–e24. [CrossRef]

50. Miranda, M.; Morici, J.F.; Zanoni, M.B.; Bekinschtein, P. Brain-Derived Neurotrophic Factor: A Key Molecule for Memory in the Healthy and the Pathological Brain. *Front. Cell. Neurosci.* **2019**, *13*, 363. Available online: https://www.frontiersin.org/articles/10.3389/fncel.2019.00363 (accessed on 18 September 2022). [CrossRef]

51. Miller, K.M.; Patterson, J.R.; Kochmanski, J.; Kemp, C.J.; Stoll, A.C.; Onyekpe, C.U.; Cole-Strauss, A.; Steece-Collier, K.; Howe, J.W.; Luk, K.C.; et al. Striatal Afferent BDNF Is Disrupted by Synucleinopathy and Partially Restored by STN DBS. *J. Neurosci.* **2021**, *41*, 2039–2052. [CrossRef]

52. Chen, Y.; Zhu, G.; Liu, D.; Zhang, X.; Liu, Y.; Yuan, T.; Du, T.; Zhang, J. Subthalamic nucleus deep brain stimulation suppresses neuroinflammation by Fractalkine pathway in Parkinson's disease rat model. *Brain Behav. Immun.* **2020**, *90*, 16–25. [CrossRef] [PubMed]

53. Herrington, T.M.; Cheng, J.J.; Eskandar, E.N. Mechanisms of deep brain stimulation. *J. Neurophysiol.* **2016**, *115*, 19–38. [CrossRef]

54. Webb, J.L.; Ravikumar, B.; Atkins, J.; Skepper, J.N.; Rubinsztein, D.C. α-Synuclein Is Degraded by Both Autophagy and the Proteasome*. *J. Biol. Chem.* **2003**, *278*, 25009–25013. [CrossRef] [PubMed]

55. Yang, N.; Li, Z.; Han, D.; Mi, X.; Tian, M.; Liu, T.; Li, Y.; He, J.; Kuang, C.; Cao, Y.; et al. Autophagy prevents hippocampal α-synuclein oligomerization and early cognitive dysfunction after anesthesia/surgery in aged rats. *Aging* **2020**, *12*, 7262–7281. [CrossRef]

56. Witt, K.; Daniels, C.; Volkmann, J. Factors associated with neuropsychiatric side effects after STN-DBS in Parkinson's disease. *Park. Relat. Disord.* **2012**, *18*, S168–S170. [CrossRef]

57. Gratwicke, J.; Zrinzo, L.; Kahan, J.; Peters, A.; Brechany, U.; McNichol, A.; Beigi, M.; Akram, H.; Hyam, J.; Oswal, A.; et al. Bilateral nucleus basalis of Meynert deep brain stimulation for dementia with Lewy bodies: A randomised clinical trial. *Brain Stimulat.* **2020**, *13*, 1031–1039. [CrossRef] [PubMed]

58. Maltête, D.; Wallon, D.; Bourilhon, J.; Lefaucheur, R.; Danaila, T.; Thobois, S.; Defebvre, L.; Dujardin, K.; Houeto, J.L.; Godefroy, O.; et al. Nucleus Basalis of Meynert Stimulation for Lewy Body Dementia: A Phase I Randomized Clinical Trial. *Neurology* **2021**, *96*, e684–e697. [CrossRef] [PubMed]

59. Thavanesan, N.; Gillies, M.; Farrell, M.; Green, A.L.; Aziz, T. Deep Brain Stimulation in Multiple System Atrophy Mimicking Idiopathic Parkinson's Disease. *Case Rep. Neurol.* **2014**, *6*, 232–237. [CrossRef]

60. Lambrecq, V.; Krim, E.; Meissner, W.; Guehl, D.; Tison, F. Deep-brain stimulation of the internal pallidum in multiple system atrophy. *Rev. Neurol.* **2008**, *164*, 398–402. [CrossRef]

61. Artusi, C.A.; Rinaldi, D.; Balestrino, R.; Lopiano, L. Deep brain stimulation for atypical parkinsonism: A systematic review on efficacy and safety. *Park. Relat. Disord.* **2022**, *96*, 109–118. [CrossRef]

62. Stankovic, I.; Krismer, F.; Jesic, A.; Antonini, A.; Benke, T.; Brown, R.G.; Burn, D.J.; Holton, J.L.; Kaufmann, H.; Kostic, V.S.; et al. Cognitive impairment in multiple system atrophy. *Mov. Disord. Off. J. Mov. Disord. Soc.* **2014**, *29*, 857–867. [CrossRef] [PubMed]

63. Meng, Y.; Hynynen, K.; Lipsman, N. Applications of focused ultrasound in the brain: From thermoablation to drug delivery. *Nat. Rev. Neurol.* **2021**, *17*, 7–22. [CrossRef] [PubMed]

64. Izadifar, Z.; Izadifar, Z.; Chapman, D.; Babyn, P. An Introduction to High Intensity Focused Ultrasound: Systematic Review on Principles, Devices, and Clinical Applications. *J. Clin. Med.* **2020**, *9*, 460. [CrossRef] [PubMed]

65. Ye, P.P.; Brown, J.R.; Pauly, K.B. Frequency Dependence of Ultrasound Neurostimulation in the Mouse Brain. *Ultrasound Med. Biol.* **2016**, *42*, 1512–1530. [CrossRef]

66. Zhou, Y.F. High intensity focused ultrasound in clinical tumor ablation. *World J. Clin. Oncol.* **2011**, *2*, 8–27. [CrossRef]

67. Stewart, E.A.; Rabinovici, J.; Tempany, C.M.C.; Inbar, Y.; Regan, L.; Gastout, B.; Hesley, G.; Kim, H.S.; Hengst, S.; Gedroye, W.M. Clinical outcomes of focused ultrasound surgery for the treatment of uterine fibroids. *Fertil. Steril.* **2006**, *85*, 22–29. [CrossRef]

68. Wu, F.; Wang, Z.B.; Chen, W.Z.; Bai, J.; Zhu, H.; Qiao, T.Y. Preliminary Experience Using High Intensity Focused Ultrasound for the Treatment of Patients with Advanced Stage Renal Malignancy. *J. Urol.* **2003**, *170*, 2237–2240. [CrossRef]

69. Chen, W.; Zhou, K. High-intensity focused ultrasound ablation: A new strategy to manage primary bone tumors. *Curr. Opin. Orthop.* **2005**, *16*, 494–500. [CrossRef]

70. Prada, F.; Kalani, M.Y.S.; Yagmurlu, K.; Norat, P.; Del Bene, M.; DiMeco, F.; Kassell, N.F. Applications of Focused Ultrasound in Cerebrovascular Diseases and Brain Tumors. *Neurotherapeutics* **2019**, *16*, 67–87. [CrossRef]

71. Samiotaki, G.; Acosta, C.; Wang, S.; Konofagou, E.E. Enhanced delivery and bioactivity of the neurturin neurotrophic factor through focused ultrasound—mediated blood–brain barrier opening in vivo. *J. Cereb. Blood Flow Metab.* **2015**, *35*, 611–622. [CrossRef]

72. Pardridge, W.M. Blood-brain barrier drug targeting: The future of brain drug development. *Mol. Interv.* **2003**, *3*, 90. [CrossRef] [PubMed]

73. Gasca-Salas, C.; Fernández-Rodríguez, B.; Pineda-Pardo, J.A.; Rodríguez-Rojas, R.; Obeso, I.; Hernández-Fernández, F.; del Álamo, M.; Mata, D.; Guida, P.; Ordás-Bandera, C.; et al. Blood-brain barrier opening with focused ultrasound in Parkinson's disease dementia. *Nat. Commun.* **2021**, *12*, 779. [CrossRef] [PubMed]

74. Zhang, H.; Sierra, C.; Kwon, N.; Karakatsani, M.E.; Jackson-Lewis, V.R.; Przedborski, S.; Konofagou, E. Focused-ultrasound Mediated Anti-Alpha-Synuclein Antibody Delivery for the Treatment of Parkinson's Disease. In Proceedings of the 2018 IEEE International Ultrasonics Symposium (IUS), Kobe, Japan, 22–25 October 2018; pp. 1–4. [CrossRef]

75. Yuan, J.; Liu, H.; Zhang, H.; Wang, T.; Zheng, Q.; Li, Z. Controlled Activation of TRPV1 Channels on Microglia to Boost Their Autophagy for Clearance of Alpha-Synuclein and Enhance Therapy of Parkinson's Disease. *Adv. Mater.* **2022**, *34*, 2108435. [CrossRef] [PubMed]

76. Moosa, S.; Martínez-Fernández, R.; Elias, W.J.; del Alamo, M.; Eisenberg, H.M.; Fishman, P.S. The role of high-intensity focused ultrasound as a symptomatic treatment for Parkinson's disease. *Mov. Disord.* **2019**, *34*, 1243–1251. [CrossRef]

77. Schlesinger, I.; Sinai, A.; Zaaroor, M. MRI-Guided Focused Ultrasound in Parkinson's Disease: A Review. *Park. Dis.* **2017**, *2017*, 8124624. [CrossRef]

78. Martínez-Fernández, R.; Máñez-Miró, J.U.; Rodríguez-Rojas, R.; del Álamo, M.; Shah, B.B.; Hernández-Fernández, F.; Pineda-Pardo, J.A.; Monje, M.H.G.; Fernández-Rodríguez, B.; Sperling, S.A.; et al. Randomized Trial of Focused Ultrasound Subthalamotomy for Parkinson's Disease. *N. Engl. J. Med.* **2020**, *383*, 2501–2513. [CrossRef]

79. Jung, N.Y.; Park, C.K.; Kim, M.; Lee, P.H.; Sohn, Y.H.; Chang, J.W. The efficacy and limits of magnetic resonance-guided focused ultrasound pallidotomy for Parkinson's disease: A Phase I clinical trial. *J. Neurosurg.* **2018**, 1853–1861. [CrossRef] [PubMed]

80. Wirth, T.; Parker, N.; Ylä-Herttuala, S. History of gene therapy. *Gene* **2013**, *525*, 162–169. [CrossRef]

81. Lin, C.Y.; Hsieh, H.Y.; Chen, C.M.; Wu, S.R.; Tsai, C.H.; Huang, C.Y.; Hua, M.Y.; Wei, K.C.; Yeh, C.K.; Liu, H.L. Non-invasive, neuron-specific gene therapy by focused ultrasound-induced blood-brain barrier opening in Parkinson's disease mouse model. *J. Control. Release* **2016**, *235*, 72–81. [CrossRef]

82. Marks, W.J.; Bartus, R.T.; Siffert, J.; Davis, C.S.; Lozano, A.; Boulis, N.; Vitek, J.; Stacy, M.; Turner, D.; Verhagen, L.; et al. Gene delivery of AAV2-neurturin for Parkinson's disease: A double-blind, randomised, controlled trial. *Lancet Neurol.* **2010**, *9*, 1164–1172. [CrossRef]

83. Kordower, J.H.; Emborg, M.E.; Bloch, J.; Ma, S.Y.; Chu, Y.; Leventhal, L.; McBride, J.; Chen, E.Y.; Palfi, S.; Roitberg, B.Z.; et al. Neurodegeneration prevented by lentiviral vector delivery of GDNF in primate models of Parkinson's disease. *Science* **2000**, *290*, 767–773. [CrossRef] [PubMed]

84. Bartus, R.T.; Johnson, E.M. Clinical tests of neurotrophic factors for human neurodegenerative diseases, part 2: Where do we stand and where must we go next? *Neurobiol. Dis.* **2017**, *97 Pt B*, 169–178. [CrossRef]

85. Price, R.J.; Fisher, D.G.; Suk, J.S.; Hanes, J.; Ko, H.S.; Kordower, J.H. Parkinson's Disease Gene Therapy: Will Focused Ultrasound and Nanovectors Be the Next Frontier? *Mov. Disord. Off. J. Mov. Disord. Soc.* **2019**, *34*, 1279–1282. [CrossRef]

86. Xhima, K.; Nabbouh, F.; Hynynen, K.; Aubert, I.; Tandon, A. Noninvasive delivery of an α-synuclein gene silencing vector with magnetic resonance–guided focused ultrasound. *Mov. Disord.* **2018**, *33*, 1567–1579. [CrossRef] [PubMed]

87. Wang, S.; Olumolade, O.O.; Sun, T.; Samiotaki, G.; Konofagou, E.E. Noninvasive, neuron-specific gene therapy can be facilitated by focused ultrasound and recombinant adeno-associated virus. *Gene Ther.* **2015**, *22*, 104–110. [CrossRef] [PubMed]

88. Mead, B.P.; Kim, N.; Miller, G.W.; Hodges, D.; Mastorakos, P.; Klibanov, A.L.; Mandell, J.W.; Hirsh, J.; Suk, J.S.; Hanes, J.; et al. Novel Focused Ultrasound Gene Therapy Approach Noninvasively Restores Dopaminergic Neuron Function in a Rat Parkinson's Disease Model. *Nano Lett.* **2017**, *17*, 3533–3542. [CrossRef]

89. Dias, V.; Junn, E.; Mouradian, M.M. The Role of Oxidative Stress in Parkinson's Disease. *J. Park. Dis.* **2013**, *3*, 461–491. [CrossRef] [PubMed]

90. Long, L.; Cai, X.; Guo, R.; Wang, P.; Wu, L.; Yin, T.; Liao, S.; Lu, Z. Treatment of Parkinson's disease in rats by Nrf2 transfection using MRI-guided focused ultrasound delivery of nanomicrobubbles. *Biochem. Biophys. Res. Commun.* **2017**, *482*, 75–80. [CrossRef]

91. Iodice, V.; Low, D.A.; Vichayanrat, E.; Mathias, C.J. Cardiovascular autonomic dysfunction in MSA and Parkinson's disease: Similarities and differences. *J. Neurol. Sci.* **2011**, *310*, 133–138. [CrossRef]

92. Tada, M.; Onodera, O.; Tada, M.; Ozawa, T.; Piao, Y.S.; Kakita, A.; Takahashi, H.; Nishizawa, M. Early Development of Autonomic Dysfunction May Predict Poor Prognosis in Patients with Multiple System Atrophy. *Arch. Neurol.* **2007**, *64*, 256–260. [CrossRef]

93. Squair, J.W.; Berney, M.; Castro Jimenez, M.; Hankov, N.; Demesmaeker, R.; Amir, S.; Paley, A.; Hernandez-Charpak, S.; Dumont, G.; Asboth, L.; et al. Implanted System for Orthostatic Hypotension in Multiple-System Atrophy. *N. Engl. J. Med.* **2022**, *386*, 1339–1344. [CrossRef] [PubMed]

94. Stagg, C.J.; Nitsche, M.A. Physiological Basis of Transcranial Direct Current Stimulation. *Neuroscientist* **2011**, *17*, 37–53. [CrossRef] [PubMed]

MDPI

St. Alban-Anlage 66

4052 Basel

Switzerland

www.mdpi.com

*Biomedicines* Editorial Office

E-mail: biomedicines@mdpi.com

www.mdpi.com/journal/biomedicines